Genome Informatics 2007

GENOME INFORMATICS SERIES (GIS)

ISSN: 0919-9454

The Genome Informatics Series publishes peer-reviewed papers presented at the International Conference on Genome Informatics (GIW) and some conferences on bioinformatics. The Genome Informatics Series is indexed in MEDLINE.

No.	Title	Year	ISBN Cl./Pa.
1	Genome Informatics Workshop I	1990	(in Japanese)
2	Genome Informatics Workshop II	1991	(in Japanese)
3	Genome Informatics Workshop III	1992	(in Japanese)
4	Genome Informatics Workshop IV	1993	4-946443-20-7
5	Genome Informatics Workshop 1994	1994	4-946443-24-X
6	Genome Informatics Workship 1995	1995	4-946443-33-9
7	Genome Informatics 1996	1996	4-946443-37-1
8	Genome Informatics 1997	1997	4-946443-47-9
9	Genome Informatics 1998	1998	4-946443-52-5
10	Genome Informatics 1999	1999	4-946443-59-2
11	Genome Informatics 2000	2000	4-946443-65-7
12	Genome Informatics 2001	2001	4-946443-72-X
13	Genome Informatics 2002	2002	4-946443-79-7
14	Genome Informatics 2003	2003	4-946443-82-7
15	Genome Informatics 2004 Vol. 15, No. 1	2004	4-946443-88-6
16	Genome Informatics 2004 Vol. 15, No. 2	2004	4-946443-91-6
17	Genome Informatics 2005 Vol. 16, No. 1	2005	4-946443-93-2
18	Genome Informatics 2005 Vol. 16, No. 2	2005	4-946443-96-7
19	Genome Informatics 2006 Vol. 17, No. 1	2006	4-946443-97-5
20	Genome Informatics 2006 Vol. 17, No. 2	2006	4-946443-99-1
21	Genome Informatics 2007 Vol. 18	2007	978-1-86094-991-3
22	Genome Informatics 2007 Vol. 19	2007	978-1-86094-984-5

Genome Informatics Series Vol. 18

ISSN: 0919-9454

Genome Informatics 2007

Proceedings of the 7th Annual International Workshop on Bioinformatics and Systems Biology (IBSB 2007)

Institute of Medical Science, University of Tokyo, Japan

31 July – 2 August 2007

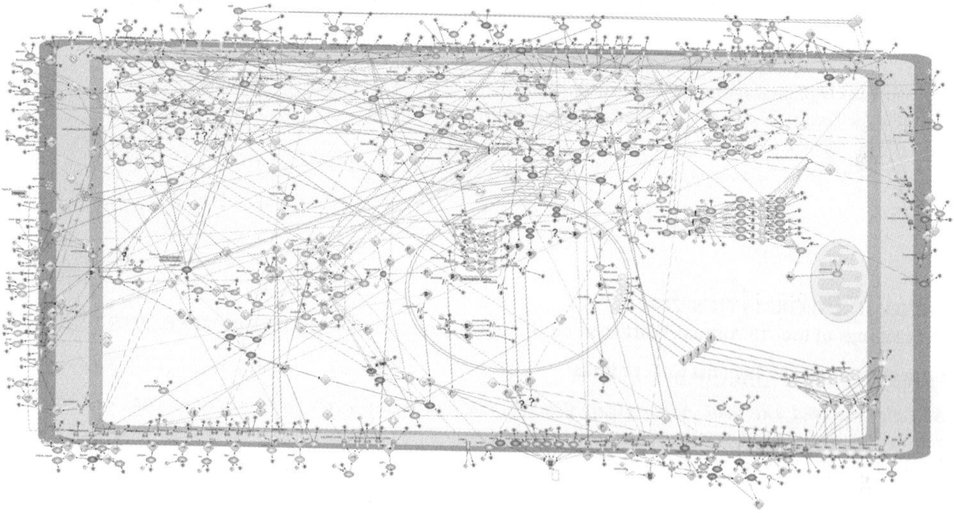

Editors

Satoru Miyano
University of Tokyo, Japan

Charles DeLisi
Boston University, USA

Herman-Georg Holzhütter
Charité–University Medicine Berlin, Germany

Minoru Kanehisa
Kyoto University, Japan

ICP

Imperial College Press

Published by

Imperial College Press
57 Shelton Street
Covent Garden
London WC2H 9HE

Distributed by

World Scientific Publishing Co. Pte. Ltd.
5 Toh Tuck Link, Singapore 596224
USA office: 27 Warren Street, Suite 401-402, Hackensack, NJ 07601
UK office: 57 Shelton Street, Covent Garden, London WC2H 9HE

British Library Cataloguing-in-Publication Data
A catalogue record for this book is available from the British Library.

GENOME INFORMATICS 2007
Proceedings of the 7th Annual Workshop on Bioinformatics and Systems Biology (IBSB 2007)

ISBN-13 978-1-86094-991-3
ISBN-10 1-86094-991-6

Printed by Fulsland Offset Printing (S) Pte Ltd, Singapore

Prof. Dr. Dr. hc. Reinhart Heinrich
(1946 ~ 2006)

This issue of "Genome Informatics" is dedicated to Prof. Dr. Dr. hc. Reinhart Heinrich, an internationally highly respected pioneer and promoter of computational systems biology, who died suddenly in October 2006 aged 60 at the height of his power.

Heinrich was one of the founders of metabolic control theory and made fundamental contributions to various fields of theoretical biophysics and biochemistry as, to name only a few, evolution and structural design of cellular networks, role of oscillations in biochemical systems and, more recently, vesicular transport in cells.

Due to Reinhart Heinrich's initiative, an international collaborative educational program was established involving the Graduate Program of the Boston University, the International Research Training Group (IRTG) "Genomics and Systems Biology of Molecular Networks" of the Humboldt University and Free University of Berlin, and the Joint Bioinformatics Education Program of Kyoto University and University of Tokyo. Reinhart Heinrich would be glad and proud to see how successful the program has developed as documented by the scientific results presented at the Seventh Annual International Workshop on Bioinformatics and Systems Biology 2007 at the University of Tokyo and published in part in this issue.

Hermann-Georg Holzhütter

PREFACE

Genome Informatics Vol. 18 contains the peer-reviewed papers presented at the Seventh Annual International Workshop on Bioinformatics and Systems Biology held on July 31–August 2 of 2007 at the Institute of Medical Science, The University of Tokyo. This workshop started in 2001 as an event for doctoral students and young researchers to present and discuss their research results and approaches in Bioinformatics and Systems Biology. The first workshop held in Berlin was organized by Prof. Dr. Reinhart Heinrich (former Professor at Humboldt University Berlin), a co-founder of this workshop. Very regretfully, he died on October 23, 2006, at the age of 60. This volume is dedicated to the memory of Prof. Dr. Dr. hc. Reinhart Heinrich.

Since 2001, the workshop has been held in Boston (2002), Berlin (2003), Kyoto (2004), Berlin (2005), and Boston (2006). In 2007, it was held in Tokyo as part of a collaborative educational program involving the leading institutions committing the following programs and partner institutions:
Programs

- Boston — Graduate Program in Bioinformatics, Boston University
- Berlin — The International Research Training Group (IRTG) "Genomics and Systems Biology of Molecular Networks"
- Kyoto/Tokyo — Joint Bioinformatics Education Program of Kyoto University and University of Tokyo

Partner Institutions

- Boston University
- Humboldt University Berlin
- Free University Berlin
- Max-Planck Institute of Molecular Genetics
- Hahn-Meitner-Institute
- Bioinformatics Center, Institute for Chemical Research, Kyoto University
- Department of Bioinformatics and Chemical Genomics, Graduate School of Pharmaceutical Sciences, Kyoto University
- Human Genome Center, Institute of Medical Science, The University of Tokyo

The submissions were pre-screened by the program committee members and each submission was reviewed by three reviewers. We have selected 31 papers after revision. These papers will be indexed in Medline, and their electronic versions are freely available from the website of the Japanese Society for Bioinformatics as *Genome Informatics Online* (http://www.jsbi.org/modules/journal/index.php/index.html). Former publications are also electronically available as Genome Informatics Vol. 15, No. 1 (2004), Vol. 16, No. 1 (2005), and Vol. 17, No. 1 (2006).

We wish to thank all of those who submitted papers and helped with the reviewing process. We also wish to thank colleagues at the Human Genome Center, The University of Tokyo, for their efforts in local arrangement, finance, and publication. In particular, Emi Ikeda, Masao Nagasaki, Hiroko Nishihata, Ayumu Saito, Asako Suzuki, and Ayako Tomiyasu.

Program Committee Chair:
Satoru Miyano

Organizers:
Charles DeLisi
Hermann-Georg Holzhütter
Minoru Kanehisa

PROGRAM COMMITTEE

Satoru Miyano	University of Tokyo, PC Chair
Tatsuya Akutsu	Kyoto University
Gary Benson	Boston University
Charles DeLisi	Boston University
Oliver Ebenhöh	Humboldt University Berlin
Susumu Goto	Kyoto University
Hermann-Georg Holzhütter	Charite-University Medicine Berlin
Seiya Imoto	University of Tokyo
Minoru Kanehisa	Kyoto University
Edda Klipp	The Max Planck Institute for Molecular Genetics
Hiroshi Mamitsuka	Kyoto University
Brandon Xia	Boston University

CONTENTS

REGULATORY ELEMENTS OF MARINE CYANOBACTERIA

SZYMON M. KIEŁBASA[1] HANSPETER HERZEL[2]
kielbasa@molgen.mpg.de h.herzel@biologie.hu-berlin.de
ILKA M. AXMANN[2]
i.axmann@biologie.hu-berlin.de

[1] *MPI MG, Max Planck Institute for Molecular Genetics, Ihnestrasse 73, D-14195 Berlin, Germany*
[2] *ITB, Institute for Theoretical Biology, Humboldt University of Berlin, Invalidenstrasse 43, D-10115 Berlin, Germany*

The free-living, oxyphototroph bacteria of the group of *Prochlorococcus* populate widely the oceans. Genomic information of nine marine cyanobacteria was used to predict signals essential for regulation. We implemented a pipeline that automatically calculates BLASTp alignments of query genomes, selects a representative subset of orthologs and predicts motifs conserved in their upstream sequences. Next, similar motifs are clustered into groups which could contain profiles recognized by different transcription factors. The phylogenetic footprinting pipeline revealed a minimal conserved set of putative transcription factors, binding sites and regulons for the chosen marine cyanobacterial genomes. DNA-binding motifs for NtcA and LexA were correctly identified. The relevance of transcriptional regulation of predicted *cis* elements was supported experimentally.

Keywords: phylogenetic footprinting; transcription factor binding sites; marine cyanobacteria.

1. Introduction

Photosynthetic bacteria such as *Prochlorococcus* and *Synechococcus* belong to the most important primary producers within the oceans. The genus *Prochlorococcus* is often present at high abundances with more than 10^5 cells per ml in nutrient-poor areas of the world's oceans splitting up in two major ecotypes – one being represented by the high-light-adapted (HL) strains such as Med4, the other by low-light-adapted (LL) strains SS120 and MIT9313 [24, 26, 36]. Nevertheless, on the basis of their ribosomal DNA similarity different ecotypes would be recognized as a single species as their rDNA sequences differ by less than 3% [15]. At the molecular level only small pieces are known about the regulatory network of marine cyanobacteria and genome-wide studies about co-regulated genes (regulons) controlled by *trans*-acting transcription factors (TFs) and their *cis* encoded DNA-binding sites do not exist. Only a few putative TF binding sites have been analysed: one for the CRP-like regulator NtcA (TGT-N_{10}-ACA) [25, 30] known to mediate nitrogen control in cyanobacteria, and a motif of putative phosphate regulator PhoB (TTAACCTT-N_3-TTAACCAT) [29]. The existance of a LexA site was suggested but not shown

by [22]. Knowledge about further *cis* elements on DNA is still rare.
To get insights into the core network of regulatory elements of multiple related species, phylogenetic footprinting is the major method. Thereby candidate regulatory elements are found by searching for conserved motifs upstream of orthologous genes from closely related species. Sequence similarity is the foundation for this computational method assuming that mutations within functional regions of genes accumulate slower than mutations in regions without sequence-specific function [35]. The phylogenetic footprinting algorithms can be divided in three parts: defining orthologous gene sequences for comparison; aligning the promoter sequences of orthologous genes; identifying segments of significant conservation. The great power of phylogenetic footprinting algorithms has been demonstrated for organisms of all kingdoms of life as the prediction of transcription regulatory sites in diverse bacterial families [23, 37], yeast [5], mouse and human [13, 20]. Reviews of methods and available resources are given by numerous articles [4, 6, 10, 31, 35]. Thereby, the initial and maybe the most difficult decision is choosing a set of genomes with the appropriate evolutionary distance of the sequences. The genomes of nine highly related but likely differentially adapted marine *Prochlorococcus* strains may represent the right distance to obtain meaningful predictions of *cis* elements.
Thus, we analyzed nine marine *Prochlorococcus* genomes and we predicted a conserved transcriptional regulatory network. For the first time, a minimal conserved core set of transcription factors, their binding sites and regulons can be suggested for the smallest known photosynthetic organism. DNA-binding motifs for NtcA and LexA were identified and several new regulatory motifs were predicted. A weak singnal corresponding to a third known motif ArsR has been observed. The importance of transcriptional regulation of two predicted *cis* elements NtcA and LexA was supported by experimental results of transcription initiation sites.

2. Materials and methods

2.1. *Computational part*

We performed a systematic intergenomic comparison to detect similar transcription factor binding sites conserved in upstream regions of orthologous genes. A pipeline implemented with BioMinerva framework [18] was used to integrate genome data and third party software tools in order to identify orthologous genes, align their promoter sequences and later to compare the alignments and interpret their similarity as a signal of regulation by a transcription factor.
Nine genomes of *Prochlorococcus* sp. were downloaded from NCBI GenBank server[a]. Tab. 1 summarizes the properties of retrieved sequences and their annotations. In order to build gene families we extracted the gene protein sequences from all studied species. Those sequences were next aligned against themselves using the BLASTp [1] algorithm run with the default parameters. The outcoming alignments can be

[a] `ftp://ftp.ncbi.nih.gov/genbank/genomes/Bacteria/`, version of February 2007

Table 1. Overview of size, annotated number of protein-coding genes (CDS), GC content, light optima (HL–high light; LL–low light) and numbers of annotated σ and transcription factors (TFs) of nine studied *Prochlorococcus* genomes.

name	accession	size	CDS	GC%	adaptation	σ	TFs
AS9601	CP000551	1669886	1921	31.32	HL	5	17
CCMP1375	AE017126	1751080	1882	36.44	LL	5	20
Med4	BX548174	1657990	1716	30.80	HL	5	23
MIT9313	BX548175	2410873	2273	50.74	LL	8	29
MIT9303	CP000554	2682675	2997	50.01	LL	8	30
MIT9312	CP000111	1709204	1809	31.21	HL	4	19
MIT9515	CP000552	1704176	1906	30.79	HL	5	19
NATL1A	CP000553	1864731	2193	34.98	LL	5	18
NATL2A	CP000095	1842899	1890	35.12	LL	5	19

understood as a graph representing evolutionary similarities between the studied genes. This graph was then processed by Markov Cluster Algorithm (MCL) [9] (an algorithm for unsupervised graph clustering based on simulation of stochastic flow in graphs) leading to a list of gene families.

We assume that genes belonging to one family are probably regulated in similar manner although they belong to different organisms. This assumption we interpret as a high chance to detect similar regulatory binding sites of the same transcription factor in corresponding promoter regions. Therefore, we predict transcription factor binding sites for each family separately in the following way. For all genes of a family we prepare their upstream DNA sequences (till the next gene or up to 300 nt of length). If an upstream region is shorter than 50 nt we assume that the gene belongs to a larger operon and we exclude it from further analysis. Afterwards, the set of upstream regions is processed by GLAM [11] – a method calculating the best possible gapless local alignment of multiple sequences with automatic determination of alignment width. If a good alignment is found it can represent a set of sites bound by a transcription factor. Since typical regulatory sites are not long we limit the maximum alignment length to 20.

As a result we obtain multiple alignments for each gene family. Next, each such alignment is converted into a position specific count matrix (PSCM) understood as a profile recognized by a potential regulating transcription factor. Since a typical transcription factor should regulate genes of more than one family we search for similar profiles calculated for different gene families. This step is carried out by a PSCMs comparison method (wmCompare [19] which bases on correlation between locations of binding sites predicted for a pair of PSCMs). The outcome of the method is an ordered list containing all matrices pairs and their similarities.

We take the most similar pairs above a chosen threshold of matrices similarity and estimate the biological significance of the choice. For this purpose we take the original matrices and shuffle their contents. In this process each position specific count matrix is converted into another one with randomly reordered positions (but still with the same size, quality, information content and GC-content). In general,

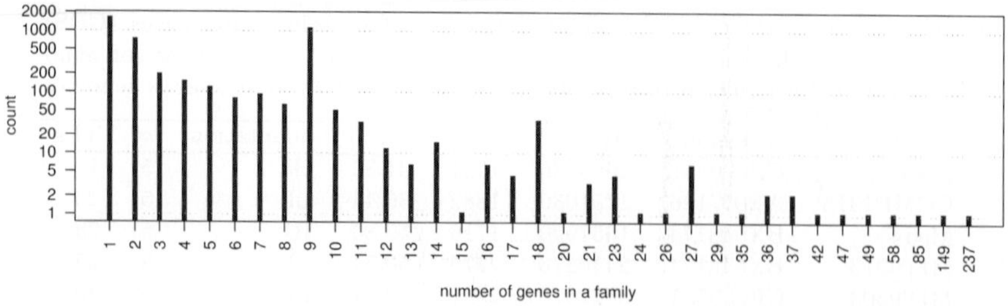

Fig. 1. 18573 coding genes have been clustered into 4072 families. For each observed family size a number of such families is shown. Single-gene families contain novel genes. Since nine genomes are analysed a peak for families containing a single gene for each genome is observed.

similar matrices are transformed into ones with lower similarity measure. Next, the shuffled matrices are compared using the same wmCompare algorithm. We repeat the shuffling 100 times and at the end we calculate the average number of pairs detected above the chosen threshold of matrices similarity. This average is a measure of the number of falsely discovered pairs of similar matrices in the original set.

We assume, that a typical transcription factor regulates more than a single gene. Therefore, we apply once again the MCL clustering algorithm to the graph constructed from the top pairs of similar matrices. This way we obtain groups of similar matrices which after alignment give us the profiles predicted to be recognized by a transcription factor. Finally, we perform a genome wide prediction of transcription factor binding sites using the obtained profiles. We use the approach proposed in [27] with parameters giving with probability 0.05 a single false positive binding site prediction for a sequence of length 500 nt.

2.2. *Experimental part*

Prochlorococcus sp. Med4 was grown in artificial seawater medium described previously [28] with a trace metal mix derived from medium Pro99 (Chisholm, personal communication). This modification resulted in the following final concentrations: 1.17 mM EDTA; 0.008 mM $ZnCl_2$; 0.005 mM $CoCl_2$; 0.09 mM $MnCl_2$; 0.003 mM Na_2MoO_4; 0.01 mM Na_2SeO_3; 0.01 mM $NiCl_2$; 1.17 mM $FeCl_3$. Cultures were kept under 10 μmol of photons $\cdot$ m^{-2} s^{-1} continuous blue light at 19 ± 1°C and harvested by centrifugation at 10 200 g for 10 min in a Dupont RC5C centrifuge.

Total RNA was isolated as previously described [12]. Transcriptional initiation sites were determined by 5'-RACE following the method of [3] with modifications outlined in detail in [33].

3. Results

3.1. *Computational analysis*

Starting point for a phylogenetic footprinting analysis is the definition of the set of orthologous protein-coding genes between the genomes of interest. From all nine

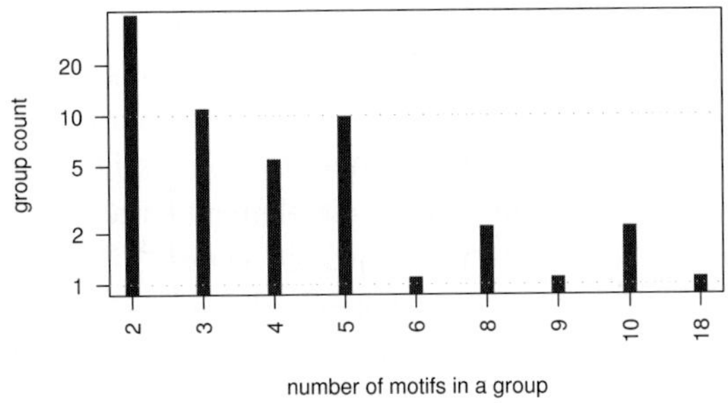

Fig. 2. Distribution of group sizes when 200 pairs of similar motifs were clustered.

genomes we could extract 18573 coding genes and after clustering we obtained 4072 orthologous gene families. Fig. 1 shows the distribution of gene families for different cluster sizes. We decided to study further approx. 35.1% of the clusters which contained at least six orthologs, to minimize problems resulting of small sample size when outcoming PSCMs are compared to each other.

For each gene family its set of gene upstream sequences was extracted and the best conserved motif in them was predicted. Then, all the obtained motifs were compared to each other. The outcoming list of motifs similarities was empirically limited to the top 200 pairs of motifs. Shuffling of motifs allowed us to estimate the average number of falsely discovered similar pairs in the top set of pairs to 132. Clusters of similar motifs were observed, suggesting existence of binding sites for *trans*-acting factors which control more than a single gene. Fig. 2 shows the distribution of number of motifs clustered into groups of high similarity. All 21 obtained clusters

Table 2. Predicted motifs and regulons identified in *Prochlorococcus* Med4 compared to all cyanobacteria motifs we could identify in literature. The ArsR motif can be assigned only manually since the most similar group contains less than four matrices.

name	known consensus	group size	predicted sequence logo	regulon
NtcA	TGT-N_{10}-ACA	10	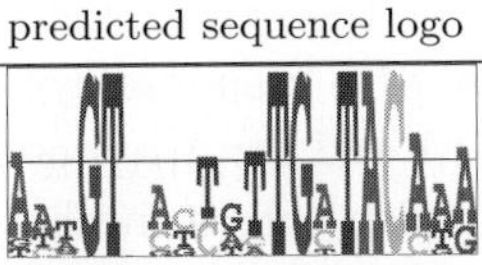	*ntcA; spt,agt; glnA; glnB; urtA*
LexA	TAGTACA-N_2-TGTACTA	6	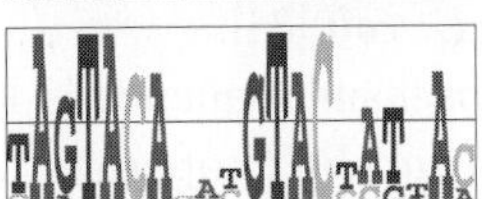	*recA*; *umuD*; *umuC*; *lexA*;
ArsR	ATCAA-N_6-TTGAT	2	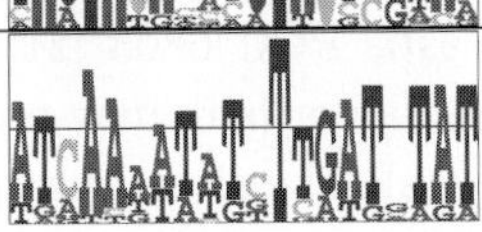	*gap*; *arsR*; *pstS*; *phoB*; *phoR*

of elements having at least four motifs were merged and in Tab. 2 we list motifs similar to all previously known from literature. The received 21 clusters of similar motifs corresponded well with the number of expected biological motifs which was estimated from the number of annotated DNA-binding proteins within marine genomes (Tab. 1). Depending on the genome we observed 4 to 8 σ factors and 17 to 30 transcription factors which can be assumed to possess DNA-binding properties. Thus, a biological meaningful number of *cis* elements was suggested between 20 and 40 compared to the information of encoded genes. Finally, these motifs were used to search for candidate regulatory elements in upstream regions of all studied genomes. The computational analysis lasted seven hours on a single-CPU typical desktop computer.

The results of the genome-wide search were analyzed in detail by assigning the downstream genes to known pathways or regulons. Towards this goal, the genome annotation as well as KEGG[b] database and an intensive literature search were informative. This final evaluation revealed three motifs with analogy to already known sites for certain cyanobacteria: NtcA, LexA and ArsR, described in detail below. Moreover, we observed several of the predicted regulons belonging to riboswitch motifs (for example THI) or other non-protein binding elements [2], which were excluded from further investigations.

NtcA is a major regulator for nitrogen control in cyanobacterial cells [16]. Those parts of the genome, which are repressed or activated by its presence, constitute the N-regulon. Here, only a small but high-scoring subset of this putative regulon was defined, including genes for major enzymes of nitrogen-metabolizing pathways such as *spt,agt* (aminotransferase) and *glnA* (glutamine synthetase) as well as important nitrogen dependent transport systems like *urtABCDE* (urea transporter). The consensus sequence, identified here, harbors additional features besides the often used TGT-N_{10}-ACA motif: The flanking A/T-rich sequences and a conserved TG (or CA) dimer [32]. Thus, our more complex motif of marine cyanobacteria corresponds partly to the profile GTA-N_8-TAC suggested recently [30].

The putative **LexA** site found for marine cyanobacteria is highly similar to the previously described consensus sequences of gram-positive and freshwater cyanobacteria [22]. Furthermore, the LexA regulon predicted here contains several genes known to be active in the SOS response system such as *umuC* and *umuD* and especially *recA* and *lexA*. RecA and LexA represent the positive and negative regulator respectively, which might indicate a mechanism surprisingly similar to the SOS system best known from *E. coli* [34].

An **ArsR**-like consensus sequence is located within the spacer region of *arsR* and *gap*. However, the *arsBHC* operon that is involved in arsenic sensing and resistance in *Synechocystis* PCC 6803 [21] was not found within the marine genomes. Thus, the ArsR-like factor here may participate in the regulation of other genes and operons. Indeed, ArsR-like sites were predicted upstream of genes like *pstS*, *phoB* (two-component response regulator, phosphate) and *phoR*, thought to be regulated by the amount of phosphate in the cell. As there is also a regulator for phosphate,

[b] `http://www.genome.jp/kegg/`

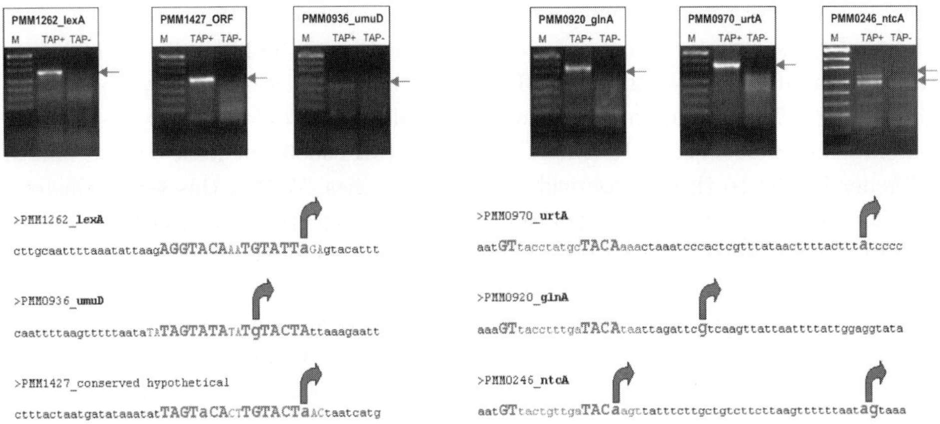

Fig. 3. Results of the PCR step during 5' RACE experiments for *lexA*, *umuD*, PMM1427 (left panel) and *urtA, glnA, ntcA* (right panel) in *Prochlorococcus* Med4. For each gene one single TIS appears except for *ntcA*, which exhibits two signals in the TAP-treated (TAP+) line. Overlay of the putative LexA recognition sequence (upper case letters) and the determined TIS (indicated by an arrow) for *lexA*, *umuD* and PMM1427 (left panel). Overlay of the predicted NtcA binding site and the mapped TIS upstream of *urtA*, *glnA*, *ntcA* (right panel).

PhoB, encoded in the marine genomes studied here, a crosstalk between both regulons might be assumed with the exception of AS9601 and MIT9515, where the ArsR-like regulator, encoded by the gene *arsR*, is missing.

3.2. *Experimental verification*

5' RACE experiments in *Prochlorococcus* Med4 were used to locate the transcription initiation site (TIS) of genes, for which putative DNA-binding sites had been predicted via our phylogenetic footprinting pipeline. For two of the best scoring motifs, NtcA and LexA, three genes were chosen, respectively. The TIS of *urtA*, *glnA*, *ntcA* as well as for *lexA*, *umuD* and PMM1427, a conserved hypothetical ORF, were mapped close to the predicted DNA-binding sites by RACE experiments. The results are shown in Fig. 3. All three predicted LexA motifs overlayed with the experimentally identified TIS as it might be assumed for LexA protein function as a repressor in bacterial gene transcription. The predicted motifs of NtcA showed different distances to the verified TIS: overlay, -10 as well as -35 distance was observed which can be easily explained by the dual function of NtcA as a repressor or activator for transcription.

4. Discussion

Phylogenetic footprinting was successfully applied to a set of sequenced marine genomes to reveal functionally relevant conservations between promoter regions of likely co-regulated genes. Thus, new information was obtained about the fundamentals of transcriptional regulation for marine cyanobacteria. In a first step, a set of orthologous coding regions was calculated resulting in 1428 families, which represents a number similar to other BLASTp comparisons of marine cyanobacterial genomes [8, 17]. Keeping in mind, that the total number of coding regions in these genomes varies between 1716 (Med4) and 2997 (MIT9303), at least around half of all genes belong to these conserved core gene families. Within this set of clusters, only 5 annotated sigma and 17 transcriptional factors were found, which likely constitute the core set of transcriptional regulatory proteins conserved between these nine marine cyanobacteria. Analyzing the orthologus upstream regions of family genes 21 motifs were detected above a chosen threshold which corresponded perfectly to an expected number of 20 to 40 DNA-binding sites. Motifs similar to previously described consensus sequences of the regulators NtcA and LexA known from freshwater cyanobacteria were identified as well as new regulatory motifs were predicted. Detailed analysis of six chosen promoters revealed that the predicted binding sites of LexA and NtcA belong to the experimentally defined promoter regions. Moreover, LexA is located exactly at the transcription initiation site for the studied genes including the *lexA* gene itself. Thus, LexA might be negatively autoregulated and could act as the repressor for several other genes. Although today, there are different functions for LexA discussed in literature [7, 14, 22] and studies about *Synechocystis* [7, 14] raised the question if all cyanobacteria possess an *E. coli*-type SOS regulon, the data obtained during this study of marine cyanobacteria give evidence for a DNA repair system surprisingly similar to the *E. coli* model.

Thus, a core set of regulons for the smallest known phototrophs is suggested here for the first time. The comparison of nine related genomes gives new insights into the minimum network of transcriptional regulation for strains within the marine ecosystem, but it does also allow drawing conclusions for cyanobacteria in general: Two known regulators, NtcA and LexA, appeared to be conserved over a wider evolutionary distance from freshwater to the group of marine cyanobacteria – from the most primitive unicellular to the filamentously growing complex species. The identification of NtcA and LexA in marine cyanobacteria illustrates how the data set might be utilized for an identification of promoters and regulatory sequences in other cyanobacterial species. In contrast, other factors like the one recognizing the ArsR-like binding site, might have evolved differentially and probably possess new functions and regulons adapted to the marine environment. Further experiments and comparisons with high throughput gene expression data will improve this initial regulatory network. Moreover, one has to remark that the computational predictions of DNA binding sites made here together with the experimentally tested examples can not serve as the entire proof of their biological function. For this purpose, additional binding studies, e.g. DNA affinity precipitation, DNase I protection or mobility shift assays, as well as detailed mutational analyses of the appropriate promoter regions might follow in the next future. Nevertheless, our global analysis represented here, could be a starting point to understand how these tiny and even so specialized organisms could dominate the oceans for millions of years although environmental conditions were and are changing.

References

[1] Altschul, S.F., Gish, W., Miller, W., Myers, E.W., and Lipman, D.J., Basic local alignment search tool, *J. Mol. Biol.*, 215(3):403–410, 1990.

[2] Axmann, I.M., Kensche, P., Vogel, J., Kohl, S., Herzel, H., and Hess, WR., Identification of cyanobacterial non-coding RNAs by comparative genome analysis, *Genome Biol.*, 6(9):R73, 2005.

[3] Bensing, B.A., Meyer, B.J., and Dunny, G.M., Sensitive detection of bacterial transcription initiation sites and differentiation from RNA processing sites in the pheromone-induced plasmid transfer system of Enterococcus faecalis, *Proc. Natl. Acad. Sci. USA*, 93(15):7794–7799, 1996.

[4] Bulyk, M.L., Computational prediction of transcription-factor binding site locations, *Genome Biol.*, 5(1):201, 2003.

[5] Cliften, P.F., Hillier, L.W., Fulton, L., Graves, T., Miner, T., Gish, W.R., Waterston, R.H., and Johnston, M., Surveying Saccharomyces genomes to identify functional elements by comparative DNA sequence analysis, *Genome Res.*, 11(7):1175–1186, 2001.

[6] Dieterich, C., Grossmann, S., Tanzer, A., Ropcke, S., Arndt, P.F., Stadler, P.F., and Vingron, M., Comparative promoter region analysis powered by CORG, *BMC Genomics*, 6(1):24, 2005. Comparative Study.

[7] Domain, F., Houot, L., Chauvat, F., and Cassier-Chauvat, C., Function and regulation of the cyanobacterial genes lexA, recA and ruvB: LexA is critical to the survival of cells facing inorganic carbon starvation, *Mol. Microbiol*, 53(1):65–80, 2004.

[8] Dufresne, A., Garczarek, L., and Partensky, F., Accelerated evolution associated with genome reduction in a free-living prokaryote, *Genome Biol.*, 6(2):R14, 2005.

[9] Enright, A.J., Van Dongen, S., and Ouzounis, C.A., An efficient algorithm for large-scale detection of protein families, *Nucleic Acids Res.*, 30(7):1575–1584, 2002.

[10] Frazer, K.A., Elnitski, L., Church, D.M., Dubchak, I., and Hardison, R.C., Cross-species sequence comparisons: a review of methods and available resources, *Genome Res.*, 13(1):1–12, 2003.

[11] Frith, M.C., Hansen, U., Spouge, J.L., and Weng, Z., Finding functional sequence elements by multiple local alignment, *Nucleic Acids Res.*, 32(1):189–200, 2004.

[12] Garcia-Fernandez, J.M., Hess, W.R., Houmard, J., and Partensky, F., Expression of the psbA gene in the marine oxyphotobacteria Prochlorococcus spp, *Arch Biochem Biophys*, 359(1):17–23, 1998.

[13] Gottgens, B., Gilbert, J.G., Barton, L.M., Grafham, D., Rogers, J., Bentley, D.R., and Green, A.R., Long-range comparison of human and mouse SCL loci: localized regions of sensitivity to restriction endonucleases correspond precisely with peaks of conserved noncoding sequences, *Genome Res.*, 11(1):87–97, 2001.

[14] Gutekunst, K., Phunpruch, S., Schwarz, C., Schuchardt, S., Schulz-Friedrich, R., and Appel, J., LexA regulates the bidirectional hydrogenase in the cyanobacterium Synechocystis sp. PCC 6803 as a transcription activator, *Mol. Microbiol*, 58(3):810–823, 2005.

[15] Hagstrom, A., Pommier, T., Rohwer, F., Simu, K., Stolte, W., Svensson, D., and Zweifel, U.L., Use of 16S ribosomal DNA for delineation of marine bacterioplankton species, *Appl. Environ Microbiol*, 68(7):3628–3633, 2002.

[16] Herrero, A., Muro-Pastor, A.M., and Flores, E., Nitrogen control in cyanobacteria, *J. Bacteriol*, 183(2):411–425, 2001.

[17] Hess, W.R., Genome analysis of marine photosynthetic microbes and their global role, *Curr. Opin. Biotechnol*, 15(3):191–198, 2004.

[18] Kielbasa, S., The BioMinerva framework (in preparation), 2007.

[19] Kielbasa, S.M., Gonze, D., and Herzel, H., Measuring similarities between transcription factor binding sites, *BMC Bioinformatics*, 6:237, 2005.

[20] Loots, G.G., Locksley, R.M., Blankespoor, C.M., Wang, Z.E., Miller, W., Rubin, E.M., and Frazer, K.A., Identification of a coordinate regulator of interleukins 4, 13, and 5 by cross-species sequence comparisons, *Science*, 288(5463):136–140, 2000.

[21] Lopez-Maury, L., Florencio, F.J., and Reyes, J.C., Arsenic sensing and resistance system in the cyanobacterium Synechocystis sp. strain PCC 6803, *J. Bacteriol*, 185(18):5363–5371, 2003.

[22] Mazon, G., Lucena, J.M., Campoy, S., Fernandez de Henestrosa, A.R., Candau, P., and Barbe, J., LexA-binding sequences in Gram-positive and cyanobacteria are closely related, *Mol. Genet. Genomics*, 271(1):40–49, 2004.

[23] McGuire, A.M., Hughes, J.D., and Church, G.M., Conservation of DNA regulatory motifs and discovery of new motifs in microbial genomes, *Genome Res.*, 10(6):744–757, 2000.

[24] Moore, L.R., Rocap, G., and Chisholm, S.W., Physiology and molecular phylogeny of coexisting Prochlorococcus ecotypes, *Nature*, 393(6684):464–467, 1998.

[25] Palinska, K.A., Jahns, T., Rippka, R.,and Tandeau De Marsac, N., Prochlorococcus marinus strain PCC 9511, a picoplanktonic cyanobacterium, synthesizes the smallest urease, *Microbiology*, 146 Pt 12:3099–3107, 2000.

[26] Partensky, F., Hess, W.R., and Vaulot, D., Prochlorococcus, a marine photosynthetic prokaryote of global significance, *Microbiol Mol. Biol. Rev.*, 63(1):106–127, 1999.

[27] Rahmann, S., Muller, T., and Vingron, M., On the power of profiles for transcription factor binding site detection, *Stat. Appl. Genet. Mol. Biol.*, 2:Article7, 2003.

[28] Rippka, R., Coursin, T., Hess, W., Lichtle, C., Scanlan, D.J., Palinska, K.A., Iteman, I., Partensky, F., Houmard, J., and Herdman, M., Prochlorococcus marinus Chisholm et al. 1992 subsp. pastoris subsp. nov. strain PCC 9511, the first axenic chlorophyll a2/b2-containing cyanobacterium (Oxyphotobacteria), *Int. J. Syst. Evol. Microbiol*, 50 Pt 5:1833–1847, 2000.

[29] Su, Z., Dam, P., Chen, X., Olman, V., Jiang, T., Palenik, B., and Xu, Y., Computational inference of regulatory pathways in microbes: an application to phosphorus assimilation pathways in Synechococcus sp. WH8102, *Genome Inform.*, 14:3–13, 2003.

[30] Su, Z., Olman, V., Mao, F., and Xu, Y., Comparative genomics analysis of NtcA regulons in cyanobacteria: regulation of nitrogen assimilation and its coupling to photosynthesis, *Nucleic Acids Res.*, 33(16):5156–5171, 2005.

[31] Ureta-Vidal, A., Ettwiller, L., and Birney, E., Comparative genomics: genome-wide analysis in metazoan eukaryotes, *Nat. Rev. Genet.*, 4(4):251–262, 2003.

[32] Vazquez-Bermudez, M.F., Flores, E., and Herrero, A., Analysis of binding sites for the nitrogen-control transcription factor NtcA in the promoters of Synechococcus nitrogen-regulated genes, *Biochim. Biophys. Acta*, 1578(1-3):95–98, 2002.

[33] Vogel, J., Axmann, I.M., Herzel, H., and Hess, W.R., Experimental and computational analysis of transcriptional start sites in the cyanobacterium Prochlorococcus MED4, *Nucleic Acids Res.*, 31(11):2890–2899, 2003.

[34] Walker, G.C., Mutagenesis and inducible responses to deoxyribonucleic acid damage in Escherichia coli, *Microbiol Rev.*, 48(1):60–93, 1984.

[35] Wasserman, W.W. and Sandelin, A., Applied bioinformatics for the identification of regulatory elements, *Nat. Rev. Genet.*, 5(4):276–287, 2004.

[36] West, N.J. and Scanlan, D.J., Niche-partitioning of Prochlorococcus populations in a

stratified water column in the eastern North Atlantic Ocean, *Appl. Environ Microbiol*, 65(6):2585–2591, 1999.

[37] Yan, B., Methe, B.A., Lovley, D.R., and Krushkal, J., Computational prediction of conserved operons and phylogenetic footprinting of transcription regulatory elements in the metal-reducing bacterial family Geobacteraceae, *J. Theor. Biol.*, 230(1):133–144, 2004.

EVOLUTIONARY CHANGES IN GENE REGULATION FROM A COMPARATIVE ANALYSIS OF MULTIPLE *DROSOPHILA* SPECIES

LAN HU[1] DANIEL SEGRÈ[1] TEMPLE F. SMITH[1,2]
hulan@bu.edu dsegre@bu.edu tsmith@darwin.bu.edu

[1]*Graduate Program in Bioinformatics, Boston University, Boston MA 02215, U.S.A.*
[2]*BioMolecular Engineering Research Center, Boston University, Boston MA 02215, U.S.A.*

Exploiting the ortholog/homolog information now available from the complete genomic sequences of twelve species of *Drosophila*, we have investigated the ability of regulatory site recognition methods to find regulatory changes for orthologs linked to chromosomal rearrangements. This has made use of the wealth of synteny information among these species. By comparing orthologs in multiple species, we found that the breakpoint of chromosomal rearrangements could have had an impact on regulatory changes of genes next to it with respect to the gene function and location. Extensions of our approach could be used to shed light on the role of gene regulation in the evolutionary adaptation to different environmental conditions.

Keywords: *Drosophila*; regulatory site; ortholog; chromosomal rearrangement.

1. Introduction

The genomes from twelve species of *Drosophila* (fruit flies) [13] have been recently sequenced, providing a wealth of data for comparative genomics analyses, and in particular for the study of how evolution may have fine-tuned the regulation of specific genes and pathways associated with different lifestyles of these species. These data include species from the two well recognized subgenera of *Drosophila*, *Sophophora* subgenus and *Drosophila* subgenus, that diverged between 40 and 60 million years ago.

While chromosomal rearrangements in *Drosophila* are common, the majority are inversions, which maintain the involved genes within the same Muller Element (ME, a chromosome arm in *D.melanogaster*). Thus few genes appear to have moved between MEs [6]. In a few cases, putative gene homologs are found to change ME, apparently via retrotransposition (i.e. mRNA is retrotranscribed to DNA and reinserted into genome at a new position). In the case of inversion, the regulatory signals may have "traveled" along with the genes, since the range of an inversion is usually large. In the case of retrotransposition, however, the genes normally would not carry along the original regulatory signals. An intriguing question is how different models of DNA rearrangement could have affected the regulatory program of a gene, especially when the upstream region of a gene has been disrupted or left behind.

Upstream small-scale deletions, insertions and point mutations are not the focus of this work. It is not that such events do not play a key role in determining gene regulation and thus expression – as a function of extent, location and/or timing – but these have been well studied, at least in micro organisms, and are generally assumed to occur more gradually. Rather, we concentrate on the less understood implications of sudden, drastic changes. Potential genetic regulation changes are particularly acute in retrotransposition, because the original regulatory region, lost during a gene transition, is unlikely to be replaced by a compatible and useful transcriptional signal. On the other hand, the fact that these genes survived indicates that, whatever change occurred, it was evolutionarily advantageous, or at least neutral. One possibility is that the moved gene has been fortuitously inserted next to a useful set of regulatory elements, however unlikely that is. Another possibility is that the gene has been inserted in an exon of another gene, regulated in a similar manner. A third possibility is that the gene is placed in a chromosomal region globally maintained at a high transcription level. In such case, given enough time, a minimal upstream region for more specialized regulation could gradually evolve. In general, if a gene is essential, this would require that a second functional copy exists. This could be realized in diploid organisms, in addition to occurring for retrotransposed genes.

Since the fruit fly is a well established model organism, its genetics and development are well studied. The availability of twelve genomes furthermore is expected to provide new comparative genomics insight on the regulation of genes that moved in different ways. In this paper, a method to characterize and compare the potential regulatory sites (PRS) of orthologs across all available species is developed. The method is applied to the central carbon metabolic genes, particularly to those genes that have disrupted upstream region by chromosomal rearrangements along the evolution. The results indicate that comparing common PRS across available species with full synteny breakpoint analysis could help to gain insights of how the breakpoint could affect the regulation of moved genes.

2. Methods

2.1. Synteny analysis

In this study, we are particularly interested in genes that "moved" at the first diverging point in the evolution of *Drosophila*, i.e. about 40~60 million years ago. The expectation would be that these orthologs keep the same neighbor context in one subgenus but not the other. The synteny analysis [5, 6] carried out hence is based on gene neighborhood comparison relative to *D.melanogaster*.

The synteny analysis is schematically illustrated in Fig. 1. Given a gene, A_mel in Fig. 1, from *D. melanogaster*, its adjacent neighbors (X_mel and Y_mel) are extracted, as well as its ortholog and neighbors (if they exist according to annotation) in another *Drosophila* species. Next, for one of the neighbors of the ortholog, N_s in Fig. 1, its ortholog and neighbors are extracted back from *D.melanogaster*. There are two possibilities, as shown in Fig. 1: (1) N_s and Y_mel are orthologs; and (2) N_s and N_mel are orthologs. The first case means that for the neighbor pair of A_mel and Y_mel in *D.melanogaster*, their orthologs (A_s and N_s respectively) are also neighbors in another species. In the second case, however, the neighbor pair relationship is not consistent, suggesting that there was a breakpoint either between A_mel and Y_mel or between A_s and N_s.

This type of synteny analysis is carried out across seven *Drosophila* species, four from the *Sophophora* subgenus (*D.melanogaster*, *D.yakuba*, *D.erecta*, and *D.ananassae*, diverged about 10~15 million years ago) and three from the *Drosophila* subgenus (*D.mojavensis*, *D.virilis*, and *D.grimshawi*, diverged about 30~35 million years ago). The genes that have possibly "moved" at the first speciation event would keep the same neighbor context in one subgenus but not the other.

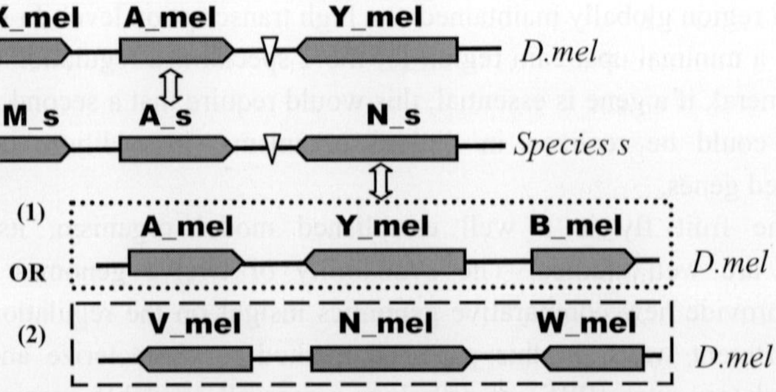

Fig. 1. Schematic illustration of synteny analysis based on gene neighbor context
Double headed arrows indicate that those two genes are orthologs. Pairs A_mel and A_s, N_s and Y_mel, N_s and N_mel are orthologs. Three genes in the same row mean that they are adjacent neighbors. Dotted line box and dashed line box show two possibilities of N's ortholog in *D.mel*: (1) N_s and Y_mel are orthologs. (2) N_s and N_mel are orthologs, showing that there is a breakpoint either between A_mel and Y_mel in *D.mel* or between G and N in species *s* (see main text). White triangles represent the possible breakpoints. *D.mel* = *D.melanogaster*. Species *s* = any other *Drosophila* species. Names with _mel suffix are genes from *D.mel*, _s suffix from Species *s*.

2.2. *Regulatory site identification*

Many different approaches for potential regulatory sites (motifs) identification have been developed. In general, motif finding falls into two categories: pattern matching to previously identified sites or *de novo* discovery. Pattern matching algorithms (e.g. MotifScanner [10] and Patser [11]) use identified patterns such as position weighted matrices (PWM) or position frequency matrices (PFM) to scan through the sequences and return the segments that have scores over some threshold. The *de novo* discovery approaches use techniques such as Gibbs Sampling (AlignACE [8]) or Expectation Maximization (MEME [3]) to detect the over-represented DNA segments in given sequences. Pattern matching approach largely depends on the patterns which ideally should come from experimentally determined sites. In fruit fly, unfortunately, the number of transcription factors whose binding sites have been characterized is still limited.

The *Drosophila* DNase I Footprint Database (FlyReg 2.0 [4]) has a collection of 1,365 DNase I footprints for *D. melanogaster* from a single experimental data type.

These data have been extracted from 201 primary references and provide a non-redundant set of high quality binding site information for 87 transcription factors. 62 motif models have been curated in the format of PWM's, and 75 in the format of PFM's.

The overall similarity of extracted upstream DNA sequences for orthologs decreases with divergence as expected. To identify the potential regulatory signals, the curated 75 alignment matrices from FlyReg 2.0 are used to scan the upstream regions (both strands) of given genes. Sites that have above threshold scores are returned as putative regulatory sites. Due to the repetitive sequence in a given DNA segment as well as the incompleteness of transcription factor binding sites, overlapping and repetitive sites could be returned by the process. To avoid that, the site with highest in the region is kept and others are discarded.

In order to compare the regulatory sites, three kinds of gene sets are constructed. They are a random gene set, a *Sophophora* ortholog set, and a *Drosophila* ortholog set. The random set(s) are generated by randomly choosing 100 genes with at least 2 kb upstream intergenic region from different species independently (which species to choose are based on individual analysis). The reason for having at least 2kb upstream intergenic region is that current annotation of gene span in species other than *D.melanogaster* does not have transcription start site but translation start site estimation, which may introduce non-transcriptional regulation information in regulatory site finding.

The other two sets are ortholog sets. One hundred genes with at least 2kb upstream intergenic region are chosen (with or without functional constraints) from *D.melanogaster* first. For those 100 genes, orthologs from four species in *Sophophora* subgenus (*D.melanogaster*, *D.yakuba*, *D.erecta*, and *D.ananassae*) constitute the *Sophophora* ortholog set; orthologs from three species in *Drosophila* subgenus (*D.mojavensis*, *D.virilis*, and *D.grimshawi*) constitute the *Drosophila* ortholog set.

In each gene set, after site scanning using Patser with p-value 10^{-3} and tiling, common potential regulatory sites (PRS) are obtained for every quadruplet (the *Sophophora* ortholog set) or triplet (the *Drosophila* ortholog set). Common PRSs are more than the intersection of PRSs from given sequences. If a PRS is detected for N ($N>1$) times in given sequences, it would be counted N times in the final common PRSs. The distribution of the common PRS then is analyzed in the relationship to moved genes.

3. Results

3.1. Functionally independent genes

Using our synteny analysis (see Methods), we identified about 1050 genes likely to have "moved" at the first speciation event relative to *D. melanogaster*. We next set to compare the upstream regulatory regions of such genes, to shed light on the potential implications of such rearrangements on transcriptional patterns.

Our first test was aimed at studying upstream region changes among functionally independent genes. For this purpose, we chose 100 *D. melanogaster* genes without functional constraints to construct two ortholog sets and applied our regulatory site identification algorithms (See Methods). For each ortholog set, we constructed a corresponding random set from the same species, to use as a baseline for comparison.

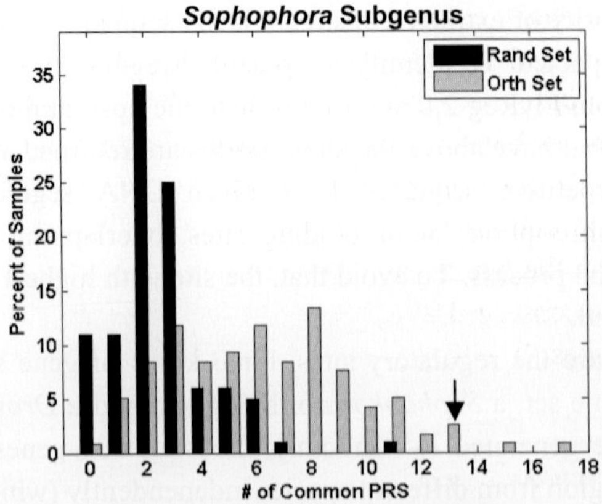

(a)

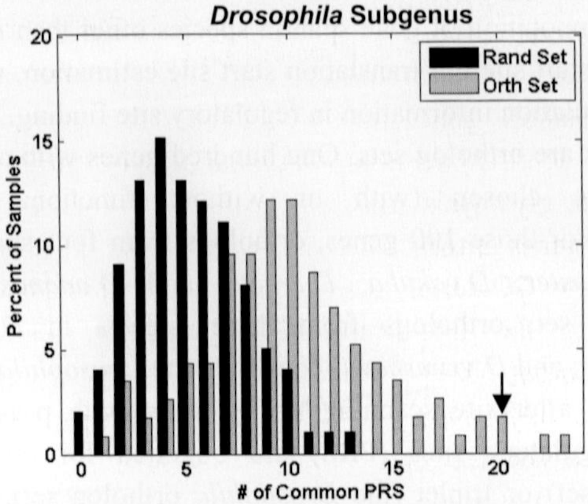

(b)

Fig. 2. Distribution of number of common PRS for functionally independent genes
(a) Between the random gene set and the *Sophophora* ortholog set. The random gene set is composed of randomly chosen genes from *D.melanogaster*, *D.yakuba*, *D.erecta*, and *D.ananassae*. (b) Between the random gene set and the *Drosophila* ortholog set. The random gene set is composed of randomly chosen genes from *D.mojavensis*, *D.virilis*, and *D.grimshawi*. In both figures, the black arrow points to where orthologs of *kek1* are binned.

As illustrated in Fig. 2, two ortholog sets have different distribution of common PRS relative to the random gene set, as expected. The orthologs, even functional independent in this case, share more common sites. The next question is whether there is intersection among the genes that have high number of common PRS in both ortholog sets. If there are genes whose orthologs share high number of common PRS in both ortholog sets, this would suggest that the regulation of those genes may not have changed throughout evolution to keep the functions of genes under the selection pressure, and some of the common PRS could be real regulatory sites.

In both ortholog sets, we singled out the top 10 genes whose orthologs share most common PRS. The resulting intersection contains CG14220, CG6621, and *kekl* (CG12283, denoted by the black arrow in Fig. 2). The synteny analysis shows that the orthologs of *kekl* in *D.mojavensis*, *D.virilis*, and *D.grimshawi* have been moved. The gene *kekl* has negative regulation of epidermal growth factor receptor activity [1, 2] and is also involved in *Drosophila* oogenesis [7]. Among all the sites detected, there are three sites that are common in both ortholog sets, two of which are binding sites of the transcription factor (TF) *apterous* and one is the binding site of TF *broad*. The TF *apterous* is involved in cell fate commitment and *broad* in cell death and oogenesis, which is consistent with the functions of *kekl*. Hence, the important functions of *kekl* in development should be conserved throughout evolution regardless of the movement of its orthologs in some species. It is likely that those three sites could be real regulatory sites (see Discussion).

3.2. Central carbon metabolic genes

We next sharpened our analysis by testing the extent of upstream region overlap for genes with known and conserved functions. Given our interest in potential correlation with lifestyle and dietary changes among species, we focused on genes coding for metabolic proteins. Metabolic genes in *D.melanogaster* thus are chosen to construct the two ortholog sets. We identified a total of 104 genes involved in *D.melanogaster* central carbon metabolism, i.e. glycolysis, pentose phosphate pathway, and tricarboxylic acid (TCA) cycle. These genes either code for enzymes or have metabolic functional annotations according to GO terms in the three pathways considered (Supplementary material). Similarly to the first test, the orthologs of these 104 genes from *D.melanogaster*, *D.yakuba*, *D.erecta*, and *D.ananassae* constitute the *Sophophora* ortholog set; orthologs from *D.mojavensis*, *D.virilis*, and *D.grimshawi* constitute the *Drosophila* ortholog set. The random gene sets are the same as those in the first test.

As shown in Fig. 3, the distribution of common PRS in both ortholog sets demonstrates a trend similar to the one found in the comparison of genes that do not necessarily share function (Fig. 2). Again, we focused on the top 10 genes that have the most common PRS from both ortholog sets. The resulting intersection includes CG5261, CG5432, and *Hex-A*. Synteny analysis shows that their orthologs in other species keep the same neighbor context. High number of common PRS in these three genes suggests that functional constraint and same gene context could keep similar gene regulation.

In addition, there are eight genes (Table 1) which have upstream region disrupted by the chromosomal rearrangement breakpoints in *Drosophila* subgenus species (*D.mojavensis*, *D.virilis*, and *D.grimshawi*) relative to *D.melanogaster*. As illustrated in Table 1, these eight genes have low number of common PRS across both subgenera, yet have more common PRS's if they are compared in either subgenus, raising the possibility that those genes may have undergone different regulation mechanisms after the first speciation event. Using the corresponding random set as background, we quantified the significance of these findings through a Z-test (Supplementary material).

Pyruvate dehydrogenate kinase (*Pdk*) has significant p-values (<0.01) in both ortholog sets, suggesting that the importance of this enzyme has kept the gene under strong selection despite the movement of its orthologs and disruption of the upstream region. Low number of common PRS in all species and high number of common PRS in either subgenus could imply that the movement of this gene may have introduced new regulatory signal to keep, and possibly fine-tune the gene's function.

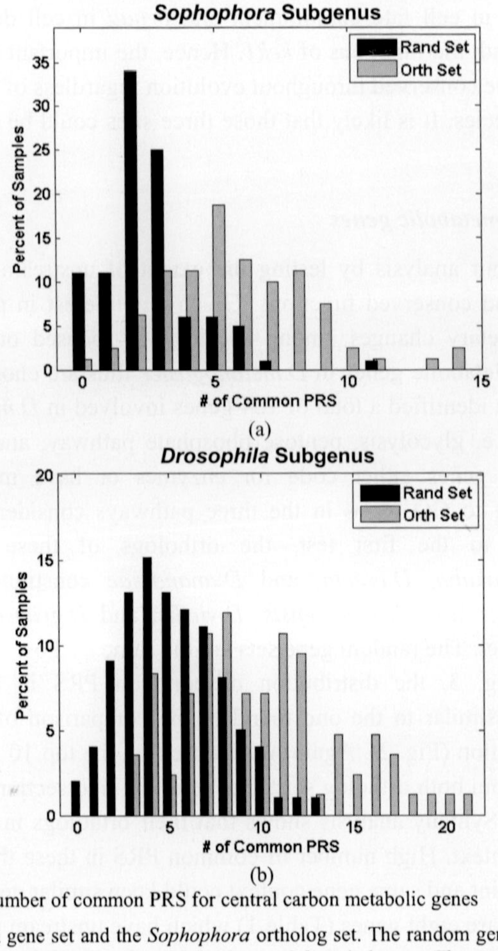

Fig. 3. Distribution of number of common PRS for central carbon metabolic genes

(a) Between the random gene set and the *Sophophora* ortholog set. The random gene set is the same as that in Fig. 2(a). (b) Between the random gene set and the *Drosophila* ortholog set. The random gene set is the same as that in Fig. 2(b).

Genes CG2964 and CG13369 have large p-values in both orthologs sets (>0.4 and >0.5 respectively), which indicates that the two genes cannot be discriminated from the random gene set. While this may potentially mean that the level of expression could have changed significantly after the rearrangement, additional evidence points to a different possible explanation. There has been evidence of expression similarity within large chromosomal domains in *D. melanogaster*: genes with unrelated function may have similar transcriptional levels merely due to their common chromosomal location [9]. The two genes under discussion, CG2964 and CG13369, turn out to fall into chromosomal domains that share similar expression profile (data not shown). This supports the possibility that "moved" genes could be fortuitously inserted next to useful regulatory elements, such as enhancers which could be far away from the gene itself. Therefore regardless of the small number of common PRS in front of these two genes, "remote" regulatory signal would still guarantee appropriate transcription level of the genes.

Table 1. Common PRS of genes with disrupted upstream region in central carbon metabolism. Numbers in parenthesis are p-values from Z-test.

Gene Name	# of Common PRS (p-value)			Molecular Function in Central Carbon Metabolism
	Sophophora ortholog set	*Drosophila* ortholog set	Both	
Pdk	8 (0.0022)	16 (0.0001)	4	pyruvate dehydrogenase kinase activity
CG2964	4 (0.4178)	7 (0.5170)	1	pyruvate kinase activity; carbohydrate kinase activity
CG13369	3 (0.8044)	7 (0.5170)	0	pentose-phosphate shunt
CG5362	9 (0.0003)	10 (0.0783)	3	L-lactate dehydrogenase activity; L-malate dehydrogenase activity
Mdh	8 (0.0022)	7 (0.5170)	1	NAD binding; malate dehydrogenase activity
CG9467	6 (0.0529)	8 (0.3083)	1	NAD binding; glyceraldehyde-3-phosphate dehydrogenase (phosphorylating) activity
CG7349	4 (0.4178)	12 (0.0123)	0	succinate dehydrogenase activity
CG6666	5 (0.1697)	8 (0.3083)	0	succinate dehydrogenase (ubiquinone) activity

Genes CG5362 and *Mdh* have significant p-value in one ortholog set but not the other (0.0003 and 0.0783 for CG5326, 0.0022 and 0.5170 for *Mdh*). This may suggest that the species from different subgenera have different metabolic requirements as a result of adaptation to different environments. The characterization of TF binding sites used in this particular measurement is solely based on *D.melanogaster*, which could introduce a bias towards species from *Sophophora* subgenus that are close to *D.melanogaster* and may not be sufficient to detect regulatory signal in more divergent species from the *Drosophila* subgenus.

4. Discussion

While our analysis shows potentially significant patterns that link genome rearrangements to regulatory adaptation of metabolic genes, one should keep in mind that the currently available set of characterized TF binding sites is rather limited, and mainly specific to *D.melanogaster* development. This inevitably introduces a bias in recognizing PRS in orthologs. The results, however, still demonstrate that comparing common PRS across available species with full synteny breakpoint analysis could help gain valuable information, for solving the puzzle of the regulation of "moved" genes.

The broad question of how DNA breakpoints affect the regulation of moved genes in *Drosophila* still remains largely unanswered. The method presented in this study takes the perspective of identifying common potential regulatory sites for sets of orthologs across different species, and attempts to elucidate how breakpoints within the upstream region of a gene could affect its regulation in light of its function.

In both tests we presented, the method shows that orthologs tend to have more common potential regulatory sites regardless of functional dependence, which is expected because the selection of function of orthologs may select the regulation as well.

In the test of functional independent gene sets (Fig. 2), the detection of gene *kek1* shows that this method could capture the common PRS in orthologs over ~50 million year span even when a gene has been "moved". This could be due to the constraint from conservation of a gene with important developmental function.

The analysis of eight genes and their orthologs that have upstream region disrupted by chromosomal rearrangement breakpoints in central carbon metabolism (Table 1) exemplifies the possible effect that the breakpoint could have on gene regulation. The importance of gene functions requires continuation of compatible regulation, despite nearby breakpoints. In the case of *Pdk*, it may also have evolved new regulatory signal to keep the gene function after the first speciation event. Some genes that "moved" without their own regulatory signal, such as CG2964 and CG13369, could have been fortuitously inserted into a highly expressed region, hence guaranteeing continuity of their functionality. For some other genes, such as CG5362 and *Mdh*, the difference in common PRS could be a sign of actual difference in gene regulation caused by the specific metabolic requirement due to adaptation to different environments. For example, *D.melanogaster* is a sympatric cosmopolitan species, feeding on necrotic fruit. *D.mojavensis* is a cactophilic species that is specifically found on the rotten arms of cacti in deserts.

Exploiting the ortholog/homolog information now available from the complete genomic sequences of twelve species of *Drosophila*, we have investigated the ability of regulatory site recognition method to find regulatory changes linked to chromosomal rearrangements, particularly the genes next to the breakpoints. We have shown that breakpoints could have multi-level effects on the regulatory changes of those genes in two subgenera of *Drosophila* with respect to the gene function and gene location. With the availability of whole genome expression data, it will give better understanding of how breakpoints change gene regulation along the evolution.

Acknowledgments

We thank AAA site [12] for genome assemblies. We also thank Douglas Smith for permission of using sequence data and Venky Iyer at Eisen Lab UC Berkeley for ortholog model. *D.erecta*, *D.ananassae*, *D.mojavensis*, *D.virilis* and *D.grimshawi* were sequenced by Agencourt Biosystems. *D.yakuba* was sequenced by Washington University. This work is sponsored by NSF grant DBI-0516000.

References

[1] Alvarado, D., Rice, A.H., and Duffy, J.B., Knockouts of *Kekkon1* define sequence elements essential for *Drosophila* epidermal growth factor receptor inhibition, *Genetics*, 166(1):201-211, 2004.

[2] Alvarado, D., Rice, A.H., and Duffy, J.B., Bipartite inhibition of *Drosophila* epidermal growth factor receptor by the extracellular and transmembrane domains of *Kekkon1*, *Genetics*, 167(1):187-202, 2004.

[3] Bailey, T.L. and Elkan, C., Fitting a mixture model by expectation maximization to discover motifs in biopolymers. *Proc. Int. Conf. Intell. Syst. Mol. Biol.*, 28-36, 1994.

[4] Bergman, C.M., Carlson, J.W., and Celniker, S.E., *Drosophila* DNase I footprint database: a systematic genome annotation of transcription factor binding sites in the fruitfly, Drosophila melanogaster, *Bioinformatics*, 21(8):1747-1749, 2005.

[5] Bhutkar A., Russo, S., Smith, T. F., and Gelbart, W.M., Techniques for Multi-Genome Synteny Analysis to Overcome Assembly Limitations, *Genome Inform.*, 17(2): 152-161, 2006.

[6] Drosophila Comparative Genome Sequencing and Analysis Consortium, Genomics on a phylogeny: Evolution of Genes and Genomes in the Genus Drosophila, Submitted

[7] Ghiglione, C., Carraway, K.L., Amundadottir, L.T., Boswell, R.E., Perrimon, N., and Duffy, J.B., The transmembrane molecule *kekkon1* acts in a feedback loop to negatively regulate the activity of the Drosophila EGF receptor during oogenesis, *Cell*, 96(6):847-856, 1999.

[8] Hughes, J.D., Estep, P.W., Tavazoie, S., and Church, G.M., Computational identification of cis-regulatory elements associated with groups of functionally related genes in *Saccharomyces cerevisiae*, *J. Mol. Biol.*, 296(5):1205-1214, 2000.

[9] Spellman, P.T. and Rubin, G.M., Evidence for large domains of similarly expressed genes in the *Drosophila* genome, *J. Biol.*, 1(1):5, 2002.

[10] Thijs, G., Marchal, K., Lescot, M., Rombauts, S., Moor, B.D., Rouzé, P., and Moreau, Y., A Gibbs sampling method to detect overrepresented motifs in the upstream regions of coexpressed genes, *J.Comput. Biol.*, 9(2):447-464, 2002.

[11] van Helden, J., Regulatory sequence analysis tools, *Nucleic Acids Res.*, 31(13):3593-3596, 2003.

[12] http://rana.lbl.gov/drosophila/

[13] http://www.flybase.org/

A STRUCTURAL GENOMICS APPROACH TO THE REGULATION OF APOPTOSIS: CHIMP VS. HUMAN

JESSICA AHMED
jessica.ahmed@charite.de

STEFAN GÜNTHER
stefan.guenther@charite.de

FRIEDRICH MÖLLER
friedrich.moeller@charite.de

ROBERT PREISSNER
robert.preissner@charite.de

Structural Bioinformatics Group, Institute of Molecular Biology and Bioinformatics, Charite-University Medicine Berlin, Arnimallee 22, 14195 Berlin, Germany

After the sequencing of the human genome, the publication of the genome of our nearest relative, the chimpanzee (Pan troglodytes) provided groundbreaking data improving the understanding of the recent human evolution. There are about forty million changes, most of them single nucleotide substitutions, which teach us about ourselves, both in terms of similarities and differences with chimpanzees. From a medical point of view differences in incidence and severity of diseases are of special importance to pinpoint novel targets and to develop innovative therapies. This analysis focuses on the cognition that chimpanzees rarely suffer from cancer. To elucidate possible reasons for this finding, we compare differences regarding apoptosis and DNA-repair on different levels of chromosome organization, gene structure, post-transcriptional and post-translational modifications to functional changes in protein structures. The result is a complex pattern of subtle variances and a few large-scale changes.

Keywords: chimpanzee; cancer; apoptosis; DNA-repair.

1. Introduction

Today, one of the most frequent causes of death is cancer. Among the malignant neoplasms, the most prominent types of cancer in Europe in 2006 are breast (30.9%), prostate (24.1%), lung (11.2%), colon, and rectum (13%) cancer [1]. Possible reasons for cancer are determined by environmental or intrinsic factors.

Nutritional differences and other ecological causes are responsible for disparity in cancer incidence, but the tenfold increase, as observed for carcinomas of breast, ovary, lung, stomach, colon and, rectum cancer, between chimpanzee and human [2] cannot be explained coherently by such arguments. This directs the focus on intrinsic factors like susceptibility, and tolerance, which depend on genetic factors. Genetic variability leads to a diversity of cancer incidence in different populations. The Japanese and non-Hispanic white men show the highest cancer risk, whereas the Alaska's men show the lowest one [3]. Especially, the prostate [4] and breast [5] types of cancer do rarely occur in the Eskimo population. However, the Indians and Aleuts show a significantly increased rate of prostate cancer compared to the native Alaska's men [5]. Well-known genetic abnormality responsible for different cancer incidence exhibits e.g. the breast cancer gene BRCA1, which possesses mutations in breast cancer patients. This protein acts as a tumor suppressor and is involved in processes of the DNA-repair. In most instances a DNA-damage leads either to DNA-repair or apoptosis. However, if these programs fail, cancer may occur. It is generally accepted that genetic variants lead to disruption of the apoptotic and/or DNA-repair pathways and therewith to a preferential cancer development [6, 7]. It was disclosed that a recurrent mutation of the BRCA1-gene occurs in cancer patients of the Chinese population [8]. Furthermore, in case of the Philadelphia chromosome a short variant of chromosome 22 exists. This short chromosome 22 leads to a transcription of a new gene, named BCR-ABL, which is involved in the Leukaemia development [9]. Currently, several Single Nucleic Polymorphisms (SNPs), which are involved in a higher cancer susceptibility, are collected in databases [10-12].

Summing up, it can be concluded, that distinct genetic differences between chimp and human exist, which are responsible for the increased cancer incidence in human, and might be associated with apoptosis or DNA-repair. An important question, which may cause even therapeutic consequences, is: How can these genetic factors, which affect cancer susceptibility, be detected? One possibility is to compare the machinery and regulation of DNA-repair and apoptosis between human and chimpanzee.

The chimpanzee, our closest living relative, has a genome wide identity of more than 98 percent [13, 14]. Nevertheless, chimpanzees rarely suffer from cancer. Especially prostate [15], breast, and lung cancer and spontaneous neoplasms [16] do rarely occur in chimpanzees population [17-19]. The sequencing of the chimpanzee genome elucidates that human and chimps diverged in about 35 million single nucleotide changes and about 5 million insertion/deletion events (indels) [13]. Previous works have shown that differences on nucleotide- and amino acid levels between chimp and human proteins, which are involved in cancer development, are existent [18]. These findings support the hypothesis that the low cancer incidence in chimpanzees has genetic reasons.

Puente et al. selected 333 cancer relevant genes and found about 1,500 changes in the amino acid sequence of the proteins [18], but their detailed analysis remains restricted to a fistful of proteins. For instance, the coding sequence for the tumor suppressor p53 has an arginine at position 73, which does only occur in the human lineage, whereas other primates like the chimpanzee, gorilla and mandrill have a proline at this position [14]. Furthermore, in the work of Nielsen et al. about 13,700 annotated genes of chimps and human were analyzed and their selectivity was calculated [20]. In this comparison, it was pointed out that several cancer related proteins, which are involved in apoptosis, tumor suppression and in the cell cycle control are positively selected. This is supported by the results of the Nature consortium [13], which determined the selectivity of groups of proteins in the GeneOntology [21]. In this case, the apoptotic pathway and the proteins, which belong to the cell cycle control, are also positively selected [13].

To pinpoint the reasons for the low cancer incidence in chimps and hopefully to deduce novel possibilities for cancer treatment, we consider an enlarged dataset of about 500 proteins which are either involved in the apoptotic or in the DNA-repair pathway. In contrast to previous studies, we will analyze these proteins on different levels:

chromosomal organization, gene structure, posttranscriptional and posttranslational modifications as well as structural and functional changes.

2. Data and Methods

Based on the assumption, that a change in the regulation of DNA-repair or apoptosis might be the reason for the decreased cancer susceptibility in chimpanzee, 493 proteins involved in these processes were selected. The protein and DNA-alignments were extracted from the Ensembl database vol. 42-44 [22]. Instead of just analyzing the number of mutations in the gene and protein sequences of both species, other topics were considered in this work as well.

These proteins were analyzed on different levels of biological organization:

- Chromosome and gene structure
- Post-transcriptional modifications and post-translational modifications
- SNPs and indels leading to structural and functional changes.

First, the chromosome organization, which means e.g. the protein distribution over the chromosomes of the 493 selected proteins, was considered for both species. Moreover, the positive selection of the group of genes and also of the single genes was calculated according the work of Nielson et al. [20]. To analyze the chromosome organization, the tool AutoGraph was used [23] to compare the positions of the genes on the chromosomes.

For the next level, the gene structure of the 493 proteins was examined and differences were recognized by changes of the number and size of introns and exons. These differences could, on the one hand, lead to different numbers of splice variants, and on the other hand to missing regulatory RNAs, which would have been transcribed from deleted intron sequences [24]. Therefore, the human proteins were selected from the Swissprot database [25] and compared with the chimpanzee and human genome of the GenBanks RefSeq [26] to find their coding region. This was carried out by use of GenomeThreader [27]. The matches for chimp and human were compared. The GenomeThreader tool marked the exons and introns to facilitate the comparison of the number of exons and the length of the introns of both species [28, 29].

Moreover, pseudogenozation, also analyzed in this work, results in a loss of genes and paralogous gene copies [14] and, therefore, leads to a decline of protein-families, which is an important event during evolution. In a previous work of Wang et al. [30], several pseudogenozations of chimp and human genes have been described. In this work the pseudogenes of the RefSeq annotation were considered. The annotation of the chimp genome is an ongoing process and will possibly lead to new results in the future. One reason for the potential of improvement of the annotation is the lack of ESTs, which usually take an important part in the annotation process. While the dbEST section of NCBI (state: July 2007) lists more than 8 million human ESTs, there are currently less than 5000 chimp ESTs available. Most annotations on *Pan troglodytes* were derived by their similarity to nucleotide and protein sequences from other species and lack support from original chimp EST and full length cDNA data.

On the RNA-level, we analyzed post-transcriptional modifications. Alternative splicing is a prerequisite after the transcription of a pre-mRNA to yield various new mRNAs. Disruptions of enhancers, which are located upstream of a gene are indispensable for alternative splicing. A single mutation in an enhancer region leads in most cases to the loss of an alternative splice site and could also cause either a new splice variant or even the loss of a splice variant. To compare the alternative splice variants, the annotated splice variants per gene, which are stored in the GenBank [26] of the NCBI database [31] were selected and compared .

Another important point we focused on in this work, is the post-translational modification after the biosynthesis. These modifications are important for the controlling of the localisation, enzyme activity and regulation of the final, native protein. For instance, the phosphorylation is a necessary step for the ubiquitination and, therefore, for the degradation of a protein. By loss or change of modification sites the biological activity of proteins could be up- or down-regulated, respectively [32]. To determine the modification sites in the protein sequences of both species, the PROSITE pattern [33] for seven different modification types was used. The number of occurrences of the seven modification types in both species was compared. The measurement of positive selection aims in the detection of genes that evolve faster than others on the basis of a selection pressure for novel forms. Changes in the nucleotide sequence that produce synonyms in the triplet code (Ks) and cause no changes in the protein sequence represent a kind of steady background noise. Non-synonymous nucleotide changes (Ka) are either advantageous for the organism or does not become accepted during evolution. A high Ka/Ks ratio indicates a strong selection pressure, while a low ratio means selection has been working to conserve the sequence. In the comparison of two species, the Ka/Ks ratio does not state which of the two species has changed the most from their common ancestor but genes that diverged the most are identified. The Ka/Ks ratio was estimated as (dA/dS)/(NA/NS) where dA and dS are the number of amino acid mutations and synonymous mutations observed in each coding sequence. NA and NS is the number of possible non-synonymous and synonymous mutations [34].

Finally, the proteins were analyzed on the structural and functional level. Structural information was used to find changes in protein-protein and protein-compound interaction sites between the chimpanzee and human lineage. Already a single amino acid change of a conserved region of a binding site could yield to a disruption of a compound binding site. The available crystal structures of 209 proteins were collected from the PDB [35] to check whether a single amino acid change between human and chimp changes a protein-protein or a protein-compound binding site. Beyond that, an occurring change of a protein-protein or protein-compound binding site, resulting of an indel, was analyzed separately, because for these cases an increased assembly error rate has to be expected [36]. For this reason 48 proteins that were excluded from the Ensembl database vol. 44 were not considered in the gene related part of this analysis.

3. Results and Discussion

3.1. Chromosome organization

The genome of the chimpanzee consists of 24 different chromosomes, whereas the human genome has 23 different chromosomes [13]. During evolution two chromosomes, which are designated 2a and 2b in chimpanzee, were merged to one larger chromosome 2 in the human genome [13]. Furthermore, nine pericentric inversions occur , in which the centromere is included [13]. One of these inversions exists on chromosome 18 and is shown in Fig. 1.

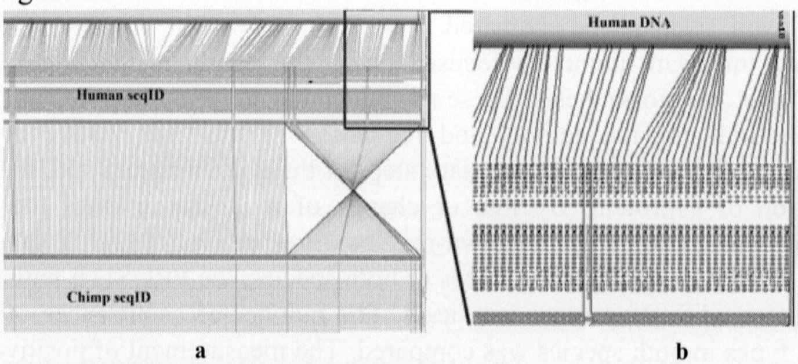

Fig. 1: Comparison of chimpanzee and human chromosome 18.
The comparison of human genes with the orthologous genes of the chimpanzee on chromosome 18 is shown in this figure. The picture was drawn with AutoGraph [23]. a) The upper part shows the human genes on chromosome 18, while the part below shows the chimpanzee genes. A pericentric inversion occurs. The black lines describe genes, which occur exclusively in the human chromosome 18 at this position. b) Magnified section.

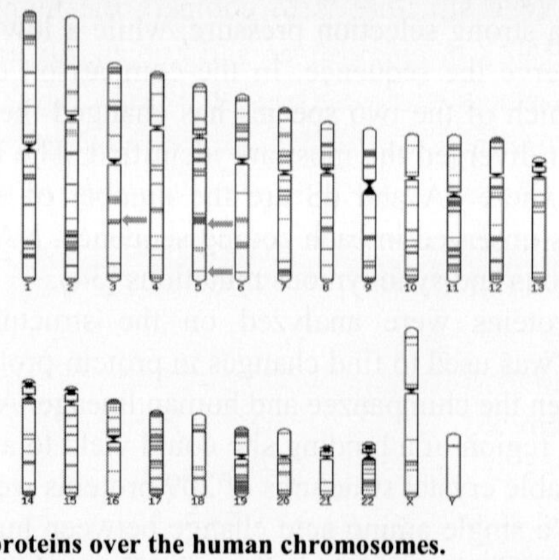

Fig. 2: Distribution of the 493 proteins over the human chromosomes.
The proteins are spread all over the chromosomes, except of chromosome y. In this respect, the y chromosome is non-coding for apoptosis and DNA-repair. The red lines present the positions of the 493 genes, which transcribe the mRNAs of the 493 proteins. The blue lines and arrows indicate changed location of corresponding chimp genes.

The 493 considered proteins, which are assumed to be responsible for the differing cancer susceptibility, are spread all over the chromosomes in the human genome (Fig. 2). This protein distribution is not identical in the chimpanzee genome. Here, we report three proteins, which have their coding sequence on different chromosomes in human and chimpanzee (Table 1). This result was attained with the GenomeThreader tool, which in a first step matched the protein sequence of the Swissprot database against the human and chimpanzee genomes and selected the annotated CDSs of GenBank.

For the RING finger protein 7 exists an annotated CDS in the GenBank for the chimpanzee and the human with 100 percent identity to the protein sequence of the Swissprot database. Nevertheless, the annotated CDS belongs to chromosome 3 in the human and to chromosome 7 in the chimpanzee genome. Moreover, a region on chromosome 3, which could also transcribe the RNA for this protein, is found with no existing annotated CDS. However, we found no matching coding region on human chromosome 7 for this protein. Similarly, for the proteins Replication protein C and the Nucleophosmin no coding regions could be detected on the corresponding chromosomes using the GenomeThreader.

Table 1: Proteins, which have its best matching coding sequence on different chromosomes.

Protein ID	Acc. number	Best match on human chromosome	Best match on chimp chromosome
RING finger protein 7	RBX2	3	7
Replication protein C	PP2AA	5	X
Nucleophosmin	NPM	5	16

3.2. Gene structure

To analyze the gene structure, the exon/intron structure has to be considered. The first step of analyzing the gene structure is to compare the numbers of exons, which are transcribed into the mRNAs in chimpanzee and human. Overall, an annotated CDS for human and chimpanzee is available for about 300 proteins with an identity of nearly 100 percent. For these CDSs the exon/intron structure was examined. About 5% of the proteins are transcribed from different numbers of exons during the human and chimpanzee biosynthesis. In one third of them the human mRNAs are transcribed of at least one more exon than the orthologous mRNA of the chimpanzee and in the two third it is conversely. These differences in the exon numbers can lead to different splice variants. To have an overview of the intron structure, the intron length of human and chimpanzee was determined and compared. Altogether, 84 proteins, which have a sequence identity of 95-100 percent and the same number of introns between human and chimpanzees have either at least one intron which is at least 50 base pairs longer than the orthologous introns. The genes of the chimpanzee have longer introns than the human genes (Fig. 3). Because introns are coding for regulatory RNAs, it might be possible that in chimpanzee more or different such RNAs exist, which might be also involved in the regulation of the apoptosis or the replication of apoptotic relevant genes.

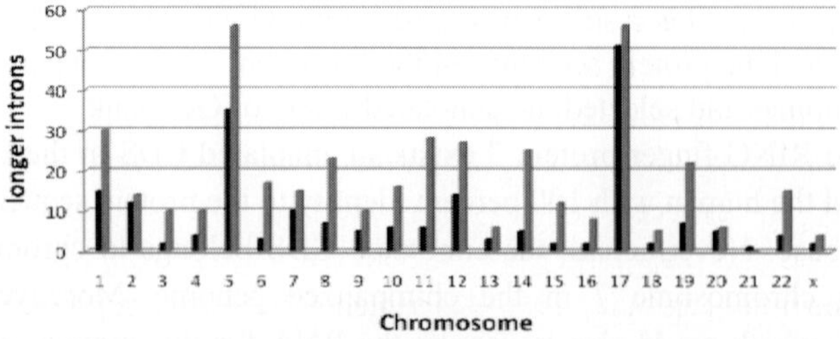

Fig. 3: Number of longer introns per human and chimpanzee chromosome for the 84 proteins. On average, the genes of the chimpanzee have longer introns than the human genes. Blue bars: number of longer introns per human chromosome, magenta: number of longer introns per chimpanzee chromosome.

On average, the apoptotic and DNA-repair proteins have about 9 mutations per gene. Nielsen et al. [20] found a very large proportion of cancer-related genes that are affected by positive selection. To detect candidates that evolve faster than others the Ka/Ks ratios of the 493 genes were calculated. To determine whether a single protein is positively selected, the Ka/Ks value has to be higher than one. Altogether, 13 proteins, including the breast cancer protein BRCA1 (Table 2) are positively selected. The positive selection and the meaning of BRCA1 for cancer development have been shown earlier [13, 20].

Table 2: Proteins with Ka/Ks value higher than 1, indicating positive selection.

Protein	Acc. number	Ka/Ks-value
Thioredoxin-like protein p46	TXND5	3,13
Immediate early protein GLY96	IEX1	2,13
Poly [ADP-ribose] polymerase 3	PARP3	2,06
dUTP pyrophosphatase	DUT	1,70
Tumor necrosis factor receptor superfamily member 10D	TR10D	1,67
Tumor necrosis factor receptor superfamily member 18	TNR18	1,43
Breast cancer type 1 susceptibility protein	BRCA1	1,28
Tumor necrosis factor receptor superfamily member 10A	TR10A	1,21
Caspase-5	CASP5	1,21
Vascular endothelial growth factor A	VEGFA	1,17
Myc proto-oncogene protein	MYC	1,14
Serpin B9	SPB9	1,02
Transcription factor E2F1	E2F1	1,00

A process, which is likely to be more important than single mutations, is the pseudogenozation [30, 37]. Pseudogenozation is a gene inactivation, which is either caused by nonsense/frameshift mutations or by the loss of paralogous gene copies that were duplicated during hominoid evolution. In the 493 considered proteins about 49 proteins are possible pseudogenes in the chimpanzee, but active genes in human. One of these proteins is the transcription factor E2F3, which is involved in the cell cycle control [38]. This protein is described as a pseudogene in the RefSeq annotation of the NCBI. In contrast, the Ensembl annotation lists this gene locus as a functional gene, but achieves this by the declaration of a cytosine as single nucleotide intron, which might be an artefact of the Ensembl annotation process. Further experimentally validation is necessary to prove if one of these annotations is right. Moreover, also an error in the existing sequence could be the cause for these annotations. The loss of the transcription factor E2F3 could lead to a decreased proliferation rate. Surprisingly, many differences in the gene structure between chimpanzee and human were found. Especially, the great number of genes, which have larger introns, could be important for the cancer development. In this process, the regulatory mRNAs as well as the great number of different splice variants play a central role. Selective siRNAs will offer the possibility to analyze their occurrence in tumor cells experimentally. Corresponding experiments are in preparation with a cooperating lab regarding Bcl-2 proteins.

3.3. Post-transcriptional and post-translational modifications

An important post-transcriptional modification is alternative splicing, which yields to a varying number of mRNAs. More than 60 percent of human genes employ alternative splicing [39]. Altogether, about 1.5 annotated splice variants per human gene and 2.5 annotated variants per chimpanzee gene exist in the 493 proteins. The increased number of alternative splice variants can be explained by the increased number of exons in chimpanzee. This could lead to a more complex regulation of the DNA-repair and apoptotic pathway.

After the translation, many proteins have to be modified. These modifications regulate for example the proteins function, activity and localisation. For the seven types of modification cAMP-, PKC- and CK2-dependent phosphorylation, myristylation, ASN-glycosilation, TYR-phosphorylation, and amidation the number of proteins, which have at least one more modification site than the orthologue are considered. Therefore, the human patterns of these modification sites were used. The result is that the human proteins have more modification sites for all seven modification types than the chimpanzee (Fig. 4). From the analysis of the frequency of post-translational modifications, it can be hypothesized that either fewer modifications in the chimpanzees proceed or, more likely, that new patterns for these modifications have evolved. Hence, the development of new PROSITE patterns for the chimpanzee will be required.

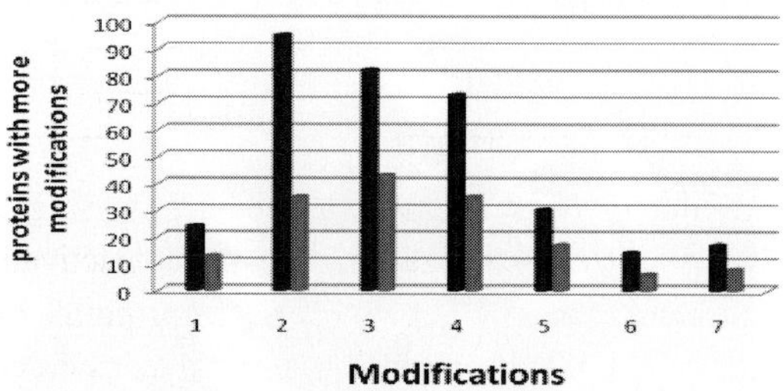

Fig. 4: Number of proteins with one minimum modification site.
Number of proteins with at least one more cAMP-phosphorylation site (1), PKC-phosphorylation site (2), CK2-phosphorylation site (3), myristyl site (4), ASN-glycosylation site (5), TYR-phosphoryaltion site (6) and/or amidation site (7). For all cases the human has more modification sites. Blue: human, magenta: chimpanzee.

3.4. SNPs and indels leading to structural and functional changes

On average, the human-chimp lineage shows two amino acid mutations per protein [40], whereas in the reduced dataset of apoptotic and DNA-repair proteins eight amino acid mutations per protein can be found. Altogether, 4,111 single amino acid changes occur and 288 insertion/deletion events. However, Puente et al. just identified 1,542 amino acid changes in 333 cancer relevant proteins [18]. Moreover, the total protein sequence identity of these 493 proteins amounts 96%. In the schedule below, recognition patterns of important domains, which are destructed by a mutation, are listed (Table 3). For instance, the BH2-motif of the Bcl-2-related protein B2La1, which retards apoptosis, is lost. To understand the genetic changes between human and chimpanzee in terms of functional differences, the analysis of the structure and function is necessary. Altogether, 209 protein structures of the 493 proteins exist in the PDB, where 59 proteins have protein-protein interaction site changes and 10 proteins have compound binding site changes.

Table 3: Proteins, in which a motif is destructed by a mutation.

Protein ID	Motif	Human	Chimp
IFNA2	Interferon α,β,δ	X*	
MIF	Macrophage migration inhibitory factor	X	
SSR3	Microbodies C-terminal targeting signal		X
TXND5	Thioredoxin family active site	X	
VEFFA	Platelet-derived growth factor family signature	X	
B2LA1	BH2 motif	X	
BRCA2	ATP-binding motif	X	
ERCC3	Helicase ATP-binding type domain profile	X	
ERN1	SER/THR protein kinases active-site signature	X	
TBB2C	Tubulin β mRNA autoregulation signal	X	

*The cross in the last two columns indicates whether the human or the chimpanzee has this motif in the protein.

An example for a change in the compound binding site is the Topoisomerase II alpha.

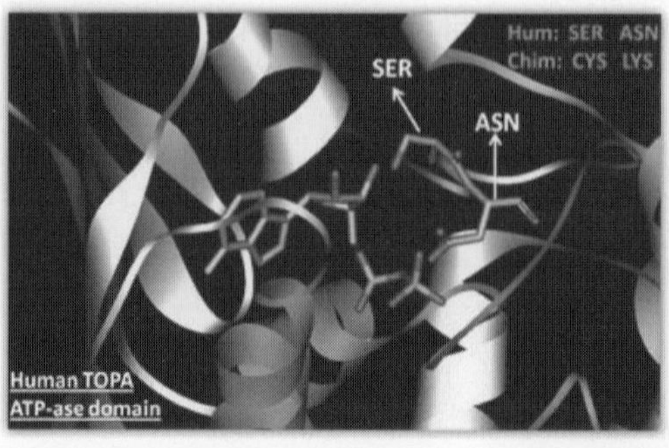

Fig. 5: ATPase domain of human Topoisomerase II alpha of pdb code: 1ZXM.
The ATPase domain with a bound ATP molecule is shown. Two residues differ in the protein sequence of the human and chimps. SER and ASN are changed to CYS and LYS. These amino acids are very important for the binding of the ATP and a change could influence the strength of the binding. In this case, the negative charge of the phosphate group and the positive charge of LYS, which is in the chimpanzee sequence, could result in a better binding of the ATP.

The crystal structure of the ATPase-domain was determined. In this case the ATP-binding site will be changed, because of a mutation of two important amino acids (Fig. 5; PDB-code: 1ZXM) [41]. The Topoisomerase II alpha is a key enzyme, which is involved in DNA replication and is frequently amplified in breast cancer, which emphasizes the meaning of this finding.

We report that the transcription factor E2F3 might be a pseudogene in the chimpanzees. A crystal structure of the E2F3 domain, which might be a pseudogene in the chimpanzee genome, but an active gene in the humans, is available in the PDB (Fig. 6; PDB-code: 1CF7) [38].

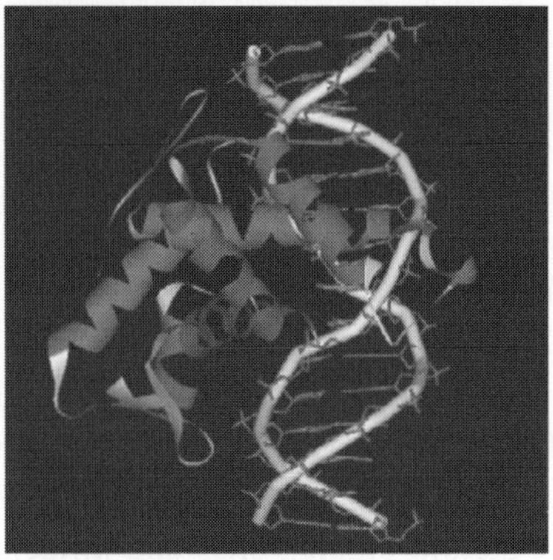

a

Fig. 6: Possible pseudogenosation of E2F3.
a) DNA-binding domain of a transcription factor E2F-family member; PDB-code: 1CF7. The E2F transcription factor is shown as ribbon and the bound DNA in stick representation. The binding of E2F family members to the promoter region of cell-cycle controlling proteins enhances their expression.

This protein belongs to the E2F-family and acts as a transcription activator, by binding the DNA at the recognition site 5'-TTTC[CG]CGC-3', which is found in the promoter region of a number of genes, which are involved in the cell cycle regulation or in the DNA replication pathway. It specifically binds the protein Rb1. Moreover, E2F3 controls the cell-cycle progression from G1 to S phase and its high concentration results in a high proliferation rate.

Further works show that this protein is highly expressed in human prostate and lung cancer type and is actually a target for cancer therapy [42, 43]. One drug which inhibits the protein E2F3 is the silybin [44]. The recent use of E2F3 as cancer target emphasizes its meaning as key regulator of apoptosis. This supports our finding that its pseudogenozation might be an important reason for the low cancer incidence in chimpanzees. Of the 159 amino acid changes occurring in the 493 proteins could be mapped onto the 3D structure. A detailed analysis regarding structural changes and their influence on the protein function will be topic of a separate publication.

Table 4: Summary.

Level	Observation
Gene structure	-49 putative pseudogenes in the chimp genome, which are active genes in human (e.g. transcription factor E2F3) -Chimps have exceeding longer introns, which may contain regulatory parts -Chimps have more exons, which transcribe for the same mRNA, which potentially leads to more splice variants -Group of proteins is positively selected (Ka/Ks = 0.4)
Post-transcriptional modification (alternative splicing)	-1.5 annotated alternative splice variants per gene in human (738 proteins) -2.5 annotated alternative splice variants per gene in chimps (1204 proteins)
Post-translational modification	-Humans have exceeding post-translational modifications or different patterns
Structure and function	-6 amino acid changes per protein -209 crystal structures of 493 proteins available -59 proteins with possible changes in protein-protein binding sites -10 proteins with changes in compound-protein binding sites (e.g. Topoisomerase II alpha)

4. Conclusion and Perspective

A major outcome of this analysis represents the pseudogenozation of a number of proteins. The increased number of exons, which leads to more splice variants, is of outstanding importance and is a new finding. Beyond that, various amino acid changes can be mapped onto 3D structures and will be analyzed elsewhere. In a next step, theirrole in the apoptotic and DNA-repair pathways will be deconstructed. Longer introns are found in the chimpanzee. Transposable elements like ALUs may be responsible for that and should be investigated circumstantially, because they add regulatory sequences, like steroid binding sites and influence the methylation status of promoters and thus, the activity of genes. This epigenetic level should be included in a further analysis.

Acknowledgments

This work was supported by the International Research Training Group Boston-Kyoto-Berlin, funded by the DFG.

References

[1] Ferlay, J., *et al.*, Estimates of the cancer incidence and mortality in Europe in 2006, *Ann. Oncol.*, 18(3):581-592, 2007.

[2] Varki, A., A chimpanzee genome project is a biomedical imperative, *Genome Res.*, 10(8):1065-1070, 2000.

[3] Miller, B.A., Scoppa, S.M., and Feuer, E.J., Racial/ethnic patterns in lifetime and age-conditional risk estimates for selected cancers, *Cancer*, 106(3):670-682, 2006.

[4] Wampler, N.S., *et al.*, Breast cancer survival of American Indian/Alaska Native women, 1973-1996., *Soz. Praventivmed.*, 50(4):230-237, 2005.

[5] Snyder, O.B., Kelly, J.J., and Lanier, A.P., Prostate cancer in Alaska Native men, 1969-2003., *Int. J. Circumpolar Health*, 65(1):8-17, 2006.

[6] Houtgraaf, J.H., Versmissen, J., and van der Giessen, W.J., A concise review of DNA damage checkpoints and repair in mammalian cells, *Cardiovasc. Revasc. Med.*, 7(3):165-172, 2006.

[7] Nakanishi, M., Shimada, M., and Niida, H., Genetic instability in cancer cells by impaired cell cycle checkpoints, *Cancer Sci.*, 97(10):984-989, 2006.

[8] Li, W.F., *et al.*, [BRCA1 1100delAT is a recurrent mutation in Chinese women with familial breast cancer], *Zhonghua Yi Xue Za Zhi*, 87(2):76-80, 2007.

[9] Balatzenko, G., *et al.*, Philadelphia variant, t(5;9;22)(q13;q34;q11), in a case with chronic myeloid leukemia, *J. BUON*, 8(1):65-67, 2003.

[10] McKusick, V.A., Mendelian Inheritance in Man and its online version, OMIM, *Am. J. Hum. Genet.*, 80(4):588-604, 2007.

[11] Shimizu, N., Ohtsubo, M., and Minoshima, S., MutationView/KMcancerDB: a database for cancer gene mutations, *Cancer Sci.*, 98(3):259-267, 2007.

[12] Bajdik, C.D., *et al.*, CGMIM: automated text-mining of Online Mendelian Inheritance in Man (OMIM) to identify genetically-associated cancers and candidate genes, *BMC Bioinformatics*, 6:78, 2005.

[13] Initial sequence of the chimpanzee genome and comparison with the human genome, *Nature*, 437(7055):69-87, 2005.

[14] Kehrer-Sawatzki, H. and Cooper, D.N., Understanding the recent evolution of the human genome: insights from human-chimpanzee genome comparisons, *Hum. Mutat.*, 28(2):99-130, 2007.

[15] Waters, D.J., *et al.*, Workgroup 4: spontaneous prostate carcinoma in dogs and nonhuman primates, *Prostate*, 36(1):64-67, 1998.

[16] Beniashvili, D.S., An overview of the world literature on spontaneous tumors in nonhuman primates, *J. Med. Primatol.*, 18(6):423-437, 1989.

[17] McClure, H.M., Tumors in nonhuman primates: observations during a six-year period in the Yerkes primate center colony, *Am. J. Phys. Anthropol*, 38(2):425-429, 1973.

[18] Puente, X.S., *et al.*, Comparative analysis of cancer genes in the human and chimpanzee genomes, *BMC Genomics*, 7(1):15, 2006.

[19] Seibold, H.R. and Wolf, R.H., Neoplasms and proliferative lesions in 1065 nonhuman primate necropsies, *Lab. Anim. Sci.*, 23(4):533-539, 1973.

[20] Nielsen, R., *et al.*, A scan for positively selected genes in the genomes of humans and chimpanzees, *PLoS Biol.*, 3(6):e170, 2005.

[21] Camon, E., *et al.*, The Gene Ontology Annotation (GOA) Database--an integrated resource of GO annotations to the UniProt Knowledgebase, *In Silico Biol.*, 4(1):5-6, 2004.

[22] Hubbard, T.J., *et al.*, Ensembl 2007, *Nucleic Acids Res.*, 35(Database issue):D610-617, 2007.

[23] Derrien, T., *et al.*, AutoGRAPH: an interactive web server for automating and visualizing comparative genome maps, *Bioinformatics*, 23(4):498-499, 2007.

[24] Mattick, J.S. and Makunin, I.V., Small regulatory RNAs in mammals, *Hum. Mol. Genet.*, 14 Spec No 1: R121-132, 2005.

[25] The Universal Protein Resource (UniProt), *Nucleic Acids Res.*, 35(Database issue):D193-197, 2007.

[26] Benson, D.A., *et al.*, GenBank, *Nucleic Acids Res.*, 35(Database issue):D21-25, 2007.

[27] Gremme, G., *et al.*, Engineering a software tool for gene structure prediction in higher organisms, *Information and Software Technology*, 47(15):965-978,2005.

[28] Usuka, J., Zhu, W., and Brendel, V., Optimal spliced alignment of homologous cDNA to a genomic DNA template, *Bioinformatics*, 16(3):203-211, 2000.

[29] Usuka, J. and Brendel, V., Gene structure prediction by spliced alignment of genomic DNA with protein sequences: increased accuracy by differential splice site scoring, *J. Mol. Biol.*, 297(5):1075-1085, 2000.

[30] Wang, X., Grus, W.E., and Zhang, J., Gene losses during human origins, *PLoS Biol.*, 4(3):e52, 2006.

[31] Maglott, D., *et al.*, Entrez Gene: gene-centered information at NCBI, *Nucleic Acids Res.*, 35(Database issue):D26-31, 2007.

[32] Basu, A., DuBois, G., and Haldar, S., Posttranslational modifications of Bcl2 family members--a potential therapeutic target for human malignancy, *Front. Biosci.*, 11:1508-1521, 2006.

[33] de Castro, E., *et al.*, ScanProsite: detection of PROSITE signature matches and ProRule-associated functional and structural residues in proteins, *Nucleic Acids Res.*, 34(Web Server issue):W362-365, 2006.

[34] Yang, Z., Balding, D., Bishop, M., and Cannings, C., *Adaptive molecular evolution, in Handbook of statistical genetics,* Wiley, London, 2001.

[35] Berman, H., *et al.*, The worldwide Protein Data Bank (wwPDB): ensuring a single, uniform archive of PDB data, *Nucleic Acids Res.*, 35(Database issue):D301-303, 2007.

[36] Overduin, B., ENSEMBL, Editor. 2007.

[37] Fairbanks, D.J. and Maughan, P.J., Evolution of the NANOG pseudogene family in the human and chimpanzee genomes, *BMC Evol. Biol.*, 6:12, 2006.

[38] Zheng, N., *et al.*, Structural basis of DNA recognition by the heterodimeric cell cycle transcription factor E2F-DP, *Genes Dev.*, 13(6):666-674, 1999.

[39] Mironov, A.A., Fickett, J.W., and Gelfand, M.S., Frequent alternative splicing of human genes, *Genome Res.*, 9(12):1288-1293, 1999.

[40] Glazko, G., *et al.*, Eighty percent of proteins are different between humans and chimpanzees, *Gene*, 346:215-219, 2005.

[41] Wei, H., *et al.*, Nucleotide-dependent domain movement in the ATPase domain of a human type IIA DNA topoisomerase, *J. Biol. Chem.*, 280(44):37041-37047, 2005.

[42] Grasemann, C., *et al.*, Gains and overexpression identify DEK and E2F3 as targets of chromosome 6p gains in retinoblastoma, *Oncogene*, 24(42):6441-6449, 2005.

[43] Oeggerli, M., *et al.*, E2F3 is the main target gene of the 6p22 amplicon with high specificity for human bladder cancer, *Oncogene*, 25(49):6538-6543, 2006.

[44] Tyagi, A., Agarwal, C., and Agarwal, R., Inhibition of retinoblastoma protein (Rb) phosphorylation at serine sites and an increase in Rb-E2F complex formation by silibinin in androgen-dependent human prostate carcinoma LNCaP cells: role in prostate cancer prevention, *Mol. Cancer Ther.*, 1(7):525-532, 2002.

GENE EXPANSION IN *TRICHOMONAS VAGINALIS*: A CASE STUDY ON TRANSMEMBRANE CYCLASES

JIKE CUI[1,2] TEMPLE F. SMITH[1]
jike@bu.edu tsmith@darwin.bu.edu

JOHN SAMUELSON[2]
jsamuels@bu.edu

[1] *Bioinformatics Program, Boston University, 44 Cummington St., Boston, MA, 02215*
[2] *Department of Molecular and Cell Biology, Boston University, 715 Albany St., Evans room 426, Boston, MA 02118*

The draft genome of *Trichomonas vaginalis* was recently published, but not much is known on why it has such a large genome. In part this size is due to many gene family expansions. For example we found over 100 members in the adenylyl cyclase family. About half are complete full length genes, and nearly half are initially confirmed to be pseudogenes, the remaining are either incomplete or the apparent result of assembly or sequencing problems. The family can be divided into two subgroups by sequence similarity. These can then be divided into functional and pseudo genes. Among all four of these sets the cyclase domain is very well conserved. We gave three possible hypotheses for that observation: a) Sequencing error or stop-codon read-through; b) Recency of duplication and mutation; c) The likelihood of functional pseudogene.

Keywords: *T. vaginalis*; cyclase; pseudogene; gene duplication.

1. Introduction

Trichomonas vaginalis is an anaerobic, parasitic flagellated protozoan. Infection with *T. vaginalis* is one of the most common sexually transmitted diseases for women with 8 million infections in North America and 180 million infections in the world each year [1-2]. The extracellular parasite resides in the urogenital tract of both sexes. *T. vaginalis* has a modified mitochondrion called the hydrogenosome, in which fermentation enzymes reside rather than enzymes of oxidative phosphorylation [3].

The recently published draft genome sequence of *T. vaginalis* by The Institute of Genomic Research (TIGR) reveals an abnormally large genome size of 160 Mb. ~60,000 protein-coding genes were identified, but only 65 were found to have introns. Two thirds of the genome consists of repeats and transposable elements [4].

Our examination of 12 animal pathogens, including fungi and protists, tells that their average genome size is ~20 Mb and ~6000 genes. These relatively small genomes provide efficiency in multiplication and infection. It is not clear why *T. vaginalis* possesses such a large genome, and how such massive gene expansion happened. The secretory pathway and signal transduction play important role in pathogenesis [28, 29]. Because cyclases are critical in eukaryotic signal transduction and have a unique structure discovered in our previous study, we propose a research project on a large family of putative transmembrane cyclase genes. And we hope the results can shed light on why and how the genome expansion happened.

2. Methods

2.1. *Identification of Cyclase Genes and Pseudogenes in T. vaginalis*

Cyclase domains were identified using NCBI Conservation Domain search tool [5]. The conserved cyclase domains and tblastn were used to search the *T. vaginalis* sequence scaffolds at the TIGR site using a cutoff E value of 1e-6 [6]. We identified 24 complete putative *T. vaginalis* transmembrane cyclases, which contained numerous transmembrane helices and a C-terminal cyclase. The length of these full-length transmembrane cyclases was ~1600 AAs (4,800 bps). In addition, we used blastx and those complete transmembrane cyclases to identify numerous other *T. vaginalis* genes encoding transmembrane cyclases, some of which were truncated or contained nonsense mutations (stop codons) or frame shifts.

2.2. *Initial Verification of Frameshift and Nonsense Mutation*

To verify each nonsense mutation, we took 30 AAs in the upstream and downstream of the stop site to make a query of 61 AAs. To verify frame shift, we took 100 bases of upstream and downstream of the end and start of each frame to make a query of 201 bases. Those queries were used to tblastn/blastn the trace from individual sequence reads of *T. vaginalis*. If there were 2 or more perfect matches, the mutation was considered real. This method actually found two assembly errors, which caused a frame shift.

2.3. *Softwares*

MUSCLE [7] and PIMA [8] are used for alignment. PAUP [9] and TREE-PUZZLE [10] were used for phylogenetic tree construction. PAML [13] was used for dN/dS analysis.

3. Results and Discussion

3.1. *Number and Topology of Cyclase Genes in T. vaginalis*

Cyclases play very important roles in eukaryotic signal transduction. The second messenger cAMP (adenosine 3',5'-cyclic monophosphate) is synthesized by an adenylyl cyclase (AC) from ATP, and cGMP (guanosine 3',5'-cyclic monophosphate) is made by a guanylyl cyclase (GC) from GTP [14]. Cyclase normally is not a big protein family, for example, we only found one cyclase in yeast. But we discovered that there are over 100 copies of the putative transmembrane cyclase genes in *T. vaginalis*. Unpublished data from our lab suggest the cyclase domains of these *T. vaginalis* transmembrane cyclases are adenylyl cyclase, and unpublished mRNA data suggest that these cyclases are expressed constitutively, that is, they are expressed at the same time.

Each full length cyclase has 12 to 16 transmembrane helixes (TMH) and a cyclase domain at the C-terminal. Some of them are truncated at the N terminal but all have at least one transmembrane helix, which means there is no cytosolic cyclase in *T. vaginalis*.

3.2. *How Did the Duplication Happen?*

Gene duplication may occur in homologous recombination, retrotransposition event, or duplication of an entire chromosome [15]. To find out more details about cyclase gene families' duplication event, we asked the following questions:

- Do the cyclase genes locate in subtelomere?
- Did the cyclase genes duplicate with other repetitive sequences?
- Were flanking sequences of the cyclase genes also duplicated?
- Were the cyclase genes duplicated by retrotransposition?

Genes in subtelomere usually have multiple copies. The cyclase family has ~100 copies, so it is reasonable to speculate that it might be in subtelomere. However Fig. 1 obviously does not support that. On average 83.7% of genes on a scaffold do not have a homolog in that scaffold, with SD of 15.7%. Although the percentage could be greatly affected by the length of scaffold, in scaffolds with more than 80 genes, the two numbers are 73.2% and 5.9%. It demonstrates that those scaffolds are not in subtelomere.

Sometimes genes could be carried by repeat elements and got duplicated when repeats move and copy themselves across the genome. We studied transposons, microsatellites, and virus-like repeat families in the upstream and downstream 5000 bases of each cyclase gene. While many repeat families were found, only one pair of genes has the same repeat at the same location. It seems cyclase genes were not duplicated with repeats.

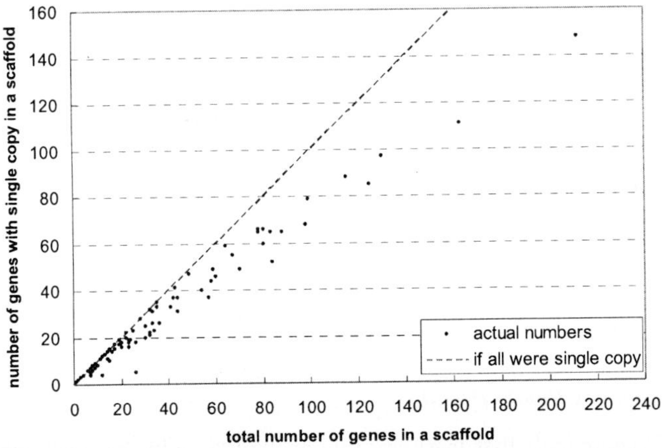

Fig. 1. Total number of genes and number of single copy genes in each of the 90 scaffolds which contains cyclase genes.

If a gene is duplicated in recombination event, its flanking sequence will probably be duplicated too. We searched ORFs in the 5000 bases region on two sides of each cyclase gene, but did not find any synteny, except some very repetitive proteins, like DNA polymerase and BspA-like surface antigen.

If cyclase genes were not duplicated with flanking sequence, they probably were copied with retrotransposons. Such a gene could have a poly-A present in its close downstream if the duplication happens very recently. And as discovered in *Entamoeba histolytica*, it may have a 9-20 bases repeat at the beginning and end of the copied region [16]. However no such evidence has been found so far among even the closest cyclase genes.

3.3. *Pseudogenes in Cyclase Genes of T. vaginalis*

Among the discovered cyclase genes, about half are complete genes ranging from 1449 to 1681 AAs. A dozen are either truncated genes that range from 340 to 1322 AAs, or incomplete genes due to the sequencing or assembly problems. The rest, which is nearly half, are pseudogenes that contain frameshifts and nonsense mutations. Fig. 2 illustrates the transmembrane cyclases in different life stages.

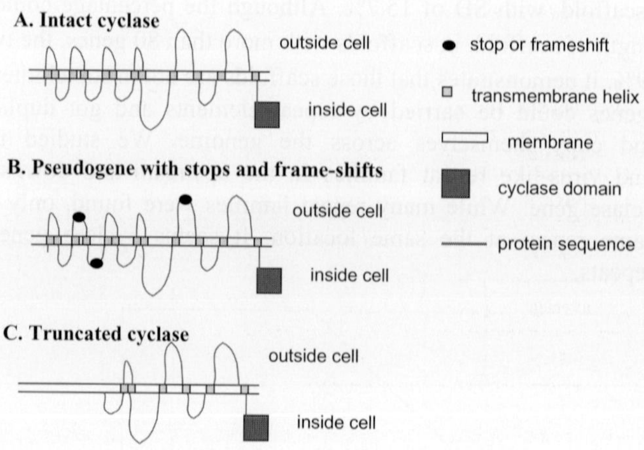

Fig. 2. Intact, pseudo, and truncated transmembrane cyclase of *T. vaginalis*.

If we refer the region before the cyclase domain as transmembrane region, an absolute majority of frameshifts and nonsense mutations happens in the transmembrane region, and very few are in the cyclase domain.

3.4. *Are Those Pseudogenes Really Pseudo?*

Alignment of the cyclases reveals that there are lots of variations in the transmembrane region, but the cyclase domain is highly conserved not only for those full length complete genes, but also in those pseudogenes. Phylogenetic analysis on the amino acid and nucleotide sequences shows that there is not a separation between pseudogenes and complete genes, and instead, they group together. Fig. 3 illustrates that grouping.

The above observation is unusual. Normally a pseudogene loses its function and is free from any selection constraint. The pseudogene then has a faster mutation clock than a functional conserved domain, and so they are not likely to group together. To study in more detail, the mutation on each codon position of the cyclase domain in 41 pairs of close full length functional genes, and 49 pairs of close functional and pseudogenes were compared. Table 1 lists the result.

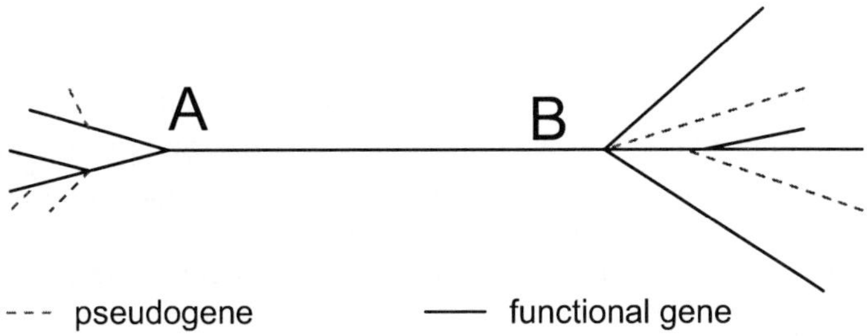

Fig. 3. A cartoon illustration of the phylogenetic grouping of the cyclase domain amino acid sequences in transmembrane cyclases using maximum likelihood.

Table 1. Number and percentage of mutation at each codon position

Pairs	statistics	1st position		2nd position		3rd position		sum
49 pairs of complete genes vs pseudogenes	average	35	20.5%	20	11.1%	99	67.1%	154
	StDev	17	6.0%	14	5.6%	24	11.2%	49
41 pairs of complete genes vs complete genes	average	42	22.1%	24	12.2%	106	64.2%	172
	StDev	23	5.7%	16	5.2%	20	10.4%	56

Surprisingly, the pseudogenes behave exactly like the functional genes. Most mutations happen at the third (wobble) codon position which is often silent, and the first codon position which is silent sometimes too. Further analysis on selection constraint using PAML program [13] gave the dN/dS ratio of 0.0436 for the functional genes, and 0.0438 for pseudogenes, suggesting that mutations in the cyclase domains of both complete genes and pseudogenes are highly selected against.

We offer the following hypotheses for the above observation,

1. Those frameshifts and nonsense mutations may be sequencing error, or they maybe real but somehow *T. vaginalis* has a mechanism to read through stop codon in translation, therefore these pseudogenes can still be expressed and be functional.
2. Those pseudogenes are real, but the frameshifts and nonsense mutations happened very recently, therefore there has not been ample time for mutations to happen in a larger scale yet.
3. Those pseudogenes cannot be expressed; however they are not treated as junk and are still being conserved. And they might serve some function in certain events.

3.5. *Evaluation of Hypothesis One and Two*

To check the sequencing error and the likelihood of *T. vaginalis* reading through nonsense mutations, PCR and sequencing of some of the critical mutation sites will be conducted using the G3 strain that is used in the TIGR sequencing project. Although most of the frameshifts and nonsense mutations have been verified by the sequencing trace file, there is not an absolute certainty until the experiment proves that. Stop-codon read-through in mRNA translation is rare, but it has been reported in yeast, human, and *E. coli* [17-20]. We will test its likelihood in *T. vaginalis* too.

Hypothesis two is in agreemenet with TIGR's conslusion that the genome expansion is recent due to the low polymorphism within high repeat protein families, and evidence of repeat expansion after *T. vaginalis* and *T. tenax* diverged [4]. However it can not explain a) Why very few frameshifts and nonsense mutations happened in the cyclase domain. b) The cyclase domains of A have much greater conservation than those of B, as shown in Fig. 2, while their TM regions have similar degree of sequence variation. Actually the TM region of group A has on average more frameshifts and stops than that of B. So the recency of the duplication in cyclase family does not seem to explain everything here.

3.6. *Pseudogenes in the Whole Genome -- Evaluation of Hypothesis Three*

Most pseudogenes come from duplicated genes because its pseudogenization is less likely to be deleterious than a singleton. Human has about 30,000 genes with 38% of duplicated genes [21], and 12,000 pseudogenes have been identified [22]. TIGR predicted that there are about 60,000 genes in *T. vaginalis* but did not mention pseudogenes. We speculate that a significant portion of the 60,000 genes might be pseudogenes. All of the pseudo and incomplete cyclase genes are included in TIGR predicted genes which start after the last frameshift or nonsense mutation. Although the amount of pseudogenes in other large gene families can not be estimated until a similar survey is conducted, our search of nonsense mutation in the whole genome has found about 3000 pseudogenes with nonsense mutation, and there could be similar or greater number of pseudogenes with frameshift. Large number of pseudogenes are present in the family of ankyrin repeat proteins, hypothetical protein, conserved hypothetical protein, adenylate cyclase, vsaA, surface antigen BspA, ANK-repeat protein, CG1651-PD-related, Dentin sialophosphoprotein precursor, ABC transporter protein, kinases, major facilitator superfamily protein, leucine rich repeat family protein, and Transmembrane amino acid transporter protein. Many of those families are involved in secretary pathway and signal transduction system, which play important role in pathogenesis.

Hartl suggests that the rate of deleting junk DNA decides genome size, and a low rate would accumulate many pseudogenes, longer introns, and intergenic regions in a genome [23-25]. However it is different in *T. vaginalis*, where there might be many pseudogenes but very little introns. Is it possible that those pseudogenes are not real junk and do serve a purpose in certain situation? That is not impossible! It was discovered that pseudogenes could act as the supplier of certain variable region to immunoglobulin gene in chicken [26], and that some pseudogenes conserve their sequences and can still be revived and functional in cow and human [27], but the mechanism is not totally understood yet.

4. Conclusion and Future Work

T. vaginalis has an enormously large genome, and many protein families underwent massive duplication. We proposed a research project to study transmembrane cyclase in *T. vaginalis*, a very important protein in cell signal transduction. Our initial results show that there are over 100 copies of genes in this family, about half are full length genes, and nearly half are pseudogenes that are initially confirmed by the sequencing trace files. The family can be roughly separated into two groups by sequence similarity based on their cyclase domains, which is very well conserved in both complete genes and surprisingly pseudogenes too.

The conservation of cyclase domain in pseudogenes is not expected. We suggested three possible reasons: a) sequencing error and stop-codon read-through which will be tested by our experiments; b) recency of duplication and mutation. It is likely to be true but can not explain some discrepancies between the TM region and cyclase domain, and between the group A and group B. c) The likelihood of functional pseudogene which is possible but does not have any evidence until we see more experimental support.

We also tried to look into how cyclase genes were duplicated. We found that they are not in subtelomere, and there is no evidence to support that they might be copied with repeats, retrotransposons, or flanking regions.

We hope that after a larger survey on other duplicated protein families and having more experimental data on the pseudogenes, we could shed light on why *T. vaginalis* possesses such a huge genome, how genes are duplicated, the quantity of its pseudogenes, and their evolution histories.

References

[1] http://www.cdc.gov/ncidod/dpd/parasites/trichomonas/factsht_trichomonas.htm
[2] Hook, E., Trichomonas vaginalis--no longer a minor STD, *Sex. Transm. Dis.*, 26(7):388-389, 1999.
[3] Upcroft, P. and Upcroft, J., Drug targets and mechanisms of resistance in the anaerobic protozoa, *Clin. Microbiol. Rev.*, 14(1):150-164, 2001.
[4] Carlton J.M., *et al.*, Draft genome sequence of the sexually transmitted pathogen Trichomonas vaginalis, *Science*, 315(5809):207-212, 2007.
[5] http://www.ncbi.nlm.nih.gov/Structure/cdd/wrpsb.cgi
[6] http://www.tigr.org/tdb/e2k1/tvg/
[7] Edgar, R.C., MUSCLE: multiple sequence alignment with high accuracy and high throughput, *Nucleic Acids Res.*, 32(5):1792-1797, 2004.
[8] Smith, R.F. and Smith, T.F., Pattern-induced multi-sequence alignment (PIMA) algorithm employing secondary structure-dependent gap penalties for use in comparative protein modeling, *Protein Eng.*, 5(1):35-41, 1992.
[9] Paup: Swofford, D. L. *PAUP*. Phylogenetic Analysis Using Parsimony (*and Other Methods).* Version 4. Sinauer Associates, Sunderland, Massachusetts, 2002.

[10] Schmidt, H.A., Strimmer, K., Vingron, M., and von Haeseler, A.., TREE-PUZZLE: maximum likelihood phylogenetic analysis using quartets and parallel computing, Bioinformatics., 18(3):502-504, 2002.

[11] Sawyer, S., Statistical tests for detecting gene conversion, *Mol. Biol. Evol.*, 6(5):526-538, 1989.

[12] Martin, D.P., Williamson, C., and Posada, D., RDP2: recombination detection and analysis from sequence alignments, *Bioinformatics,* 21(2):260-262, 2005.

[13] http://abacus.gene.ucl.ac.uk/software/paml.html

[14] Roelofs, J., Smith, J. L., and Van Haastert, P. J. M., cGMP signalling: different ways to create a pathway, *TRENDS Genet.,* 19(3):132-134, 2003.

[15] Zhang, J., Evolution by gene duplication: an update, *Trends in Ecology & Evolutio,* 18(6):292-298, 2003.

[16] Van Dellen, K., Field, J., Wang, Z., Loftus, B., and Samuelson, J., LINEs and SINE-like elements of the protist Entamoeba histolytica, *Gene.*, 297(1-2):229-239, 2002.

[17] Williams, I., Richardson, J., Starkey, A., and Stansfield, I., Genome-wide prediction of stop codon readthrough during translation in the yeast Saccharomyces cerevisiae, *Nucleic Acids Res.,* 32(22):6605-6616, 2004.

[18] Namy, O., Duchateau-Nguyen, G., Hatin, I., Hermann-Le Denmat, S., Termier, M., and Rousset, J.P., Identification of stop codon readthrough genes in Saccharomyces cerevisiae, *Nucleic Acids Res.*, 31(9):2289-2296, 2003.

[19] Lai, C.H., Chun, H.H., Nahas, S.A., Mitui, M., Gamo, K.M., Du, L., and Gatti, R.A., Correction of ATM gene function by aminoglycoside-induced read-through of premature termination codons, *Proc. Natl. Acad. Sci. USA*, 101(44):15676-15681, 2004.

[20] Engelberg-Kulka, H. and Schoulaker-Schwarz, R., Stop is not the end: physiological implications of translational readthrough, *J. Theor. Biol.*, 131(4):477-485, 1988.

[21] Li, W.H., Gu, Z., Wang, H., and Nekrutenko, A., Evolutionary analyses of the human genome, *Nature*, 409(6822):847-849, 2001.

[22] Zhang, Z., Harrison, P.M., Liu, Y., and Gerstein, M., Millions of years of evolution preserved: a comprehensive catalog of the processed pseudogenes in the human genome, *Genome Res.*, 13(12):2541-2558, 2003.

[23] Hartl, D.L. and Wirth, D.F., Duplication, gene conversion, and genetic diversity in the species-specific acyl-CoA synthetase gene family of Plasmodium falciparum, *Mol. Biochem. Parasitol.*, 150(1):10-24, 2006.

[24] Petrov, D.A. and Hartl, D.L., High rate of DNA loss in the Drosophila melanogaster and Drosophila virilis species groups, *Mol. Biol. Evol.,* 15(3):293-302, 1998.

[25] Lozovskaya, E.R., Nurminsky, D.I., Petrov, D.A., and Hartl, D.L., Genome size as a mutation-selection-drift process, *Genes. Genet. Syst.,* 74(5):201-207, 1999.

[26] Ota, T. and Nei, M., Evolution of immunoglobulin VH pseudogenes in chickens, *Mol. Biol. Evol.*, 12(1):94–102, 1995.

[27] Kleineidam, R.G., Jekel, P.A., Beintema, J.J., and Situmorang, P., Seminal-type ribonuclease genes in ruminants, sequence conservation without protein expression?, *Gene.*, 231(1-2):147-153, 1999.

[28] Lopez, L.B., Braga, M.B., Lopez, J.O., Arroyo, R., and Costa e Silva Filho, F., Strategies by which some pathogenic trichomonads integrate diverse signals in the decision-making process, *An. Acad. Bras. Cienc.*, 72(2):173-186, 2000.
[29] Kucknoor, A.S., Mundodi, V., and Alderete, J.F., The Proteins Secreted by Trichomonas vaginalis and Vaginal Epithelial Cell Response to Secreted and Episomally-Expressed AP65, *Cell. Microbiol.,* 2007.

STATISTICAL PROPERTIES AND INFORMATION CONTENT OF CALCIUM OSCILLATIONS

ALEXANDER SKUPIN MARTIN FALCKE
alexander.skupin@hmi.de falcke@hmi.de

Hahn-Meitner-Institut, Department of Theoretical Physics, Glienicker Str. 100, 14109 Berlin, Germany

Calcium is the most important second messenger in living cells serving as a critical link between a large variety of extracellular stimuli and the intracellular target. Often, the Ca^{2+} signal is carried by $[Ca^{2+}]$ oscillations. Our recent studies have demonstrated that in contrast to traditional ideas Ca^{2+} oscillations do not occur by simple synchronization of channel clusters opening and closing in an oscillatory fashion but originate from microscopic fluctuation caused by the stochastic binding of the ligands Ca^{2+} and IP_3 to the receptor's binding sites. They are orchestrated spatially on the cell level by wave nucleation. In this paper we analyze the stochastic data and show how internal properties can be determined from global observations. Further, we analyze the information content of spontaneous and stimulated oscillations.

Keywords: cell signaling; calcium oscillations; noise; information estimation.

1. Introduction

Calcium is a ubiquitous second messenger which regulates multiple cellular functions like gene expression, secretion, muscle contraction or synaptic plasticity. The Ca^{2+} signal employed by such a variety of processes is a transient increase of the intracellular concentration [2, 4, 7, 11, 15]. This $[Ca^{2+}]$ increase is due to entry through the cell membrane and to Ca^{2+} release from internal storage compartments especially the endoplasmic reticulum (ER) and the sarcoplasmic reticulum. Release from these stores is a nonlinear process. It leads in many cells to the formation of spatio-temporal signals in form of waves of high Ca^{2+} concentration travelling across the cell and global oscillations [7]. The information transmitted by these signals arrives as a stimulus at the plasma membrane and is translated into intracellular Ca^{2+} oscillations.

Ca^{2+} release from storage compartments is controlled by channels. A channel type present in the ER membrane of many cells is the inositol 1,4,5-trisphosphate (IP_3) receptor channel (IP_3R). The open probability of the IP_3R depends on the IP_3 concentration and the calcium concentration in the cytosol (see [7, 11, 14] for reviews). It increases with increasing Ca^{2+} concentration, i.e. Ca^{2+} release is self-amplifying. That is called Calcium Induced Calcium Release (CICR). Ca^{2+} released by one channel diffuses in the cytosol and increases the open probability of neighboring channels (see Fig. 1A). This coupling of channels by Ca^{2+} diffusion causes the spatial spread of release. Very high Ca^{2+} concentrations inhibit the channel and terminate release. Activation and inhibition together create a bell-shaped dependence of the open probability on cytosolic $[Ca^{2+}]$. There is some structure in the spatial arrangement of channels: They are grouped into clusters on the ER membrane containing 1-40 channels. Clusters are randomly scattered across the ER membrane. Due to their importance and their frequent appearance, Ca^{2+} oscillations were the representative of cellular oscillators most prominent in mathematical physiology and modeling for the last twenty years or so. Most traditional models assume cells to behave like stirred chemical reactors, i.e. neglect concentration gradients. They describe Ca^{2+} dynamics by deterministic ordinary differential equations [12]. Our recent studies have shown, that intracellular Ca^{2+} oscillations using IP_3 receptor channels are sequences of random spikes. Spikes are initiated by the local stochastic behavior of ion channels and transformed into a global Ca^{2+} transient by wave nucleation, i.e. by a spatial phenomenon.

Wave nucleation can be explained by the hierarchical picture of intracellular Ca^{2+} dynamics. It arose upon observation of random release events from single channel clusters - called puffs [3, 5, 7, 9, 10]. They can be considered as the elemental events of intracellular Ca^{2+} dynamics. According to the hierarchical concept, random opening of a single channel in a cluster causes the other channels of the same cluster to open thus generating a puff. That puff may cause neighboring clusters to open, too. If a supercritical number of puffs arises - a supercritical wave nucleus - release spreads through the whole cell. The existence of such a critical nucleus was shown experimentally and theoretically [6, 9, 10]. It introduces the dependence of the triggering probability of global events on the strength of spatial coupling by Ca^{2+} diffusion [6].

We have shown recently [13] that intracellular Ca^{2+} oscillations exhibit interspike-interval (ISI) distributions with an approximately linear relation between the average ISI and the standard deviation. We have measured the time course of Ca^{2+} concentrations during oscillations in four different cell types. Here, we analyze the data with respect to information on the properties of the specific cell and with respect to the information content of stimulated and spontaneous spike trains using the Kullback entropy as measure for information.

2. Methods and Results

2.1. *Experimental Methods*

We measured Ca^{2+} oscillations in four different cell types: spontaneous oscillations in astrocytes, microglia, in from fat tissue developed human stem cells (PLA) and by carbamyl choline (CCh) induced oscillations in human embryonic kidney cells (HEK). Cells were loaded with distinct fluorescence dyes and illuminated with monochromatic light to measure the global cytosolic Ca^{2+} concentration. The resulting fluorescence light was recorded after filtering with an appropriated bandpass

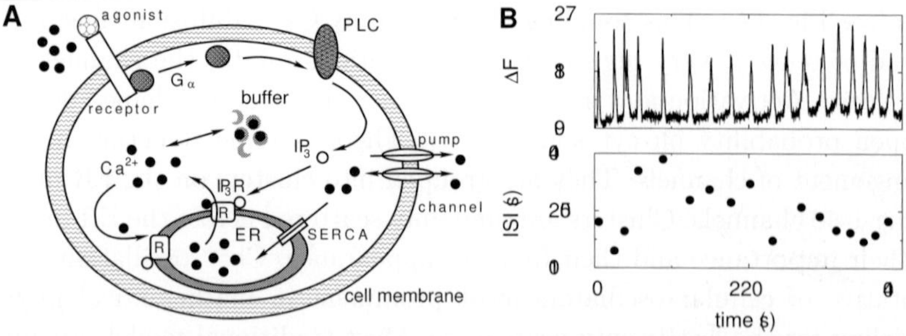

Fig. 1. **A**: Model of CICR. At stimulation with an agonist receptors in the cell membrane induce IP_3 (circles) production by a G-protein (G_α) activated PLC. If IP_3 and Ca^{2+} (dots) are bound to the IP_3R (R) it opens and Ca^{2+} diffuses from the ER into the cytosol. There it is caught by buffers and pumped back into the ER by SERCA pumps or into the extracellular space by pumps and can activate adjoining IP_3Rs. Some cell types express channels (gap junctions) to transmit the signal to neighboring cells. **B**: A typical fluorescence signal ΔF of a PLA cell is shown in the upper panel. In the lower panel for each Ca^{2+} spike the following ISI is shown indicating that Ca^{2+} oscillations have a stochastic character since the ISIs vary from about 100 s to 400 s.

filter by a digital camera with sample rates between 0.2 and 0.5 Hz.
The fluorescence signals F were scaled to the initial fluorescence level F_0 leading to $\Delta F = F/F_0$ shown in Fig. 1B in the upper panel. We determined the interspike interval (ISI) as the time between consecutive fluorescence maxima. The lower panel of Fig. 1B shows the ISI following each spike at the time of the spike. We obtained the characteristics of the Ca^{2+} oscillations as the averaged ISI T_{av} and their standard deviation σ from these series.
In the buffer experiments, we used BAPTA or EGTA which were added to cells by loading for 5 to 10 min. For more experimental details see [13].
The study conforms to the Declaration of Helsinki and all cell donors gave their informed written consent to use part of their fat tissue for the generation of processed lipoaspirate (PLA) cells.

2.2. *Characterizing the oscillations*

We characterize the statistical properties of these oscillations by means of the average interspike interval T_{av} and its standard deviation σ and their relation. The data of distinct cells and cell types are shown in Fig. 2, where σ is plotted over T_{av} for individual cells, each data point representing one cell. The data for astrocytes (black) and HEK cells (red) are shown in panel A and panel B contains those of microglia (black) and PLA cells (red). To quantify the different populations and for further analysis we also plotted the linear regression lines for each cell type within the corresponding colors having slopes $m_{astro} = 0.93$, $m_{HEK} = 0.58$, $m_{mic} = 1.01$ and $m_{PLA} = 0.7$.
There are three prominent features of the data. We observe a minimal ISI as there are no oscillations with T_{av} shorter than about 40 s, the standard deviation σ

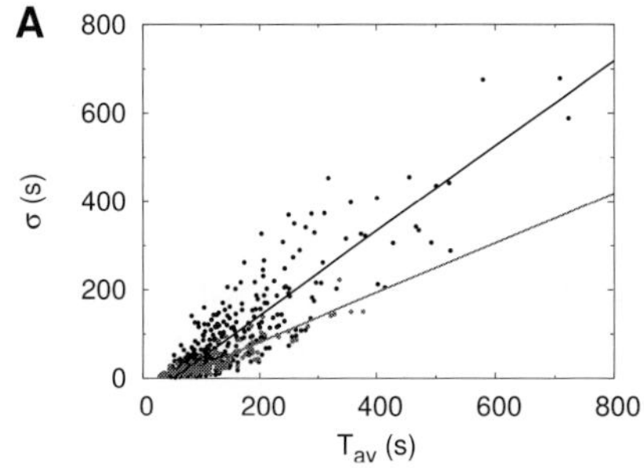

Fig. 2. Ca^{2+} spikes occur randomly. The standard deviation σ of the Ca^{2+} oscillation are in the same range as T_{av} shown for astrocytes (black) and stimulated HEK cells (red) in panel **A** and for microglia (black) and PLA cells (red) in panel **B** with the corresponding linear regression lines.

increases with T_{av} and σ reaches the same order of magnitude as T_{av} [13]. The minimal ISI arises from the existence of a deterministic part of the ISI. It is the minimal time required for the cell to recover from a concentration spike and reflects processes like store depletion, recovery from inhibition or negative feed back of Ca^{2+} to IP_3 production [13]. This deterministic time is specific for individual cells and thus we will refer to it as T_{cell}.

The fact that the standard deviations σ are in the same range as the T_{av} demonstrates the stochastic character of spike generation [13]. Moreover, the data structure indicates a linear dependence of σ on T_{av} which is in accordance with the hypothesized wave nucleation process [13]. The nucleation probability sets the stochastic part t of the interspike interval. It depends on the spatial organization of cells.

We conclude from these findings that a single interspike interval is given by the cell specific recovery time and a random time t as

$$\mathrm{ISI} = T_{cell} + t \,. \tag{1}$$

We denote the distribution density of t by $p(t)$. The average ISI of an individual cell has a contribution T_{cell} and a contribution arising from $p(t)$. The latter part specifies the shape of the σ-T_{av} relation

$$T_{av} = T_{cell} + \int_0^\infty t p(t) dt = T_{cell} + T_{stoch} \,. \tag{2}$$

The spread of data points in Fig. 2 represents properties of individual cells. They differ in T_{cell} generating spread in the direction of T_{av} and second in the properties of the stochastic process, shifting them in the σ-T_{av} plane along curves parameterized by characteristics of the stochastic process. In the following we will try to decompose the measured T_{av} according to expression (2) into its parts T_{stoch} and T_{cell} to get information about the processes shaping the data.

We start with the simplifying assumption that the cell recovers completely before a spike is set off and that the probability for a nucleation event is 0 during recovery and constant after recovery. Therefore, wave nucleation is expected to obey a Poisson process with its characteristic waiting time distribution density

$$p_{\text{poi}}(t) = \lambda e^{-\lambda t} , \tag{3}$$

where λ denotes the wave nucleation rate. A common feature of the Poisson process is the equality of its average and standard deviation. We therefore expect that T_{stoch} equals σ and is given by $T_{\text{stoch}} = \sigma = 1/\lambda$, i.e. the data of cells with the same T_{cell} should lie on a straight line with slope one. Consequently for diverse finite T_{cell} the data structure should be explainable by an array of lines with slope one. For a finite number of cells and with the reasonable assumption of an upper limit for T_{cell} the resulting population slope might be an appropriate approximation for that relation. Indeed, we can identify in Fig. 2 such dependence for the astrocytes (panel A black) and for the microglia (panel B black) by both, inspection by eye and by consideration of the population slopes that are around one.

Hence, we can determine the cell specific time T_{cell} by eq. (2) by setting $T_{\text{stoch}} = \sigma$ as

$$T_{\text{cell}} = T_{\text{av}} - \sigma . \tag{4}$$

On the other hand the red dots in Fig. 2 seem to disobey this interpretation by a Poisson process since the population slopes for PLA cells and stimulated HEK cells are different from one.

We extend our assumptions on the relation between recovery and nucleation to explain these data. We assume that the spike generation rate is 0 only at the time immediately after a spike and relaxes to a constant during the recovery process. A spike might be set off before recovery is complete. Note that this matches well with ideas that the recovery time is set by recovery from channel inhibition, store refilling or similar processes. We study a nucleation rate $\Lambda(t)$ increasing in time $\Lambda(t) = \lambda(1 - e^{-\rho t})$ where ρ can be interpreted as regeneration rate. Therewith the probability density to observe a Ca^{2+} transient at time t takes the form

$$p_\rho(t) = \lambda\left(1 - e^{-\rho t}\right) \exp\left[-\int_0^t \lambda\left(1 - e^{-\rho t'}\right) dt'\right] , \tag{5}$$

illustrated in Fig. 3A. With this definition we can calculate the first two moments as

$$T_{\text{av}} = \frac{e^{\frac{\lambda}{\rho}}\left(\frac{\lambda}{\rho}\right)^{1-\frac{\lambda}{\rho}}}{\lambda}\left[\Gamma\left(\frac{\lambda}{\rho}\right) - \Gamma\left(\frac{\lambda}{\rho}, \frac{\lambda}{\rho}\right)\right] \tag{6}$$

$$\langle T^2 \rangle = \frac{2e^{\frac{\lambda}{\rho}}}{\lambda^2}\mathcal{F}\left[\left(\frac{\lambda}{\rho}, \frac{\lambda}{\rho}\right), \left(1 + \frac{\lambda}{\rho}, 1 + \frac{\lambda}{\rho}\right), -\frac{\lambda}{\rho}\right] \tag{7}$$

where $\Gamma(x)$ denotes the Euler Γ-function, $\Gamma(x, y)$ the incomplete Gamma function and $\mathcal{F}[\mathbf{x}]$ is the generalized hypergeometric function ${}_2F_2(\mathbf{x})$ [1]. The standard deviation given by

$$\sigma = \sqrt{\langle T^2 \rangle - {T_{\text{av}}}^2} \tag{8}$$

can be calculated and plotted over T_{av} in dependence of λ and ρ as shown in Fig. 3B, where lines with constant λ are obtained by varying ρ and vice versa for lines with constant ρ. Note that we did not include a deterministic part T_{cell}. That would

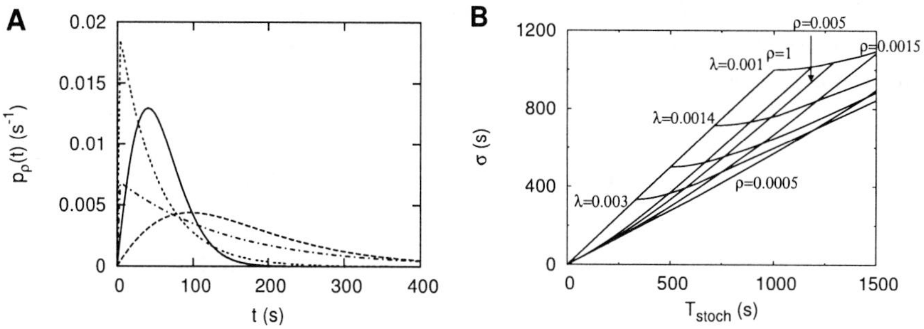

Fig. 3. Influence of a time dependent nucleation rate. **A**: The density (5) for different parameters. For high values of ρ $p_\rho(t)$ merge into a pure Poisson process as shown for $\rho = 1$ and $\lambda = 0.02$ (dotted line) and $\lambda = 0.007$ (dash dotted line). With decreasing ρ the maximum of the distribution is shifted to higher values of t. For fixed $\rho = 0.01$ the width increases with decreasing λ as for the examples $\lambda = 0.06$ (solid line) and $\lambda = 0.01$ (dashed line). **B**: The σ-T_{av} relation in dependence of the nucleation rate λ and the regeneration rate ρ given by eqs. (6) and (7). By comparison with the data in Fig. 2A we conclude that stimulated HEK cells have high λ and small ρ.

shift the curves to the right. The analysis shows that for small regeneration rates the slope of the σ-T_{av} relation decreases and can be approximated for large T_{av} and σ by the coefficient of variation CV. By inserting (6) and (7) into equation (8) we obtain the functional dependence of the form

$$m \approx \mathrm{CV} = \frac{\sigma}{T_{av}} = f\left(\frac{\lambda}{\rho}\right) \tag{9}$$

which depends on the ratio $r = \lambda/\rho$ only. From this we conclude that the stimulated HEK cells have high λ and small ρ. A possible physiological interpretation of this result is the following. The intracellular IP_3 concentration is high due to the stimulation so that the IP_3Rs are sensitized permanently leading to Ca^{2+} signals arising before cells are in their rest states again.

Nevertheless one might explain the distinct data structure by a large diversity of T_{cell}. Additionally, we relied on the slope of the linear regression (m_{pop}) averaging over the cell populations for information on the slope of the σ-T_{av} relation. To clarify these issues we performed experiments providing approximations for the slope of the σ-T_{av} relation. The stochastic part T_{stoch} depends on the nucleation rate λ and we can consequently characterize the stochastic process by changing λ and study the repercussion on σ and T_{av}. λ depends strongly on the concentration of Ca^{2+} binding proteins (buffers) in the cytosol as they determine the diffusion length of free Ca^{2+} and hence the strength of spatial coupling between channel clusters [7]. Higher buffer concentrations yield in lower nucleation rates. On the other hand we do not expect the small buffer concentrations used to have an influence on the regeneration rate ρ or on the deterministic time T_{cell} since both are not very sensitive to the cytosolic buffering capacity. Hence, we predict cells to be shifted on lines corresponding to $\rho = constant$ in Fig. 3B by lowering λ. We denote the slope of the σ-T_{av} relation estimated from the data points before and after buffer addition as m_{shift}. If the population slopes before and after adding additional buffer are in the same range as m_{shift}, we can take them as estimate of the slope of the σ-T_{av} relation.

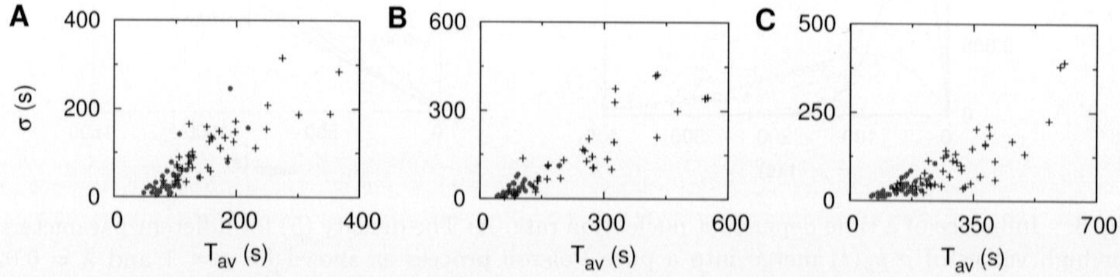

Fig. 4. Ca^{2+} buffers influence both σ and T_{av} in astrocytes (**A**, **B**) and in HEK cells (**C**). For experiments with astrocytes we loaded cells with 20 nM BAPTA (**A**) as well as with 1 μM EGTA (**B**) showing a similar effect. HEK cells were loaded with 1 μM BAPTA (**C**). For all cell types and buffer concentrations cell seems to be shifted in the direction of the population.

In the experimental realization we measured oscillations in astrocytes and HEK cells for several tens of minutes, loaded additional Ca^{2+} buffer (BAPTA as well as EGTA) and restarted the measurements. The resulting data are shown in Fig. 4, where again the standard deviation σ is plotted over T_{av} for astrocytes (A, B) and HEK cells (C) before (red dots) and after (blue crosses) buffer loading. In the astrocytes experiments we used two distinct buffers, BAPTA (A) and EGTA (B), with different concentrations, namely 20 nM and 1 μM respectively, leading to a similar effect,what is in accordance with theoretical predictions [7].
We determined the population slopes m_{pop} by linear regression and the average shifting slope m_{shift} for the analysis. We find that the population slopes are quite self consistent and are in good agreement with the shift slopes. Despite the variability of m_{shift} shown in table 1, we nevertheless find a separation between the astrocytes and the HEK cells indicating the different dynamical regimes of both populations.

Table 1. Comparison between the population slopes m^b_{pop} before and m^a_{pop} after buffer application and the average shifting slope m_{shift}.

kind of experiment	m_{shift}	m^b_{pop}	m^a_{pop}
astrocytes with 10 nM BATA	1.01 ± 0.16	0.84	0.91
astrocytes with 20 nM BATA	0.94 ± 0.23	1.04	0.89
astrocytes with 1 μM EGTA	0.88 ± 0.19	0.86	0.79
HEK cells with 1 μM EGTA	0.49 ± 0.21	0.56	0.58

We now return to the decomposition of T_{av}. From the results above we expect that relation (4) should work for astrocytes, i.e. that the distributions $p(T_{cell})$ obtained for data without and with additional buffer should coincide. In addition they should be comparable with the distribution calculated from the intersection of the individual shifting lines with the T_{av} axis. The verification is shown in Fig. 5A, in which the three distribution have a similar shape and posture of their maxima. The naive use of eq. (4) for HEK cells will lead to failure as can be seen in Fig. 5B caused by different parameters for HEK cells in relation (5). Rescaling of σ by the inverse of the corresponding m_{pop} leads again to a nice consistency of the distribution as shown in the inset of 5B. The small contributions to $p(T_{cell})$ with negative T_{cell} probably occur due to spike trains which are not yet equilibrated.

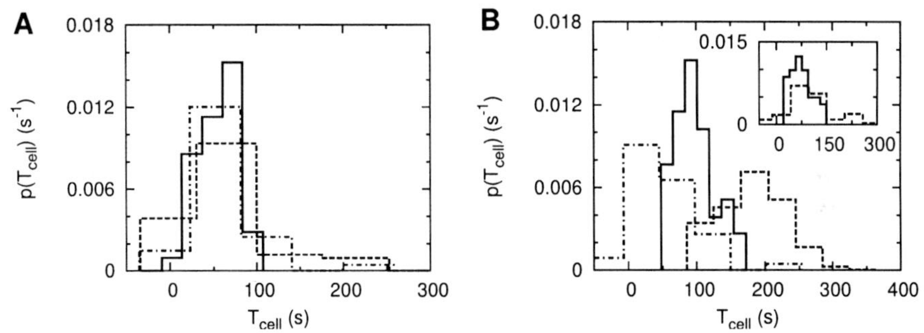

Fig. 5. Comparison of distribution densities of $p(T_{cell})$ obtained by (4) before and after buffer application (solid and dashed lines respectively) and the density determined by shifting lines (dash dotted lines) for astrocytes (**A**) loaded with 20 μM BAPTA and HEK cells (**B**) loaded with 1 μM BAPTA. The distributions coincide very well for astrocytes, whereas diverge for HEK cells caused by the fact that latter have a time dependent nucleation rate. The inset shows the corrected distributions taking that into account.

We have thus demonstrated that stimulated HEK cells have distinct stochastic properties than spontaneously oscillating astrocytes and that this difference can be estimated by the structure of the population data. Based on that we presented how to extract the intrinsic property T_{cell} of a cell out of the global observations σ and T_{av}.

2.3. *Information theory*

In order to interpret the experimental results we are interested in the information distance of a pure (eq. 3) and a time dependent (eq. 5) Poisson process. Therefore we look at the information difference by switching from one process to another one given by the Kullback entropy [8]

$$\mathcal{K}(p_0, p_{st}) = k \int_0^\infty p_0(t) \log \frac{p_0(t)}{p_{st}(t)} dt. \tag{10}$$

By plugging in $p_0(t) = p_{\text{poi}}(t)$ and $p_{\text{st}}(t) = p_\rho(t)$ into equation (10) we find for the information gain the analytical solution

$$\mathcal{K} = k\left[H\left(\frac{\lambda}{\rho}\right) + \frac{1}{\left(1+\frac{\lambda}{\rho}\right)} - 1\right] \tag{11}$$

where $H(\lambda/\rho)$ denotes the continuous harmonic series given by

$$H(x) = \gamma + \frac{d}{dx}\log\Gamma(x+1) \tag{12}$$

with the Euler-Mascheroni constant γ and the Gamma function $\Gamma(x+1)$. That means for $\rho \to \infty$, i.e. $p_\rho \to p_{\text{poi}}$, $\mathcal{K}$ goes to zero as expected and for $\rho \ll \lambda$ it turns out that $\mathcal{K} \to \log(\lambda/\rho)$. Note that in this description the nucleation rates λ for both processes were equated. The analysis in the previous section has shown that HEK cells act at higher λs what additionally increases the information gain and thus expression (11) is a lower estimate.

The expression (11) for $\mathcal{K}$ depends only on the ratio $r = \lambda/\rho$ as the relation for the slope m in eq. (9) does, thus we can estimate the information contained in Ca^{2+} oscillations by the slope m. In Fig. 6A the relations (9) and (11) are shown in dependence on r where we used natural units leading to $k = 1$. Fig. 6B displays the dependence of $\mathcal{K}$ on m.

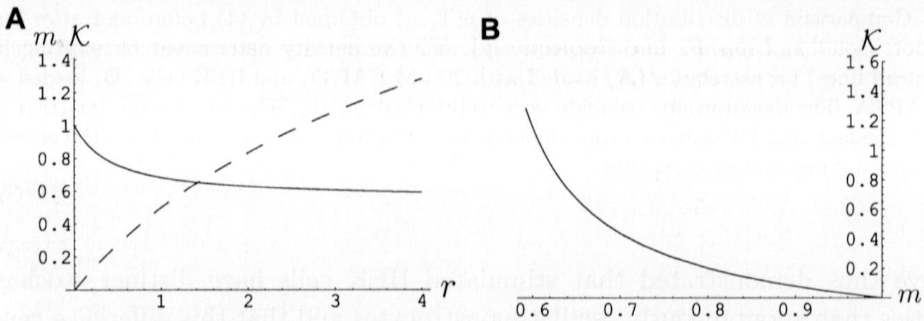

Fig. 6. **A**: The slope $m \approx$ CV (solid line) and the information gain $\mathcal{K}$ (dashed line) in dependence of the ratio $r = \lambda/\rho$. **B**: Due to that relation we can estimate the detectable information in a signal by the experimental population slope m_{pop} of the σ-T_{av} relation. Values for $\mathcal{K}$ were calculated within natural units, i.e. $k = 1$.

The experimental findings make sense in this context. The spontaneous oscillations in astrocytes and microglia might demonstrate the structure of the internal spatial signaling network within cells. The organization of IP_3R clusters is an optimized design from the information theoretical point of view, since it corresponds to a Poisson process. That is the process with the maximal entropy for processes which are restricted to positive stochastic values with mean μ. Therefore this arrangement is able to detect an eventually occurring signal most efficiently in terms of entropy export. Further we can deduce from the data of HEK cells that these cells perform information processing. In Fig. 2B we found for the PLA cells a slope $m_{\mathrm{PLA}} = 0.7$, which is between the stimulated HEKs and the spontaneous glia cells. That result is suggestive, since these stem cells are in the phase of cell differentiation in which Ca^{2+} oscillations might control gene expression.

3. Discussion

Based on the observation that Ca^{2+} oscillations occur randomly [13] we here analyzed that stochastic process. The ISI consists of a deterministic part T_{cell} and a stochastic part T_{stoch}. We could extract both of them from global observations. We showed that the stochastic part of spontaneous oscillations obeys a Poisson process, whereas stimulated oscillations can be described by a time dependent nucleation rate. Variability of the deterministic part causes about 25% of the spread of T_{av} of a cell population. The remaining 75% arise from the stochastic part T_{stoch}. We have shown by an information theoretical approach that stimulated oscillations indeed carry a signal in comparison to spontaneous behavior.

References

[1] Abramowitz, M. and Stegun, A., *Handbook of Mathematical Functions*, 9th edition, Dover Publication, New York, 1970.

[2] Berridge, M., Inositol trisphosphate and calcium signalling, *Nature*, 361: 315–325, 1993.

[3] Berridge, M., Elementary and global aspects of calcium signalling, *J. Physiol.*, 499, 291–306, 1997.

[4] Berridge, M., Lipp, P., and Bootman, M., The versatility and universality of calcium signalling, *Nature Rev. Mol. Cell Biol.*, 1, 11–22, 2000.

[5] Bootman, M., Niggli, E., Berridge, M., and Lipp, P., Imaging the hierarchical Ca^{2+} signalling in HeLa cells, *J. Physiol*, 499, 307–314, 1997.

[6] Falcke, M., On the role of stochastic channel behavior in intracellular Ca^{2+} dynamics, *Biophys. J.*, 84, 42–56, 2003.

[7] Falcke, M., Reading the patterns in living cells - the Physics of Ca^{2+} signaling, *Advances in Physics*, 53, 255–440, 2004.

[8] Goychuk, I. and Hänggi, P., Stochastic resonance in ion channels characterized by information theory, *Phys. Rev. E*, 61, 4272–4280, 1999.

[9] Marchant, J., Callamaras, N., and Parker, I., Initiation of IP_3-mediated Ca^{2+} waves in Xenopus oocytes, *The EMBO J.*, 18, 5285–5299, 1999.

[10] Marchant, J. and Parker, I., Role of elementary Ca^{2+} puffs in generating repetitive Ca^{2+} oscillations, *The EMBO Journal*, 20, 65–76, 2001.

[11] Putney, J. and Bird, G., The inositolphosphate-calcium signaling system in nonexcitable cells, *Endocrine Reviews*, 14: 610–631, 1993.

[12] Schuster, S., Marhl, M., and Höfer, T., Modelling of simple and complex calcium oscillations, *Eur. J. Biochem.*, 269, 1333–1355, 2001.

[13] Skupin, A., *et al.*, Constructive use of noise leads to intracellular Ca^{2+} oscillations, *submitted*, 2007.

[14] Taylor, C., Inositol trisphosphate receptors: Ca^{2+}-modulated intracellular Ca^{2+} channels, *Biochimica and Biophysica Acta*, 1436, 19–33, 1998.

[15] Tsien, R. and Tsien, R., Calcium channels, stores and oscillations, *Annu. Rev. Cell Biol.*, 6, 715–760, 1990.

A MINIMAL CIRCADIAN CLOCK MODEL

ILKA M. AXMANN
i.axmann@biologie.hu-berlin.de

STEFAN LEGEWIE
s.legewie@biologie.hu-berlin.de

HANSPETER HERZEL
h.herzel@biologie.hu-berlin.de

ITB, Institute for Theoretical Biology, Humboldt University of Berlin, Invalidenstrasse 43, D-10115 Berlin, Germany

The coordination of biological activities into daily cycles provides an important advantage for the fitness of diverse organisms. Thereby, an internal circadian oscillator drives gene expression in an approximate 24 hours rhythm. Circadian clocks are found in most eukaryotes. In prokaryotes only cyanobacteria are known to regulate their activities in a circadian rhythm. *In vitro* experiments showed that three cyanobacterial proteins KaiA, KaiB and KaiC together with ATP are sufficient to generate temperature-compensated circadian oscillations of KaiC protein phosphorylation. Thus, in contrast to eukaryotic clock models the cyanobacterial core oscillator operates independently of transcription and translation processes. Most previous models of the bacterial circadian clock used complex mathematical descriptions. Here, we suggest a minimal and manageable heuristic system. Even though only four reaction steps were assumed, our model exhibited sustained oscillations of KaiC phosphorylation. A simulation of known experimental data was successful as well as oscillations maintained even for a concerted increase of Kai protein concentration. Thus, we provided a useful minimal system of differential equations which might serve as a core module of the holistic cyanobacterial clockwork in the future.

Keywords: circadian clock; oscillation; feedback; cyanobacteria; mathematical model.

1. Introduction

Numerous biological activities follow rhythmic cycles with a circa 24 hour period. Even in the absence of environmental stimuli (light, temperature) this behavior preserves a so-called free-running rhythm. The underlying core oscillator is termed circadian clock, which is ticking in most eukaryotes. The 24 h period of the free-running rhythm is not affected by normal changes in the natural environment. For several examples it is shown that the clock can compensate for perturbations within a physiological range. On the other hand, a circadian clock is able to be entrained to an environmental cycle as light [7] or temperature pulses [21].
Circadian clocks were longtime thought to be restricted to the eukaryotic kingdom. However, the optimal temporal coordination of biological processes and adaptation to daily fluctuations play an important role for diverse organisms. Green plants and phototrophic bacteria are dependent on a 12 h light - 12 h dark rhythm be-

cause of their photosynthetic activity. An internal circadian clock has been found in these photoautotrophic organisms as well. In particular, for the cyanobacterium, *Synechococcus elongatus*, a robust circadian cycling has been observed even under constant darkness. In contrast to eukaryotic clock models the cyanobacterial clock keeps operating in the presence of transcription and translation inhibitors [30]. Thus, compared to former clock models this core oscillator operates independently of transcription and translation processes. Moreover, only three different cyanobacterial proteins (KaiA, KaiB, KaiC) together with ATP are sufficient to achieve temperature-compensated 24 h rhythms of KaiC phosphorylation *in vitro* [23].

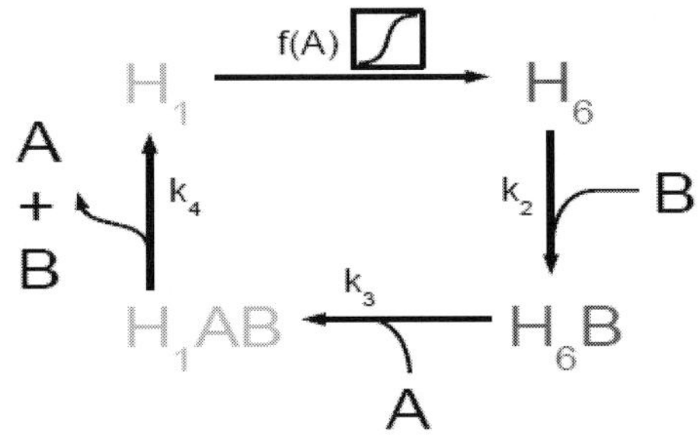

Fig. 1. Minimal model of the cyanobacterial circadian core clock comprising KaiA, KaiB and KaiC proteins. (1) Depending on the available amount of free KaiA proteins, KaiC hexamers switch between a de-(H_1) and a phosphorylated (H_6) state. (2) KaiB proteins bind to phosphorylated KaiC hexamers resulting in a conformational change of the KaiBC protein complex (H_6B). (3) KaiA proteins join the KaiBC complex. Conformational change and sequestration of free KaiA by complexation favor KaiC dephosphorylation. (4) The dephosphorylated complexes (H_1AB) break-up and release KaiA, KaiB and dephosphorylated KaiC hexamers (H_1) which restarts the circle. Default parameters of the reaction kinetics are given in Table 1.

Over the past 10 years numerous experimental investigations gave insights into molecular details of the cyanobacterial clock. Nevertheless, it remained unclear how a biochemical mechanism missing protein synthesis and degradation can keep time so precisely over long periods of several days. Now various modeling methods have been applied to the KaiABC system to simulate the chemical network that is able to generate self-sustained oscillations. In summary, multiple phosphorylation states [19], allosteric rearrangement [33], KaiA sequestration [1], monomer exchange [2, 21] or different combinations of them [29, 38] have been suggested to serve as basic mechanisms producing synchronized oscillations. Other theoretical approaches introduced a positive feedback on KaiC phosphorylation [17] or multiple states of KaiA and KaiB [15], but these models appear to be inconsistent to latest experiments [13, 21].

In order to study the recently proposed mechanism of KaiA sequestration [1] leading to oscillation of the KaiABC clock we applied a heuristic approach. Only four reaction steps were assumed which sufficed to generate sustained oscillations. Known experimental data were simulated successfully exhibiting robust oscillatory behavior even for the concerted increase of Kai protein concentration. Thus, we defined a useful minimal model of the core circadian oscillator which might serve as a basic module to design more complex networks of the holistic cyanobacterial clockwork.

2. Model assumptions

Extensive experimental studies on cyanobacterial cells and their clock proteins revealed details about the molecular background of circadian oscillations. Here, the experimental observations were translated into a condensed mathematical model (Fig. 1). The 6 variables of the system, A, B, H_1, H_6, H_6B and H_1AB represent the concentrations of clock proteins, KaiA, KaiB, KaiC and their complexes. The dynamics of these 6 variables is described by the following system of differential equations:

$$\frac{dH_1}{dt} = -f(A) \cdot H_1 + k_4 \cdot H_1AB \quad (1)$$

$$\frac{dH_6}{dt} = f(A) \cdot H_1 - k_2 \cdot B \cdot H_6 \quad (2)$$

$$\frac{dH_6B}{dt} = k_2 \cdot B \cdot H_6 - k_3 \cdot A \cdot H_6B \quad (3)$$

$$\frac{dH_1AB}{dt} = k_3 \cdot A \cdot H_6B - k_4 \cdot H_1AB \quad (4)$$

$$\frac{dB}{dt} = k_4 \cdot H_1AB - k_2 \cdot B \cdot H_6 \quad (5)$$

$$\frac{dA}{dt} = k_4 \cdot H_1AB - k_3 \cdot A \cdot H_6B \quad (6)$$

KaiC hexamers: For our model, we considered KaiC proteins as stable hexamers (termed H) because experiments showed that this seems to represent the normal situation within the living cell [10, 20, 25]. KaiC possesses an auto-kinase and -phosphatase activity [25, 30]. Two main phosphorylation sites (T432 and S341) were described for the KaiC monomer [35], which resulted in 12 phosphorylatable sites per KaiC hexamer. **Phosphorylation:** KaiA proteins (A) were demonstrated to interact with KaiC hexamers and to enhance KaiC phosphorylation [14, 26, 31, 32, 37]. For simplification, we lumped multiple phosphorylation steps to a single reaction which is assumed to be catalyzed by KaiA. Thereby, depending on the available amount of free KaiA proteins, KaiC hexamers switch between a dephosphorylated (H_1) and a phosphorylated (H_6) state. **Conformational change:** Further it has been described that KaiB attenuates KaiA-enhanced phosphorylation of KaiC [14]. In the model, we assumed that KaiB proteins bind to highly phosphorylated KaiC hexamers. This might mediate a conformational change of the KaiBC protein complex (H_6B) so that KaiC dephosphorylation is initiated because the enhancer KaiA no longer stimulates KaiC phosphorylation. **Dephosphorylation:** As described by

different authors [6, 12], during the subjective night all three Kai proteins, KaiA, KaiB and KaiC, form stable complexes *in vivo* with yet unknown stoichiometry. In our model we assumed that KaiA joins the KaiBC complex (H_6AB) which decreases the concentration of free KaiA proteins which in turn leads to a decreased rate of KaiC phosphorylation. Thus, the sequestration of KaiA to the KaiBC complex constitutes the necessary feedback, which is indispensable for biological oscillations [1, 4]. **Break-up:** The dephosphorylated complexes (H_1AB) were assumed to become instable and to break-up. This last step led to the release of KaiA, KaiB and dephosphorylated KaiC hexamers (H_1), thus, re-starting the circle.
We chose linear and bilinear kinetics for all steps except for the KaiA-mediated KaiC phosphorylation (Fig.1). This step is assumed to depend on the concentration of free KaiA in a non-linear fashion. Our assumption seems to be justified because previous theoretical studies demonstrated [8, 9, 28] that multiple phosphorylation states result in a switch-like behavior. Thus, we modeled phosphorylation of 12 sites in the KaiC hexamer using a Hill function:

$$f(A) = \frac{f_{max} \cdot A^h}{{K_1}^h + A^h} \tag{7}$$

Using a Runge-Kutta algorithm the system of differential equations was solved numerically. MATLAB (The MathWorks, Natick, MA) implementation was helpful to analyze robustness of our model and to perform the bifurcation analysis. ASAMIN, a MATLAB gateway routine to ASA (Adaptive Simulated Annealing; www.ingber.com), was applied to fit our model to quantitative experimental data.

3. Simulations and Results

The numerical solution of our minimal model (Fig. 2) was able to generate sustained 24 hour oscillations for a chosen set of default parameters, listed in Tab. 1. The initial concentrations of Kai proteins were chosen from literature [13], which resemble the default protein concentration of the *in vitro* experiment. The simulation of the KaiC phosphorylation cycle is shown in Fig. 2 (upper graph) in the way it can be measured experimentally by time-resolved measurements of KaiC phosphorylation. The temporal development of the 6 variables of our system, A, B, H_1, H_6, H_6B and H_1AB is plotted below (Fig. 2 (lower graph)). Surprisingly, the simulated time course of KaiC phosphorylation is qualitatively very similar to experimental observations: The initial high amplitude until the system is tuned, can be found in diverse *in vitro* experiments [13, 23]. A detailed analysis of all model species (Fig. 2 (lower graph)) demonstrates that H_1 and H_6 exhibited the largest changes in concentration – dephosphorylated and phosphorylated KaiC hexamers, whereas KaiA, KaiB as well as all complexes, H_6B and H_1AB, show only small amplitude ranges (Fig. 2 (lower graph)).
Although our first simulations exhibited sustained oscillations, the sensitivity of

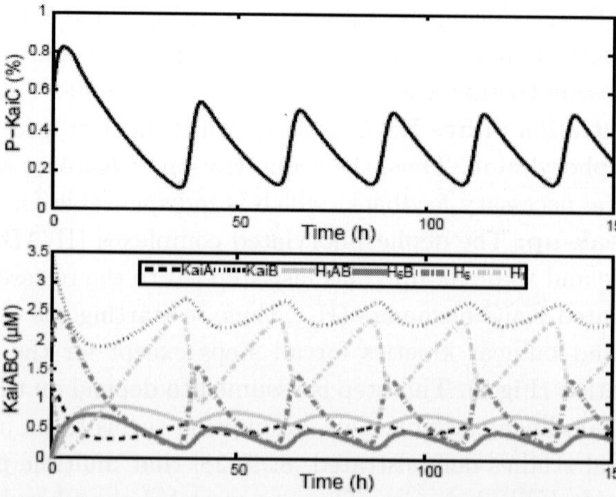

Fig. 2. The numerical solution of our minimal model using the default parameters, listed in Tab. 1. The simulation of the phosphorylated amount of KaiC (P-KaiC) in the system is shown (upper graph) exhibiting variables H_6 and H_6B normalized to the total concentration of KaiC ($KaiC_T$). This time course is qualitatively similar to *in vitro* experiments. The temporal development of the 6 variables of the system, A, B, H_1, H_6, H_6B and H_1AB is plotted below.

the system with respect to parameter changes remained to be tested. We performed a bifurcation analysis and analyzed 10-fold changes in both directions for each parameter. As main characteristics for oscillatory behavior amplitude and period were calculated (Fig. 3 left). Here, our model appeared to be robust over a 20-fold range with respect to parameters k_2 and k_4. The dephosphorylation rate, k_3, was observed to be more sensitive because oscillation were restricted to a 10-fold parameter range. We also analyzed the sensitivity of the oscillations towards the Hill coefficient in Eq. 7. These simulations clearly showed that a lower limit of $h = 11$ exists. The observed change in period was small for all parameters except for k_4,

Table 1. Model parameters

parameter	default value	interpretation
h	38	Hill coefficient
K_1	0.6 μM	ratio of phosphorylation and dephosphorylation rates
f_{max}	0.6 h^{-1}	maximal stimulatory effect
k_2	0.04 $h^{-1}\mu M^{-1}$	complex formation rate
k_3	0.4 $h^{-1}\mu M^{-1}$	dephosphorylation rate
k_4	0.1 h^{-1}	break-up rate of complexes
A_T	1.2 μM	concentration of KaiA
B_T	3.5 μM	concentration of KaiB
C_T	3.5 μM	concentration of KaiC

the break-up rate (Fig. 3 right). Here, our system exhibited the highest sensitivity to variations. This observation might indicate that complex break-up is the slowest and, therefore, the rate limiting step of our system.

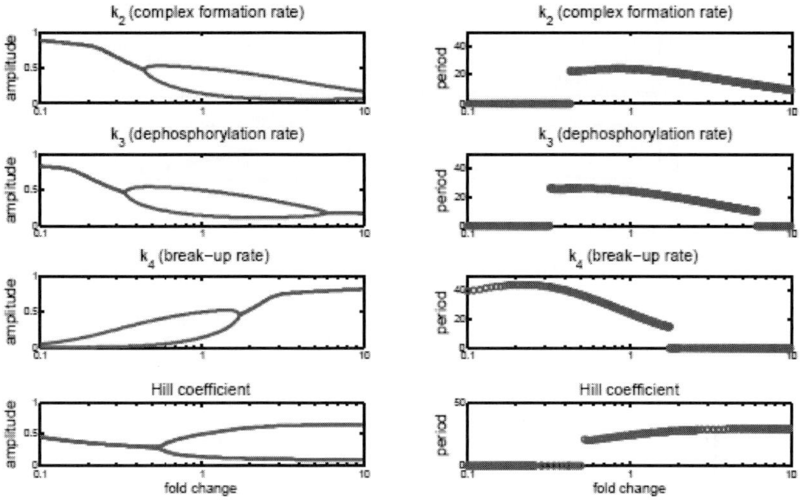

Fig. 3. Default parameters, k_2, k_3, k_4, and Hill coefficient are changed within a +/- 10-fold range from the default values given in Table 1. The observed amplitudes (left) and periods (right) are plotted visualising sensitivities and oscillation ranges.

Additionally, we tested model behavior with respect to changes of the protein concentrations. According to our bifurcation analysis shown in Fig. 4, we observed sustained oscillations for an at least 10-fold range of KaiB and KaiC concentrations as well as for a concerted variation of all three proteins. Further, it turned out that variation of KaiA concentration is the most critical element of our system. Here, amplitude and period were lost due to less than 5-fold change. The sensitivity to KaiA concentration might be explained by the fact that we assumed a KaiA-dependent Hill function for the first reaction (Eq. 7). More specifically, the concentration of free KaiA needs to cycle around the threshold parameter K_1 in order to switch phosphorylation on and off.

In summary, our results were in agreement with observations by Kageyama and colleagues [13]. They varied protein concentrations individually and showed that P-KaiC oscillations are slightly more sensitive to variation of KaiA than to KaiB which is in accordance with our simulations (compare Fig. 4 top left and 2nd row left). One still missing experiment would be the increase of KaiC protein amount for the *in vitro* experiment. Here, our model predicts oscillatory behavior even if KaiC is increased more than 10-fold, see Fig. 4 3rd row. Further, Kageyama's experiments showed that a concerted 5-fold increase in all protein concentration led to phosphorylation rhythms nearly identical to those seen under standard conditions, whereas decreased concentrations of proteins led to non-oscillatory behavior as seen

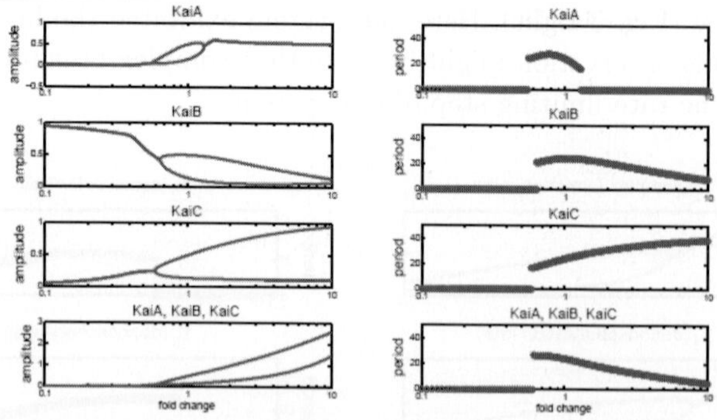

Fig. 4. Default concentrations of Kai proteins, KaiA, KaiB and KaiC, are changed 10-fold individually and for all three proteins. The corresponding amplitudes and periods are plotted visualising sensitivities and oscillation ranges.

in the model (Fig. 4 bottom left). Our model failed to reproduce Kageyama's observation that the amplitude and period remained unchanged upon a 5-fold increase of protein concentration. Using bimolecular reactions and a strong non-linearity (Eq. 7) the model appeared to be more sensitive to the variation of protein levels.

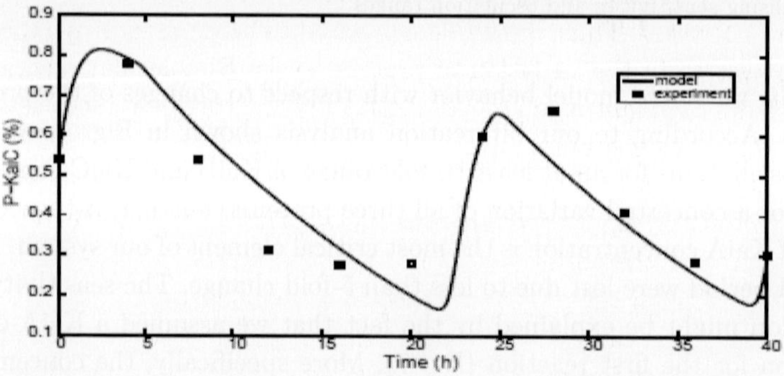

Fig. 5. Fit of our minimal circadian clock model (solid line) to quantitative experimental data (open squares) [13] using ASAMIN optimization algorithm. P-KaiC resembles the phosphorylated amount of KaiC in the system exhibiting variables H_6 and H_6B normalized to the total concentration of KaiC ($KaiC_T$). The optimized values of parameters are h = 100.0, $K_1 = 0.5838\ \mu M$, $f_{max} = 0.7033\ h^{-1}$, $k_2 = 0.0594\ h^{-1}\mu M^{-1}$, $k_3 = 0.4299\ h^{-1}\mu M^{-1}$ and $k_4 = 0.17\ h^{-1}$.

To test whether our minimal model of a circadian clock might be in quantitative agreement with experiments, we used an optimization procedure to fit the model to experimental data. Using the optimized parameter set, again our minimal model generated sustained P-KaiC oscillations. Moreover, the experimentally observed large amplitude of KaiC phosphorylation was achieved [13]; see Fig. 5.

4. Discussion and outlook

Here, we introduced a minimal model of the circadian protein oscillator of a cyanobacterial cell using a system of ordinary differential equations. The system comprised only 6 variables which represent the concentrations of clock proteins and their complexes. Thus, a system of differential equations was obtained which can be interpreted directly in biological terms because each experimentally observed protein and protein complex is given explicitly by a variable. For further simplification we focused on linear and bilinear kinetics for all steps of our model except the KaiA-mediated KaiC phosphorylation. This reaction is assumed to depend in a highly non-linear manner on the concentration of free KaiA. The assumption seems to be justified by several theoretical approaches [8, 9, 28] demonstrating that multiple phosphorylation states can result in a switch-like behavior. Thus, we included KaiA-mediated phosphorylation of KaiC by using a Hill function. Surprisingly, this minimal model was able to simulate sustained 24 hour oscillations despite the numerous simplifications. Even the large amplitude between KaiC phosphorylation and dephosphorylation level was achieved by fitting experimental data. One has to state here that a reduced model is always just a cartoon of biological reality whose predictions are expected to fail in details. For example, our cartoon of the protein clock was less successful to simulate a robust behavior of oscillation period and amplitude against concerted increase of protein concentration. When we simulated this scenario, oscillations were sustained but the period increased and the amplidude was lowered. Thus, our model appeared to be more sensitive than the real biochemical system to the variation of protein levels. Bimolecular reactions and a strong non-linearity might mediate this sensitivity to protein variations. To achieve a model as robust as the biochemical system, one or more additional steps might be included in the future. For example, an additional rate-limiting, monomolecular step (which might, e.g., represent KaiC dephosphorylation) between reactions 2 and 3 might improve the performance of our model. Much more extensive modeling approaches focusing on the cyanobacterial clock demonstrated that the robustness of circadian oscillation can be simulated. In these models, many more reaction steps were required and often unknown or less-explored states were assumed by the investigators which improved robust oscillatory behavior. To date, essential kinetic data are still not available, which makes it difficult to define a comprehensive mathematical description. Our minimal model can be extended stepwise by different kinetic mechanisms in order to get insights into the core mechanisms underlying robustness. For an oscillatory system at least one feedback and a certain delay is needed [1, 4]. In our minimal model, the sequestration of KaiA to the KaiBC complex constitutes the necessary feedback. In general, high Hill coefficients, an explicit delay or Michaelis-Menten kinetics can reduce the number of reaction steps that are needed to obtain oscillations. Our small system based on only four reactions including one non-linear step required a high Hill coefficient which kept the number of parameters in the system low. It was shown for other oscillatory systems that a necessary delay can be caused by several processes, e.g. posttranslational modification, degradation, complex formation or nuclear import and export [27, 34]. In our case – the *in vitro* clock of Kai proteins – degradation and transport processes can be neglected. Thus, a delay might be caused by multiple phosphorylations on the KaiC hexamer modeled as a switch or by formation and break-up of Kai protein complexes. Our bifurcation analysis revealed that oscillation period was most sensitive towards the break-up rate, k_4, exhibited the highest period range. This might indicate that the break-up of complexes is the rate limiting step and, therefore, the cause of a significant delay in our system. Previous minimal models of biological oscillations such as the Goodwin model have proven to be useful tools, as they provide insights into the basic mechanism of oscillations. Also, these minimal models can easily be extended in order to understand the role of additional regulatory loops. We hope that our minimal circadian clock model in the same way might serve as a simple module to be integrated into more complex models. For example, it has been suggested that circadian timing in cyanobacteria might be additionally regulated via transcriptional / translational feedback loops [11, 16, 22, 29]. Such TTO-loops could simply be added to our core model in order to understand their impact on time keeping. The cyanobacterial Kai proteins studied here do not share sequence similarity to any known eukaryotic clock component. Nevertheless, phospho-proteins involved in circadian timing had been found in diverse organisms. Therefore, it is conceivable that a clock which is solely based on posttranslational modifications ('phoscillator' [18]) might be a general mechanism mediating circadian rhythms. Today, even for higher eukaryotes a circadian phoscillator is in discussion. Accordingly, it was shown that sustained oscillations of PER protein, a key mediator of the transcriptional feedback loop, were maintained even if its mRNA was constitutively expressed such that its coding mRNA levels were no longer under circadian control [3, 5, 24, 36]. Thus, even for eukaryotic systems transcription-translation feedback might not be the core of the circadian mechanism. Here, we suggested a useful circadian clock model which might serve as a module of holistic clockworks in the future. Compared to more complex mathematical models, in our minimal system one can comprehensively study the dynamic behavior. Our simulations confirm that KaiA sequestration can lead to self-sustained oscillations as observed experimentally.

References

[1] Clodong, S., Duhring, U., Kronk, L., Wilde, A., Axmann, I., Herzel, H., and Kollmann, M., Functioning and robustness of a bacterial circadian clock, *Mol. Syst. Biol.*, 3:90, 2007.

[2] Emberly, E., and Wingreen, N.S., Hourglass model for a protein-based circadian oscillator. *Phys. Rev. Lett.*, 96(3):038303, 2006.

[3] Fan, Y., Hida, A., Anderson, D.A., Izumo, M., and Johnson, C.H., Cycling of cryptochrome proteins is not necessary for circadian-clock function in mammalian fibroblasts, *Curr. Biol.*, 17(13):1091–1100, 2007.

[4] Friesen, W.O. and Block, G.D., What is a biological oscillator?, *Am. J. Physiol*, 246(6 Pt 2):R847–853, 1984.

[5] Fujimoto, Y., Yagita, K., and Okamura, H., Does mPER2 protein oscillate without

its coding mRNA cycling?: post-transcriptional regulation by cell clock, *Genes Cells*, 11(5):525–530, 2006.

[6] Garces, R.G., Wu, N., Gillon, W., and Pai, E.F., Anabaena circadian clock proteins KaiA and KaiB reveal a potential common binding site to their partner KaiC, *EMBO J.*, 23(8):1688–1698, 2004.

[7] Geier, F., Becker-Weimann, S., Kramer, A., and Herzel, H., Entrainment in a model of the mammalian circadian oscillator, *J. Biol. Rhythms*, 20(1):83–93, 2005.

[8] Goldbeter, A. and Koshland, D.E. Jr, An amplified sensitivity arising from covalent modification in biological systems, *Proc. Natl. Acad. Sci. USA*, 78(11):6840–6844, 1981.

[9] Gunawardena, J., Multisite protein phosphorylation makes a good threshold but can be a poor switch, *Proc. Natl. Acad. Sci. USA*, 102(41):14617–14622, 2005.

[10] Hayashi, F., Suzuki, H., Iwase, R., Uzumaki, T., Miyake, A., Shen, J.R., Imada, K., Furukawa, Y., Yonekura, K., Namba, K., and Ishiura, M., ATP-induced hexameric ring structure of the cyanobacterial circadian clock protein KaiC, *Genes Cells*, 8(3):287–296, 2003.

[11] Iwasaki, H., Nishiwaki, T., Kitayama, Y., Nakajima, M., and Kondo, T., KaiA-stimulated KaiC phosphorylation in circadian timing loops in cyanobacteria, *Proc. Natl. Acad. Sci. USA*, 99(24):15788–15793, 2002.

[12] Kageyama, H., Kondo, T., and Iwasaki, H., Circadian formation of clock protein complexes by KaiA, KaiB, KaiC, and SasA in cyanobacteria, *J. Biol. Chem.*, 278(4):2388–2395, 2003.

[13] Kageyama, H., Nishiwaki, T., Nakajima, M., Iwasaki, H., Oyama, T., and Kondo, T., Cyanobacterial Circadian Pacemaker: Kai Protein Complex Dynamics in the KaiC Phosphorylation Cycle In Vitro, *Mol. Cell*, 23(2):161–171, 2006.

[14] Kitayama, Y., Iwasaki, H., Nishiwaki, T., and Kondo, T., KaiB functions as an attenuator of KaiC phosphorylation in the cyanobacterial circadian clock system, *EMBO J.*, 22(9):2127–2134, 2003.

[15] Kurosawa, G., Aihara, K., and Iwasa, Y., A model for the circadian rhythm of cyanobacteria that maintains oscillation without gene expression, *Biophys J.*, 91(6):2015–2023, Jun 2006.

[16] Kutsuna, S., Nakahira, Y., Katayama, M., Ishiura, M., and Kondo, T., Transcriptional regulation of the circadian clock operon kaiBC by upstream regions in cyanobacteria, *Mol. Microbiol*, 57(5):1474–1484, 2005.

[17] Mehra, A., Hong, C.I., Shi, M., Loros, J.J., Dunlap, J.C., and Ruoff, P., Circadian rhythmicity by autocatalysis, *PLoS Comput. Biol.*, 2(7):e96, 2006.

[18] Merrow, M., Mazzotta, G., Chen, Z., and Roenneberg, T., The right place at the right time: regulation of daily timing by phosphorylation, *Genes Dev.*, 20(19):2629–2633, 2006.

[19] Miyoshi, F., Nakayama, Y., Kaizu, K., Iwasaki, H., and Tomita, M., A mathematical model for the Kai-protein-based chemical oscillator and clock gene expression rhythms in cyanobacteria, *J. Biol. Rhythms*, 22(1):69–80, 2007.

[20] Mori, T., Saveliev, S.V., Xu, Y., Stafford, W.F., Cox, M.M., Inman, R.B., and Johnson, C.H., Circadian clock protein KaiC forms ATP-dependent hexameric rings and binds DNA, *Proc. Natl. Acad. Sci. USA*, 99(26):17203–17208, 2002.

[21] Mori, T., Williams, D.R., Byrne, M.O., Qin, X., Egli, M., McHaourab, H.S., Stewart, P.L., and Johnson, C.H., Elucidating the Ticking of an In Vitro Circadian Clockwork. *PLoS Biol.*, 5(4):e93, 2007.

[22] Nair, U., Ditty, J.L., Min, H., and Golden, S.S., Roles for sigma factors in global circadian regulation of the cyanobacterial genome, *J. Bacteriol*, 184(13):3530–3538,

2002.

[23] Nakajima, M., Imai, K., Ito, H., Nishiwaki, T., Murayama, Y., Iwasaki, H., Oyama, T., and Kondo, T., Reconstitution of circadian oscillation of cyanobacterial KaiC phosphorylation in vitro, *Science*, 308(5720):414–415, 2005.

[24] Nishii, K., Yamanaka, I., Yasuda, M., Kiyohara, Y.B., Kitayama, Y., Kondo, T., and Yagita, K., Rhythmic post-transcriptional regulation of the circadian clock protein mPER2 in mammalian cells: a real-time analysis, *Neurosci Lett.*, 401(1-2):44–48, 2006.

[25] Nishiwaki, T., Iwasaki, H., Ishiura, M., and Kondo, T., Nucleotide binding and autophosphorylation of the clock protein KaiC as a circadian timing process of cyanobacteria, *Proc. Natl. Acad. Sci. USA*, 97(1):495–499, 2000.

[26] Pattanayek, R., Williams, D.R., Pattanayek, S., Xu, Y., Mori, T., Johnson, C.H., Stewart, P.L., and Egli, M., Analysis of KaiA-KaiC protein interactions in the cyanobacterial circadian clock using hybrid structural methods, *EMBO J.*, 25(9):2017–2028, 2006.

[27] Reppert, S.M. and Weaver, D.R., Coordination of circadian timing in mammals, *Nature*, 418(6901):935–941, 2002.

[28] Salazar, C. and Hofer, T., Versatile regulation of multisite protein phosphorylation by the order of phosphate processing and protein-protein interactions, *FEBS J.*, 274(4):1046–1061, 2007.

[29] Takigawa-Imamura, H. and Mochizuki, A., Transcriptional autoregulation by phosphorylated and non-phosphorylated KaiC in cyanobacterial circadian rhythms, *J. Theor. Biol.*, 241(2):178–192, 2006.

[30] Tomita, J., Nakajima, M., Kondo, T., and Iwasaki, H., No transcription-translation feedback in circadian rhythm of KaiC phosphorylation, *Science*, 307(5707):251–254, 2005.

[31] Uzumaki, T., Fujita, M., Nakatsu, T., Hayashi, F., Shibata, H., Itoh, N., Kato, H., and Ishiura, M., Crystal structure of the C-terminal clock-oscillator domain of the cyanobacterial KaiA protein, *Nat. Struct. Mol. Biol.*, 11(7):623–631, 2004.

[32] Vakonakis, I. and LiWang, A.C., Structure of the C-terminal domain of the clock protein KaiA in complex with a KaiC-derived peptide: implications for KaiC regulation, *Proc. Natl. Acad. Sci. USA*, 101(30):10925–10930, 2004.

[33] van Zon, J.S., Lubensky, D.K., Altena, P.R., and Ten Wolde, P.R., An allosteric model of circadian KaiC phosphorylation, *Proc. Natl. Acad. Sci. USA*, 104(18):7420–7425, 2007.

[34] Vanselow, K., Vanselow, J.T., Westermark, P.O., Reischl, S., Maier, B., Korte, T., Herrmann, A., Herzel, H., Schlosser, A., and Kramer, A., Differential effects of PER2 phosphorylation: molecular basis for the human familial advanced sleep phase syndrome (FASPS), *Genes Dev.*, 20(19):2660–2672, 2006.

[35] Xu, Y., Mori, T., Pattanayek, R., Pattanayek, S., Egli, M., and Johnson, C.H., Identification of key phosphorylation sites in the circadian clock protein KaiC by crystallographic and mutagenetic analyses, *Proc. Natl. Acad. Sci. USA*, 101(38):13933–13938, 2004.

[36] Yang, Z. and Sehgal, A., Role of molecular oscillations in generating behavioral rhythms in Drosophila, *Neuron*, 29(2):453–467, 2001.

[37] Ye, S., Vakonakis, I., Ioerger, T.R., LiWang, A.C., and Sacchettini, J.C., Crystal structure of circadian clock protein KaiA from Synechococcus elongatus, *J. Biol. Chem.*, 279(19):20511–20518, 2004.

[38] Yoda, M., Eguchi, K., Terada, T.P., and Sasai, M., Monomer-Shuffling and Allosteric Transition in KaiC Circadian Oscillation, *PLoS ONE*, 2:e408, 2007.

PROMOTER ANALYSIS OF MAMMALIAN CLOCK CONTROLLED GENES

KATARZYNA BOŻEK[1]
k.bozek@biologie.hu-berlin.de

SZYMON M. KIEŁBASA[2]
kielbasa@molgen.mpg.de

ACHIM KRAMER[3]
achim.kramer@charite.de

HANSPETER HERZEL[1]
h.herzel@biologie.hu-berlin.de

[1] *Institute for Theoretical Biology, Humboldt University, Invalidenstraße 43, D-10115 Berlin, Germany*
[2] *Max Planck Institute for Molecular Genetics, Ihnestrasse 63–73, 14195 Berlin, Germany*
[3] *Laboratory of Chronobiology, Institute of Medical Immunology, Charité Universitätsmedizin Berlin, Hessische Str. 3–4, 10115 Berlin, Germany*

The circadian clock is a biological system providing an internal self-sustained temporal framework and adaptation mechanisms to the daily environmental rhythm. One of its behavioral implication in humans is the sleep-wake cycle. The core mammalian circadian clock is a system composed of interacting regulatory feedback loops present in many tissues throughout the body. The core set of circadian clock genes codes for proteins feeding back to regulate not only their own expression, but also that of clock output genes and regulatory pathways. Still, however, our understanding of processes regulated in a circadian fashion and the linkage between the molecular system and behavioral or physiological outputs is poor. Our work aims at identification of clock-controlled genes (CCGs) and their regulatory motifs. We analyzed several microarray measurements of genes with a daily oscillating expression and extracted 2065 of them together with their peak expression phases and oscillation amplitudes. For an in-depth analysis we selected a subset of 167 genes reported by multiple microarray experiments. Gene promoters were scanned in the search for known regulatory motifs of clock genes (E-Box, RRE, D-Box, CRE) as well as other over-represented regulatory motifs. We found an over-representation of the E-boxes and D-boxes in the selected subset of 167 CCGs. This over-representation is smaller when the list of 2065 genes is analyzed. The search for other regulatory motifs contained in the TRANSFAC database revealed a strong over-representation of some of them such as Sp1, AP-2, STAT1, HIF-1 and E2F. The signals found in the promoter sequences indicate possible regulatory mechanisms important for the coordination of circadian rhythms.

Keywords: circadian clock; clock controlled genes; DNA arrays; promoter; transcription factor binding site (TFBS); position specific count matrix (PSCM).

1. Introduction

The circadian clock, an internal oscillating system with a period of about 24 hours, is one of the most ubiquitous and preserved biological timing systems. It provides an internal temporal framework and regulates the activities of an organism in relation to environmental cycles. Some of its implication on the human behavior, physiology or metabolism are the sleep-wake cycle, hormone rhythms and core body temperature fluctuations.

On the molecular level the mammalian circadian system is composed of interacting positive and negative gene regulatory feedback loops. The key components are the basic helix-loop-helix transcription factors CLOCK and BMAL1 forming a complex that activates the transcription of Per, Cry and RevErbα genes. The PER-CRY protein complex abrogates the transcriptional activity of CLOCK-BMAL1 after translocation to the nucleus. This way the PER and CRY proteins inhibit their own expression. On the other hand, by acting on CLOCK-BMAL1, the complex of PER-CRY indirectly increases its own expression through abolishing the expression of RevErbα — an inhibitor of Bmal1 [1]. Such a system of interconnected positive and negative feedback loops of transcription, translation, protein-protein interaction, phosphorylation, nuclear translocation and protein degradation contributes to delays that create a coordinated molecular cycle approximating the 24h environmental period.

The molecular circadian clock is present in many mammalian tissues throughout the body. The core set of circadian clock genes codes for proteins feeding back to regulate not only their own expression, but also that of clock output genes. The clock transcription factors either directly control expression of target genes or indirectly through regulation of other transcription factors.

Studies suggest a hierarchical organization of the circadian regulation in a body with the central pacemaker located in the suprachiasmatic nucleus (SCN) in brain [2]. The master clock is self-sustained and entrained to the daily light/dark cycle. It transmits synchronizing signals to local circadian oscillators in peripheral tissues to achieve and maintain adaptive phase control. Hence, this model of hierarchical circadian oscillators is responsible for regulating the rhythmic outputs observable, e.g., in behavior and physiology.

Although the core molecular pacemaker generating circadian rhythms has been defined both in the SCN and peripheral organs, the molecular outputs that ultimately regulate circadian control of cellular physiology, organ function and behavior are poorly understood. The inter-tissue synchronization mechanism, sustainment of peripheral oscillators or regulation of output pathways by the circadian clock remain an open question. To decipher the interactions among the oscillators and the link between circadian transcriptional output and physiology it is important to study clock-controlled genes, i.e. genes that are under the direct or indirect transcriptional control of the clock transcription factors.

This work aims at identification of *cis* regulatory motifs of clock-controlled genes. Numerous gene-expression analyses have shown that the mRNA levels of many clock-related genes have high-amplitude circadian oscillations in the SCN or peripheral tissues. Those genes that are not an essential part of the core clock system but reveal circadian oscillations in their mRNA levels are named clock-controlled genes. Recent studies show their importance in mediating local processes within a specific tissue or in such fundamental processes as cell-cycle or metabolism [1]. Investiga-

tion of the genes and their regulation is a way to come closer to the understanding of the circadian regulation of output pathways in different tissues, synchronization of peripheral circadian oscillators and adaptation to the environmental day-night rhythm.
First, through a literature search, a set of microarray experiments has been found and a list of genes expressed in a circadian manner has been assembled. Next, a promoter analysis of these genes has been performed. We analyze a subset of best annotated genes as well as the full list of genes in the search for known clock-regulated motifs. Lastly, a complete search for any over-represented motif from the TRANSFAC database has been completed. We confirm known regulatory motifs such as E-boxes and predict novel binding sites including AP-2, E2F, STAT1, HIF-1 and Sp1.

2. Method and Results

2.1. *Meta-analysis of microarray experiments*

We performed an extensive literature search in order to find published microarray experiments reporting genes expressed in a circadian manner in mammalian tissues. An initial list of publications [3–13] has been limited to those containing complete gene annotation and full information on gene expression levels and phases.
From the selected papers [4, 5, 8, 10, 12, 13] a gene list has been assembled. After unification of the overlapping genes and removal of inconsistencies of annotations the list contains 2065 genes expressed in several mouse tissues such as liver, heart, SCN, skeletal muscle and in rat fibroblasts. The overlap among the tissues is limited: 77 genes are expressed in 2 tissues, 23 in 3 and more. The mean deviation of measured peak expression phases of the same gene expressed in two different tissues is striking (e.g. 5.8 h for the liver–SCN genes, 4.5 h for liver–heart genes) which is consistent with the observations from other studies [12].
In the first analysis we selected a subset of 167 top-scoring genes that appear in at least three published gene lists. The fact that these genes have been detected by independent experiments indicates that their circadian expression is robust.

2.2. *Promoter analysis*

Following this data search we extracted sequences ranging over 3 kbp upstream and 2 kbp downstream of the transcription start site (TSS) of each gene. The choice of the region has been motivated by previous promoter studies [14–17] detecting clock-related *cis* elements within several hundred base pairs upstream of a gene TSS up to its first intron. The sequences have been downloaded from EnsEMBL 43 mouse genome [18].
In the search of transcription factor binding sites we used a method for the background model computation and the cut-off threshold estimation for the predicted sites proposed by Rahmann *et al.* [19] implemented with BioMinerva framework [20].

In addition to the basic search in the mouse genome, we performed homology tests by reducing the search space to the conserved parts of promoter sequences of other species orthologous genes (rat, human) as reported by BLASTZ [21].

2.3. *Search for known motifs*

In the first stage of this study we have chosen four *cis* elements that are known to participate in the circadian regulation: E-boxes, DBP/E4BP4 binding elements (D-boxes), RevErbα/ROR binding elements (RREs) and cyclic AMP responsive elements (CRE) [15, 22]. Position Specific Count Matrices (PSCMs) of the three last motifs have been taken from TRANSFAC version 10.4 [23], whereas the E-box matrix has been constructed separately basing on eight published experiments on vertebrate tissues [14–17, 24–27]. As suggested in [25] the binding affinity of specific transcription factors to the short 6 bp E-box might be supported by the flanking nucleotides. Hence in the construction of the E-box PSCM 4 bp upstream and downstream of the E-box have been aligned. The sequence logos [28] of the profiles recognized by the four transcription factors are depicted in Fig. 1.

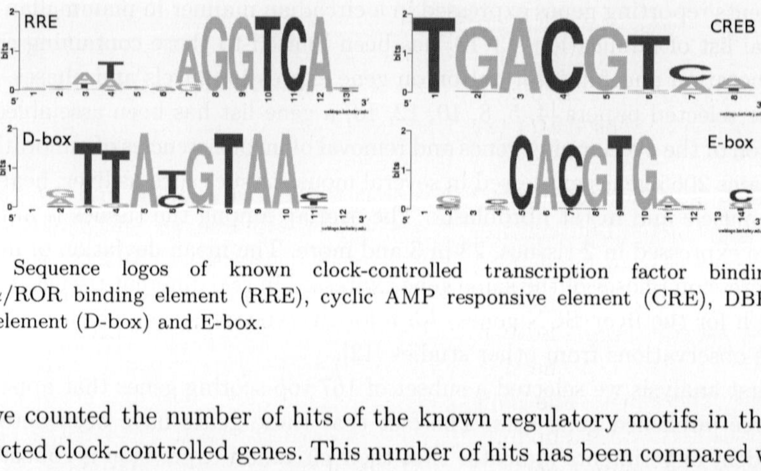

Fig. 1. Sequence logos of known clock-controlled transcription factor binding sites: RevErbα/ROR binding element (RRE), cyclic AMP responsive element (CRE), DBP/E4BP4 binding element (D-box) and E-box.

First, we counted the number of hits of the known regulatory motifs in the list of 167 selected clock-controlled genes. This number of hits has been compared with the average number of hits obtained for the promoters of sets of genes of the same size (167) sampled 100 times randomly from a subset of 5000 mouse genes not containing those reported as clock-controlled. The promoter region is defined as discussed above. The average number of hits of each motif and the standard deviation of samplings has been calculated. The same sampling of background sequences has been repeated on the full list of 2065 clock-controlled genes. The results of these tests are presented in Table 1.

An over-representation of E-box and D-box motifs can be observed in the promoters of the selected clock-controlled genes as compared to random mouse genes. The differences between the CCGs and random genes exceed 3 standard deviations of the random sampling in the case of E-boxes and nearly 2 standard deviations in the case of D-boxes. The number of motifs in the random samples is approximately

Table 1. Known clock-controlled transcription factor binding sites predicted in the promoters of 167 selected genes. D-boxes and E-boxes are over-represented in the selected promoters when compared to the predictions done on different background models — a random subset of 5000 mouse genes not containing the CCGs and the full list of 2065 CCGs. Standard deviations of the background samples are given in parentheses.

Motif	Consensus sequence	Hits in CCGs	Random EnsEMBL promoters	Full list of CCGs promoters
RRE	DNWWNDAGGTCAH	235	227 (31.7)	221 (14.2)
CRE	TGACGTMW	41	39 (11.8)	50 (12.6)
D-box	NRTTAYGTAAYN	**162**	**135 (17.7)**	**127 (14.2)**
E-box	NSNMCACGTGWNNS	**166**	**101 (19.3)**	**135 (18.5)**

normally distributed. Thus an excess of motifs in the clock controlled genes of more than 2 standard deviations can be considered as significant. The smaller number of hits in the case of sampling from the full list of CCGs justifies our selection of the subset of 167 genes. In other words, the meta-analysis of multiple microarray experiments enhances the over-representation of E-boxes. Therefore the smaller set of well selected genes allows an easier detection of signals in the promoter analysis. The fact that CRE and RRE do not appear as over-represented regulatory motifs of the analyzed CCGs might be due to the selection of the genes for this analysis. These motifs as well as D-boxes are known to be relevant for the gene regulatory network of the core oscillator [22] whereas we analyze mainly large sets of output genes.

To validate the relevance of the E-box construction we counted and compared the number of predicted binding sites of another E-box matrix proposed by Kiełbasa *et al.* [29] as well as the number of canonical E-box sequences (`CACGTG`) in the CCG and random promoters. Since both approaches lead to equally high scores (data not shown) we consider our PSCM to be well defined and use it in the following tests. Likewise, we performed the site search and counting procedure on the full list of clock-controlled genes. The number of hits was compared to the average number of hits obtained from sampling the same number of genes (2065) from randomly chosen EnsEMBL mouse genes. The results of this test are presented in Table 2.

Table 2. Known clock-controlled transcription factor binding sites predicted in the set of 2065 clock-controlled genes. E-boxes and CRE motifs are considerably over-represented when compared to the predictions done on a random subset of 5000 mouse genes not containing the CCGs. Standard deviations of the background samples are given in parentheses.

Motif	Consensus sequence	Hits in CCGs	Random EnsEMBL promoters
RRE	DNWWNDAGGTCAH	2626	2872 (104.9)
CRE	TGACGTMW	**619**	**533 (35.2)**
D-box	NRTTAYGTAAYN	1566	1730 (35.2)
E-box	NSNMCACGTGWNNS	**1637**	**1303 (58.4)**

The search for the E-boxes confirms our previous results, however the D-box signal disappears. This might be due to noise in the data as well as different regulatory mechanism of the selected subset of 167 clock-controlled genes.
Fig. 2 contains a schematic illustration of the predicted transcription factor binding sites (TFBSs) and their conservation on the promoter of Per1. All predicted CRE sites and 4 out of 5 predicted E-box sites are confirmed by other studies [16]. The unconfirmed E-box lies outside of the region analyzed by Hida *et al.* [16].

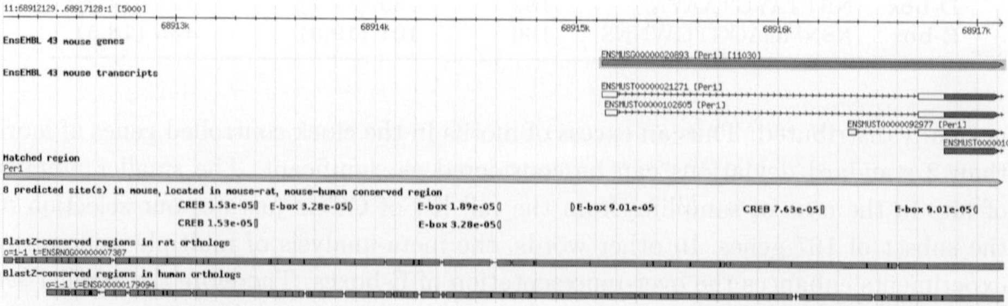

Fig. 2. Predicted conserved TFBSs in the region 3 kbp upstream and 2 kbp downstream of the TSS of Per1 mouse gene.

Another example shown in Fig. 3 is the binding site prediction of the RevErbα, a gene encoding a member of the nuclear receptor superfamily. This gene is known to be activated by CLOCK-BMAL1 heterodimer binding to E-box sequences and to be regulated through D-boxes and RRE motifs [22, 30, 31]. The E-box predictions in the first intron are in accordance with other studies [31], the co-occurrence of D-boxes could be an indication of a combinatorial regulatory mechanism.

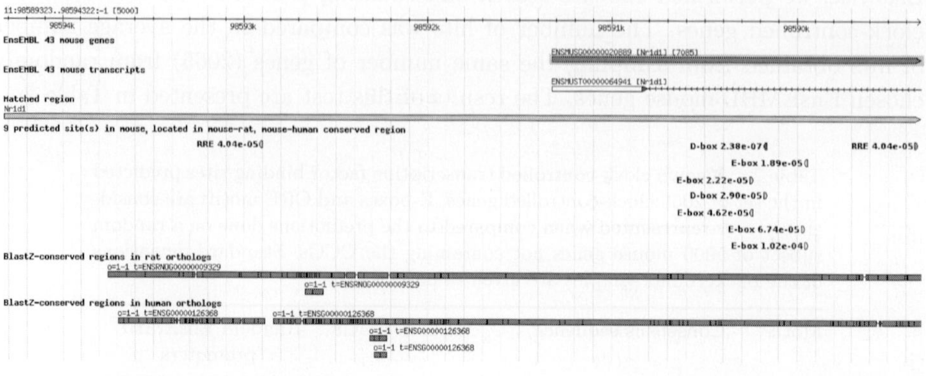

Fig. 3. Predicted conserved TFBSs in the region 3 kbp upstream and 2 kbp downstream of the TSS of RevErbα (Nr1d1) mouse gene.

2.4. *TRANSFAC motifs search*

Having detected an over-representation of some of the known clock-related transcription factor binding sites, we continued the analysis with the search for an over-representation of any TRANSFAC regulatory motif in the clock-controlled gene promoters. All 815 position specific count matrices from the TRANSFAC version 10.4 have been downloaded and the search of the binding sites has been performed on the 5 kbp promoter regions of the subset of 167 CCGs. The number of hits of each of the motifs has been calculated and compared to the average number of hits of the given motif in sampled random mouse gene promoters.

Since this method resulted in a large number of highly over-represented motifs, in the further analysis we considered only those constructed from mammalian experiments and with the number of hits greater than 3 standard deviations above the average number of hits in random genes. The limited list has been clustered according to their sequence similarity [29]. The results of this analysis are shown in the Table 3.

Table 3. A selection of discovered over-represented TRANSFAC binding sites in the 167 selected clock-controlled genes. Genes presented in the table have the number of hits greater than 3 standard deviations above the average number in random genes. The random genes are sampled from a set of 5000 mouse genes not containing the CCGs. Standard deviations of the background samples are given in parentheses.

Name	Hits in CCGs	Random EnsEMBL promoters	Name	Hits in CCGs	Random EnsEMBL promoters
DEAF1	149	96 (17.4)	E2F	78	39 (7.8)
HIC1	377	214 (21.7)	E2F-1	64	46 (6.0)
Pax-4	559	420 (33.3)	E2F-1:DP-2	64	39 (5.5)
VDR	682	534 (43.6)	GC box	885	616 (61.7)
WT1	1098	788 (81.5)	Lmo2 complex	359	279 (22.5)
HES1	188	136 (16.1)	LRF	328	235 (30.5)
KROX	1074	721 (70.8)	MAZ	983	728 (52.3)
Nrf-1	229	139 (25.7)	Muscle initiator	687	440 (44.0)
NF-kappaB (p50)	254	162 (17.3)	MZF	659	540 (34.9)
STAT1	159	112 (14.6)	NF-Y	210	153 (16.1)
HNF4	1062	674 (53.1)	NRSF	495	391 (24.6)
HIF-1	85	49 (10.3)	Sp1	1148	729 (63.9)
AP-2	415	247 (31.8)	SREBP-1	314	236 (17.6)
AP-2alpha	426	269 (31.7)	USF2	326	233 (26.8)
AP-2gamma	450	299 (37.4)	ZF5	245	123 (21.9)
CLOCK:BMAL	165	112 (13.3)	ZNF219	1015	674 (58.1)
CP2	318	250 (20.9)			

The same comparison test has been done on the promoters of orthologous genes of rat and human. We found a big overlap in the over-representation of motifs in the orthologous gene promoters. Most of the clustered motifs (AP-2, E2F, DEAF1, HES1, HIC1, KROX, MAZ, Muscle initiator, MZF, NF-kappaB, Nrf-1, NRSF, Pax-4, Sp1, VDR, WT1, ZF5) show high representation signals in the promoters of clock-controlled genes of rat and human as well.

The over-representation of the CLOCK:BMAL binding site in spite of the high cut-

off value criteria confirms our previous results and validates our construction of the E-box matrix. Another confirmation is USF2, a motif belonging to the Myc-Max matrix cluster, showing a strong similarity to the E-box sequence [29]. When a cut-off threshold of 2 standard deviations is considered other members of the Myc-Max cluster appear to be over-represented (N-Myc, USF, c-Myc:Max).
To study if the enhancement of certain regulatory motifs is due to the GC content similarity of promoters with the motifs, we calculated the GC ratio of the analyzed promoter regions. The GC content of the regulatory motifs has been obtained from their position count matrices by counting the ratio of G and C nucleotides to all nucleotides of sequences used in the matrix construction. Most of the over-represented motifs have a high GC ratio of the composing sequences (e.g. Sp1 – 0.79, GC box – 0.68), significantly above the average ratio of the TRANSFAC matrices (0.48). However even for such a GC-rich motif as Sp1 its over-representation in the predictions is still higher than expected from the ratio of the GC composition of the CCG promoters (0.49) to the one of background sequences (0.46).

3. Discussion

The identification of clock-controlled genes and understanding of their regulation is a step towards linking the molecular circadian system with a phenotype. In this study we generated a list of clock-controlled genes based on a set of microarray experiments. We performed promoter analysis in the search for known clock transcription factor binding sites in a selected subset as well as in the full list of CCGs. Having found a strong over-representation of E-boxes and D-boxes we continued with the search for any over-represented regulatory motif included in the TRANSFAC database. We found a set of motifs with a number of hits over 3 standard deviations bigger than in the sampled sets of random mouse genes (Table 3).
The predicted participation of the Sp1 transcription factor in the circadian regulation is consistent with recent biological results. It was shown by electrophoretic mobility shift assays that Sp1 binds certain DNA sequences in a circadian fashion (Hans Reinke and Ueli Schibler, personal communication).
Most of the predicted clock-related transcription factor binding sites have a high GC content. The affinity of the GC-rich *cis* elements to clock-controlled gene promoters could be therefore a feature specific to the clock-controlled gene regulation.
The approach used can help to identify statistically observable regulatory motifs of this specific gene set. However, due to potential false positives in the TFBS search the predictions on the level of individual genes should be additionally validated.
However, if the predicted transcription factors are a part of the circadian system and show an oscillating transcriptional activity, their analysis can help to reveal biological processes regulated by the circadian system. As an example, the E2F transcription factor family is involved in the control of cell cycle dependent expression of genes that are essential for cellular proliferation. E2F genes may regulate cell cycle progression, apoptosis and suppress cell proliferation [32] which plays an important role in the tumor progression. Revealing such a connection of the circa-

dian clock to cell cycle and apoptosis might be an important contribution to the chronotherapy of cancer.

Other over-represented binding sites, such as NF-kappaB or STAT1, point to a link between the circadian clock and the immune system. NF-kappaB transcription factors control the expression of multiple genes essential for the immunogenic response and apoptosis [33]. Decoding its relation to the circadian clock might therefore have implications on the therapeutics of a wide variety of human diseases, including immunodeficiency, inflammation, arthritis or cancer.

Another interesting result of our study relates to the transcription factor hypoxia inducible factor 1 (HIF-1). It is a member of the family of transcription factors that respond to changes in available oxygen in the cellular environment. The gene coding for HIF-1 is found to be clock controlled [10, 12]. As shown in Table 3, binding sites of this factor are significantly over-represented in the clock controlled genes. Moreover, 9 of 14 target genes included in TRANSFAC are reported to be clock controlled.

Taken together these data may contribute to the understanding of the circadian rhythms in mammals and help to explain regulation of different biological processes. Our further work aims at investigating the specificity of regulation within tissues, and the correlation of the transcription factor binding sites over-representation to the phase of expression peak.

References

[1] Schibler, U., The daily rhythms of genes, cells and organs. Biological clocks and circadian timing in cells, *EMBO Rep.*, 6:S9–13, 2005.

[2] Yamazaki, S., *et al.*, Resetting Central and Peripheral Circadian Oscillators in Transgenic Rats, *Science*, 288(5466):682–685, 2000.

[3] Akhtar, R.A., *et al.*, Circadian Cycling of the Mouse Liver Transcriptome, as Revealed by cDNA Microarray, Is Driven by the Suprachiasmatic Nucleus, *Curr. Biol.*, 12(7):540–550, 2002.

[4] Duffield, G., *et al.*, Circadian Programs of Transcriptional Activation, Signaling, and Protein Turnover Revealed by Microarray Analysis of Mammalian Cells, *Curr. Biol.*, 12(7):551–557, 2002.

[5] Grundschober, C., *et al.*, Circadian Regulation of Diverse Gene Products Revealed by mRNA Expression Profiling of Synchronized Fibroblasts, *J. Biol. Chem.*, 276(50):46751–46758, 2001.

[6] Humphries, A., *et al.*, cDNA Array Analysis of Pineal Gene Expression Reveals Circadian Rhythmicity of the Dominant Negative Helix-Loop-Helix Protein-Encoding Gene, Id-1, *J. Neuroendocrinol*, 14(2):101–108, 2002.

[7] Kita, Y., *et al.*, Implications of circadian gene expression in kidney, liver and the effects of fasting on pharmacogenomic studies, *Pharmacogenetics*, 12(1):55–65, 2002.

[8] Miller, B., *et al.*, Circadian and CLOCK-controlled regulation of the mouse transcriptome and cell proliferation, *Proc. Natl. Acad. Sci. USA*, 104(9):3342–3347, 2007.

[9] Oishi, K., *et al.*, Genome-wide Expression Analysis of Mouse Liver Reveals CLOCK-regulated Circadian Output Genes, *J. Biol. Chem.*, 278(42):41519–41527, 2003.

[10] Panda, S., *et al.*, Coordinated Transcription of Key Pathways in the Mouse by the Circadian Clock, *Cell*, 109(3):307–320, 2002.

[11] Resuehr, D., Sikes, H., and Olcese, J., Exploratory Investigation of the Effect of Mela-

tonin and Caloric Restriction on the Temporal Expression of Murine Hypothalamic Transcripts, *J. Neuroendocrinol*, 18(4):279–289, 2006.

[12] Storch, K.F., *et al.*, Extensive and divergent circadian gene expression in liver and heart, *Nature*, 417(6884):78–82, 2002.

[13] Ueda, H., *et al.*, A transcription factor response element for gene expression during circadian night, *Nature*, 418(6897):534–539, 2002.

[14] Leclerc, G. and Boockfor, F., Pulses of Prolactin Promoter Activity Depend on a Noncanonical E-Box that Can Bind the Circadian Proteins CLOCK and BMAL1, *Endocrinology*, 146(6):2782–2790, 2005.

[15] Chen, W. and Baler, R., The rat arylalkylamine N-acetyltransferase E-box: differential use in a master vs. a slave oscillator, *Brain Res. Mol. Brain Res.*, 81(1-2):43–50, 2000.

[16] Hida, A., *et al.*, The Human and Mouse Period1 Genes: Five Well-Conserved E-Boxes Additively Contribute to the Enhancement of mPer1 Transcription, *Genomics*, 65(3):224–233, 2000.

[17] Yamaguchi, S., *et al.*, Role of DBP in the Circadian Oscillatory Mechanism, *Mol. Cell Biol.*, 20(13):4773–4781, 2000.

[18] `http://www.ensembl.org/Mus\_musculus/index.html`

[19] Rahmann, S., Möller, T., and Vingron, M., On the Power of Profiles for Transcription Factor Binding Site Detection, *Stat. Appl. Genet. Mol. Biol.*, 2:Article7, 2003.

[20] Kiełbasa, S.M., The BioMinerva framework, *in preparation*, 2007.

[21] Schwartz, S., *et al.*, Human-Mouse Alignments with BLASTZ, *Genome Res.*, 13(1):103–107, 2003.

[22] Ueda, H., *et al.*, System-level identification of transcriptional circuits underlying mammalian circadian clocks, *Nat. Genet.*, 37(2):187–192, 2005.

[23] `http://www.gene-regulation.com/pub/databases.html\#transfac`

[24] Hogenesch, J., *et al.*, The basic-helix-loop-helix-PAS orphan MOP3 forms transcriptionally active complexes with circadian and hypoxia factors, *Proc. Natl. Acad. Sci. USA*, 95(10):5474–5479, 1998.

[25] Muñoz, E., Brewer, M., and Baler R., Circadian Transcription: thinking outside the E-box, *J. Biol. Chem.*, 277(39):36009-36017, 2002.

[26] Muñoz, E., Brewer M., and Baler, R., Modulation of BMAL/CLOCK/E-Box complex activity by a CT-rich cis-acting element, *Mol. Cell Endocrinol*, 252(1-2):74–81, 2006.

[27] Jin, X., *et al.*, A Molecular Mechanism Regulating Rhythmic Output from the Suprachiasmatic Circadian Clock, *Cell*, 96(1):57–68, 1999.

[28] Crooks, G.E., *et al.*, WebLogo: A sequence logo generator, *Genome Res.*, 14(6):1188–1190, 2004.

[29] Kiełbasa, S.M., Gonze, D., and Herzel, H., Measuring similarities between transcription factor binding sites, *BMC Bioinformatics*, 6:237–248, 2005.

[30] Chopin-Delannoy, S., *et al.*, A specific and unusual nuclear localization signal in the DNA binding domain of the Rev-erb orphan receptors, *J. Mol. Endocrinol*, 30(2):197–211, 2003.

[31] Triqueneaux, G., *et al.*, The orphan receptor Rev-erbalpha gene is a target of the circadian clock pacemaker, *J. Mol. Endocrinol*, 33(3):585–608, 2004.

[32] DeGregori, J. and Johnson, D.G., Distinct and Overlapping Roles for E2F Family Members in Transcription, Proliferation and Apopt, *Curr. Mol. Med.*, 6(7):739–748, 2006.

[33] Liang, Y., Zhou, Y., and Shen, P., NF-kappaB inhibitors for the treatment of inflammatory diseases and cancer, *Cell Mol. Immunol.*, 1(5):343–350, 2004.

MODELING DEVELOPMENT: SPIKES OF THE SEA URCHIN

CLEMENS KÜHN[1] ALEXANDER KÜHN[1]
kuehn@molgen.mpg.de kuehn_a@molgen.mpg.de
ALBERT J. POUSTKA[1] EDDA KLIPP[1,2]
poustka@molgen.mpg.de klipp@molgen.mpg.de

[1] *Max-Planck-Institute for Molecular Genetics, Ihnestr 63-73, 14195 Berlin, Germany*
[2] *Humboldt Universität zu Berlin, Institute for Biology, Invalidenstr 42, 10115 Berlin, Germany*

Modeling of specification events during development poses new challenges to biochemical modeling. These include data limitations and a notorious absence of homeostasis in developing systems. The sea urchin is one of the best studied model organisms concerning development and a network, the Endomesoderm Network, has been proposed that is presumed to control endoderm and mesoderm specification in the embryo of *Strongylocentrotus purpuratus*. We have constructed a dynamic model of a subnetwork of the Endomesoderm Network. In constructing the model, we had to resolve the following issues: choice of appropriate subsystem, assignment of embryonic data to cellular model, choice of appropriate kinetics. Although the resulting model is capable of reproducing fractions of the experimental data, it falls short of reproducing specification of cell types. These findings can facilitate the refinement of the Endomesoderm Network.

Keywords: modeling; development; sea urchin.

1. Introduction

Development of a complex organism begins with the fertilized egg. Through a series of differential cleavages, specification events and morphological changes, the adult organism is formed. This specific and complex sequence of interconnected events requires a hard-wired plan or program, located in the genome. Together with maternal transcription factors (TFs), the genome contains all information necessary to develop to an adult organism . The single events in development are mainly mediated by differential gene expression to establish extracellular gradients in the embryo or discriminate certain cells from others [3]. Sea urchins like *Strongylocentrotus purpuratus* have been used since the end of the 19th century to study developmental processes [4]. Using modern microbiological techniques, the understanding of development of *S.purpuratus* has increased dramatically. Besides the sequence of the genome of *S.pur.* [20], a complex gene regulatory network, the Endomesoderm Network has been established that is presumed to control mesoderm and endoderm specification [2, 26]. This network, based on experimental data, is available as a graphic representation, but not as a more complex mathematical model.

A model capable of reproducing all major events and interactions in the developing sea urchin needs to be comprised of a growing number of cells in order to enable the establishment of gradients and and reproduce morphological changes. A prerequisite for such a large-scale model, though, is the existence of a working small-scale model capturing the events in one single cell.

We will show the construction of such a cellular model based on the Endomesoderm Network. Since experimental data concerning development like mRNA concentrations is, in most cases, not measured on a cellular basis but for the whole embryo and distinction between not fully specified cell types is not trivial, this modeling involves careful recalculation of experimental values. Furthermore, experimental data is very sparse so that the necessary parameter estimation becomes computationally demanding. To perform this estimation efficiently, we partition the model to minimize the number of parameters estimated simultaneously.

The result is a dynamic model that, because of its shortcomings, gives important indications for the refinement of the Endomesoderm Network.

2. Materials and Methods

The goal of this investigation was the establishment of a mathematical modeling of the processes outlined in the Endomesoderm Network. Therefore, we chose to construct a model of ordinary differential equations (ODEs) that refers to a single cell. By emulating external inputs caused by cell-cell interactions or extracellular gradients, this model should be able to reproduce experimental data. As the Endomesoderm Network focuses on the specification of different cell types, endoderm, mesoderm and primary mesenchyme cells (PMC), the model should be able to generate three distinct expression patterns.

Because of sparse available data, we chose a subset of the Endomesoderm Network for which quantitative timecourse data generally exists. This subset includes the genes *Wnt8* [23], *Otx* [10], *Blimp1* [11], *Brn* [24], *Bra* [17], *FoxA* [16], *Hox* [5], *GataE* [9], *Eve*, *Pmar1* [15] and *Notch*. We attempted to correct obvious shortcomings of the network by incorporating a model of the canonical Wnt-Pathway [7, 8].

Upon close examination, the time course data proved inadequate, since the experimental time courses are determined using all cells of the embryo. Thus, increased expression in a rather small region of the embryo following basal expression in the entire embryo would not be accounted for by the raw data. We resolved this problem by using fate maps and data from modern imaging techniques.

2.1. *Recalculation of Experimental Data*

The expression data available for the genes used in this study is determined as transcripts per embryo. It seems fairly obvious that, in a growing, developing organism, transcript number per embryo is not necessarily equal transcript number per cell. For the model constructed here, the number of transcripts per cell for each cell type is essential.

To calculate the expression of any gene per cell of a given cell type, we need information on the expression in this cell type relative to the expression in the entire embryo. This can be obtained from whole mount in situ hybridization(WISH) data, as available at [18, 27]. WISH data qualitatively determines the localization of transcripts of a given gene in the embryo.

The simplest case is that WISH data shows that a given gene is expressed in only one cell type at a given time point. The only additional information needed to calculate transcript number per cell is the number of cells comprising the given cell type. For early stages of the embryo, this can be obtained from fate maps [1, 14, 22]. For later stages, advanced imaging techniques are necessary to infer the number of cells expressing a certain marker [13, 21].

In this simple case, the number of transcripts per cell can be determined by dividing the number of transcripts at a given time point by the number of cells in the territory expressing the gene. Trivial as it sounds, this approach requires knowledge of the amount of cells expressing and transcript abundance at the same time point, which is usually not obtainable. Therefore, the number of cells in a given territory at a certain timepoint is inferred by assuming linear growth between different experimental measurements here.

In more difficult cases, i.e. a gene is expressed in multiple territories with different rates, these rates have to be approximated as good as possible from WISH data. As WISH data is rather qualitative, we assume here that genes expressed in more than one territory are expressed equally among these territories, thus the number of transcripts per embryo is divided by the number of cells of all territories expressing the gene in question.

A comparison between embryonic expression rates and recalculated expression is given in Fig.1.

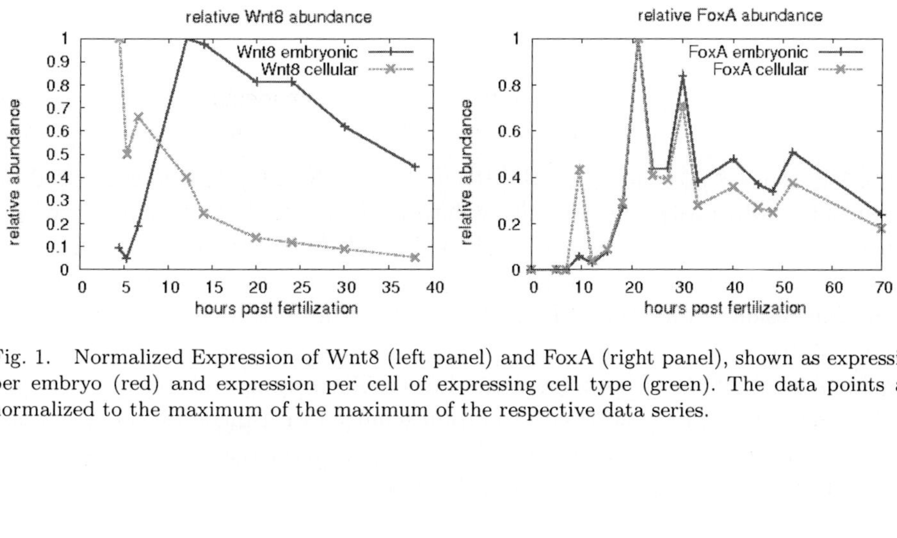

Fig. 1. Normalized Expression of Wnt8 (left panel) and FoxA (right panel), shown as expression per embryo (red) and expression per cell of expressing cell type (green). The data points are normalized to the maximum of the maximum of the respective data series.

2.2. *Details of the ODE Model*

The model was formulated as a set of ODEs. It was implemented in Systems Biology Markup Language (SBML) [6], in order to use available parameter estimation tools. To focus solely on the regulatory interactions and omit any mechanisms for which experimental data is missing, compartimentation of the model as well as transport processes are omitted. Translation and degradation reactions are modelled using first-order reaction kinetics of the form

$$v_x = k_{rt} \cdot [Y] \tag{1}$$

where x is the identifier of the reaction, k_{rt} is a constant for a given reaction type and $[Y]$ is the concentration of substrate. Transcription kinetics are formulated using a modular approach. This approach facilitates the transfer of Boolean functions to ODE models: A given TF can have an activating or an inhibiting influence on its target gene's expression. An activatory influence is of TF A on gene G is defined as

$$\varphi_{G_A} = \frac{k_{A_G} \cdot [A]}{c_{A_G} + [A]} \tag{2}$$

whereas an inhibitory input of TF I on G is given by

$$\varphi_{G_I} = \frac{k_{I_G} \cdot c_{I_G}}{c_{I_G} + [I]} \tag{3}$$

with k_{A_G}, c_{A_G}, k_{I_G} and c_{I_G} parameters specific for each distinct combination of TF and effected gene.

To further exemplify, consider the case of some gene G, activated by TF *A1* as well as *A2*, both acting additively. The formalism above combined with a degradation term yields

$$\frac{d[G_{mRNA}]}{dt} = \frac{k_{A1_G} \cdot [A1]}{c_{A1_G} + [A1]} + \frac{k_{A2_G} \cdot [A2]}{c_{A2_G} + [A2]} - k_{deg\_mRNA} \cdot [G_{mRNA}] \tag{4}$$

Using k_{A1_G},c_{A1_G},k_{A2_G} and c_{A2_G}, the strength of each input can be controlled independently. Obviously, this formalism introduces a rather large number of parameters, but these are necessary to allow for different activatory and inhibitory strengths of one TF on different genes and different contributions of multiple TFs to the expression of one gene. The rate laws of this formed used in this model are given in Table 1.

Any number of inputs of both types can be combined using multiplication in case the influence of all TFs is necessary to produce output or using addition in the case that the influence of either one of the TFs is required. Such a combination is then used as the velocity of a given transcription reaction.

Most parameters in the resulting set of ODEs are undetermined and cannot be obtained from literature, necessitating parameter estimation. We chose SBML-PET [25] to estimate these parameters.
Since most parameters depend, directly or indirectly, on other parameters, estimation of all parameters simultaneously is computationally nearly infeasible. We therefore partitioned the model into submodels, emulating necessary inputs according to experimental data, and estimated the parameters in small groups.
Therefore and to mimic transcriptional regulation arising from extracellular gradients, we made use of event structures provided in SBML. Event structures do have the unfavorable characteristic to introduce discontinuous elements in the system of ODEs. We therefore constructed a formula that consists of the sum of an activatory Hill-Kinetic and an inhibitory Hill-Kinetic. Only one of the two summands is active at a time, depending on whether the modeled concentration is rising (activatory term) or declining (inhibitory term). The activity of either term and the parameters involved are controlled using events. The change in concentration of x is given by

$$\frac{d[x]}{dt} = S_1 \cdot k \cdot \frac{t^h}{\Theta^h + t^h} + (1 - S_1) \cdot k \cdot (1 - \frac{t^h}{\Theta^h + t^h}) - k_{deg} \cdot [x] \quad (5)$$

Here, k is used to control the maximum concentration of x, t is simulation time, Θ equals the value of t where x reaches half of its maximal value, h is the hill coefficient controlling the steepness of the slope and $-k_{deg} \cdot [x]$ is a degradation term.
S_1 controls whether the concentration of x is rising or falling. Requiring $S_1 \in \{0, 1\}$, either the first or the second summand in Eq. 5 are non-zero. Reseting S_1 and fine-tuning of the other parameters using events enables the reproduction of complex temporal patterns without discontinuities.

3. Results

As quantitative time courses of mRNA concentrations for most genes in the Endomesoderm Network are very sparse, we selected genes for which detailed experimental data is available. These genes are also the presumed key genes of the network. A graph representation of the resulting network is given in Fig.2.
As explained before, the difference between different genes arises from differences in transcriptional regulation alone. Hence, the rate laws controling transcription are given in Table 1 along with the general kinetics used for the other reactions. To drive differential expression in the different cell types, the helper variables $Notch$, TCF_REP, $meso\_REP$, $PMC\_repressor$ and Otx_REP are used. Their activities are shown in Table 2. The activities are given at the mRNA level and as protein degradation is slower than mRNA degradation, protein abundance lasts longer.
Further details of the ODE model, like those concerning the Wnt-Pathway model included in the model can be obtained from the SBML file in supplemental data.
Simulation results of the model were compared to experimentally determined and recalculated time courses. The two sets are compared in Fig.3. In general, the simu-

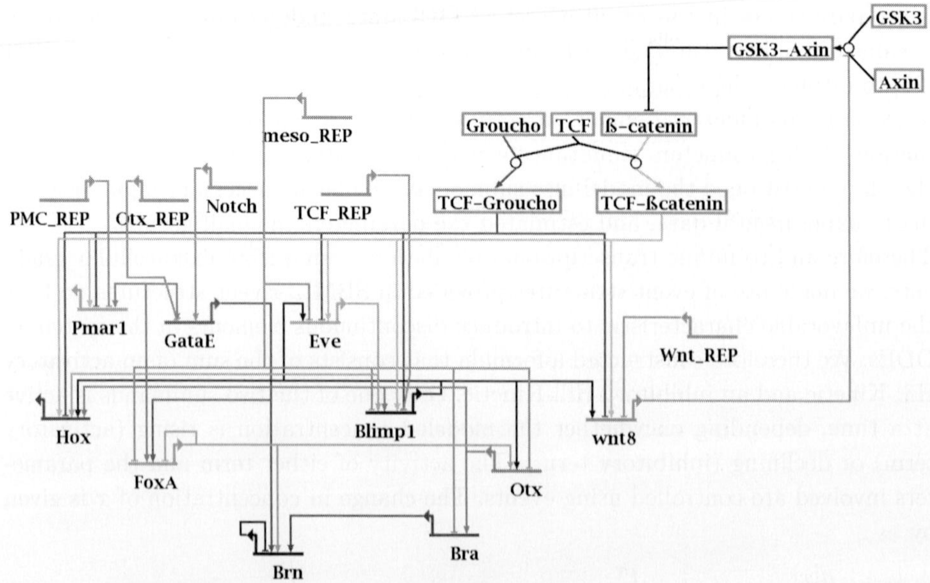

Fig. 2. Topology of the Boolean model underlying the model analyzed here. Arrows indicate activatory interactions, barred lines indicate inhibitory interactions. Genes are represented as horizontal lines with arrows, proteins and protein complexes as rectangles. Helper variables (grey genes) represent influences necessary for the boolean network to reproduce experimnetal data that are not based on any experimental data or assumptions. Note all helper variables have inhibitory effects. Figure created using Biotapestry [12]

Table 1. Rate laws for transcriptional regulation and general kinetics

Gene/Reaction	Rate Law
Blimp1	$(\frac{3.506\cdot[bcat\_TCF]}{1.118+[bcat\_TCF]}+\frac{1.0\cdot[Otx]}{(0.5397+[Otx])}+\frac{1.0\cdot[Eve]}{100.0+[Eve]})\cdot(\frac{2.073\cdot 1.0}{1.0+[Gro\_TCF]}+\frac{1.0\cdot[TCF\_REP]}{100+[TCF\_REP]})\cdot\frac{1.0\cdot 1.0}{1.0+[meso\_REP]}$
Bra	$\frac{7.366\cdot[Otx]}{100.0+[Otx]}\cdot\frac{14.89\cdot 0.13}{0.13+[Otx\_REP]}\cdot\frac{5.475\cdot 71.91}{71.91+[meso\_REP]}$
Brn	$\frac{1.942\cdot[GataE]}{1.942+[GataE]}\cdot\frac{25.0\cdot[FoxA]}{700.0+[FoxA]}\cdot\frac{25.0\cdot[Bra]}{700.0+[Bra]}\cdot\frac{2.0\cdot 1.0}{2.0+[Brn]}$
Eve	$\frac{3.22\cdot[Blimp1]}{0.1+[Blimp1]}\cdot\frac{6.586\cdot[bcat\_TCF]}{100.0+[bcat\_TCF]}\cdot\frac{34.29\cdot 0.1}{0.1+[Gro\_TCF]}\cdot\frac{1.0\cdot 1.0}{1.0+[meso\_REP]}$
FoxA	$((\frac{2.035\cdot[Otx]}{0.1+[Otx]}\cdot\frac{4.534\cdot 5.804}{5.804+[Otx\_REP]}+\frac{43.49\cdot[GataE]}{10.61+[GataE]})\cdot\frac{1.001\cdot 5.0}{5.0+[meso\_REP]}+(\frac{2.035\cdot[Otx]}{0.1+[Otx]}\cdot\frac{4.534\cdot 5.804}{5.804+[Otx\_REP]}))\cdot\frac{1.001\cdot 100.0}{100.0+[FoxA]}$
GataE	$\frac{4.958\cdot[Otx]}{0.1+[Otx]}\cdot\frac{1.0\cdot 1.384}{1.384+[Otx\_REP]}+\frac{2.982\cdot[Notch]}{100.0+[Notch]}$
Hox	$(\frac{100.0\cdot[bcat\_TCF]}{13.98+[bcat\_TCF]}+\frac{1.0\cdot[Eve]}{100.0+[EvepMC]}+\frac{1.0\cdot[Blimp1]}{100.0+[Blimp1]})\cdot\frac{1.0\cdot 0.1}{0.1+[Gro\_TCF]}\cdot\frac{1.0\cdot[Otx]}{0.2236+[Otx]}$
Otx	$(\frac{14.62\cdot[GataE]}{0.1752+[GataE]}\cdot\frac{1.0\cdot[FoxA]}{1.0+[FoxA]}+\frac{11.06\cdot[Blimp1]}{7.275+[Blimp1]})\cdot\frac{1.0\cdot 1.0}{1.0+[meso\_REP]}$
Pmar1	$\frac{3.408\cdot[Otx]}{0.1+[Otx]}\cdot\frac{37.24\cdot[bcat\_TCF]}{100.0+[bcat\_TCF]}\cdot\frac{29.08\cdot 0.1}{0.1+[Gro\_TCF]}\cdot\frac{1.0\cdot 1.0}{1.0+[pmc\_REP]}$
Wnt8	$(\frac{2.644\cdot[bcat\_TCF]}{99.92+[bcat\_TCF]}+\frac{29.64\cdot[Blimp1]}{0.3815+[Blimp1]})\cdot\frac{7.164\cdot 0.1}{0.1+[Wnt\_REP]}\cdot\frac{10.4\cdot 0.1}{0.1+[Gro\_TCF]}$
$degradation_{mRNA}$	$0.2\cdot[X_{mRNA}]$
translation	$2.0\cdot[X_{mRNA}]$
$degradation_{X_{protein}}$	$0.10193\cdot[X]$

Table 2. Period of activity of helper variables in different territories. Cells values pertain to the values of $\Theta_{activatory}$ and $\Theta_{inhibitory}$ in Equation 5

Variable	Endoderm	Mesoderm	PMC
PMC_{REP}	OFF	$START : END$	$START : END$
$meso_{REP}$	$8 : END$	$15 : END$	OFF
TCF_{REP}	OFF	OFF	$24 : END$
Otx_{REP}	$START : 10$	$Start : 17$	$Start : 17$
Wnt_{REP}	$10 : END$	$10 : END$	$10 : END$

lation results are –at least– qualitatively similar to the experimental data, although the time scale is not equivalent among all genes. One exception is FoxA, which shows oscillating behavior not reproduced by the model.

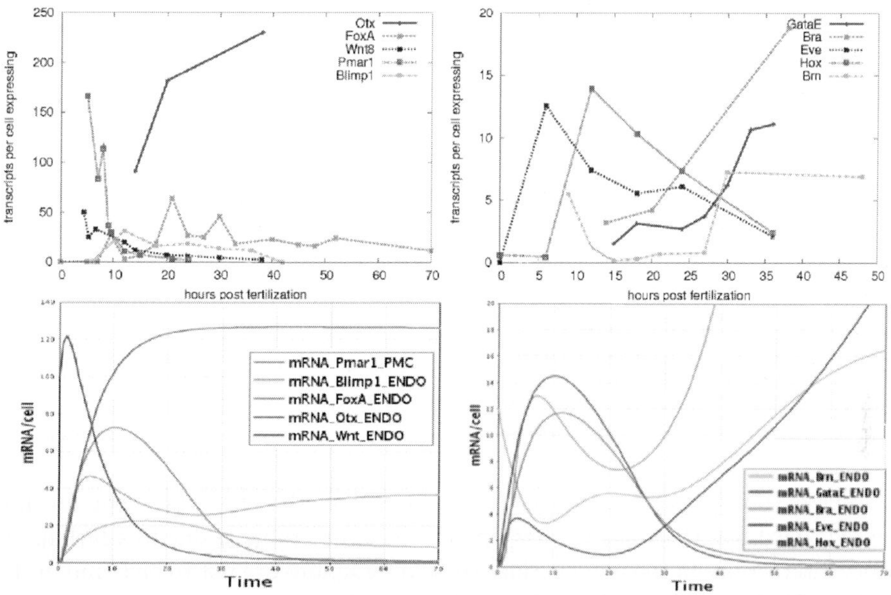

Fig. 3. Time courses as determined from experimental data (top row) and simulation results (bottom row). To determine transcripts/cell, we assumed that each gene is expressed in one cell type only but equally throughout this cell type. Therefore, the expression of each gene is shown for only one cell type (Endoderm for Blimp1, FoxA, Otx, Wnt, Brn, GataE, Bra, Eve, Hox; PMC for Pmar1). In the recalculated experimental data, transcript abundance in other cell types is assumed to be 0. In simulation results, this clear distinction between cell types could not be obtained.

In general, the designed model reproduces the available experimental data to a satisfying extend.

4. Discussion

The present analysis highlights a few issues arising when modeling developmental GRNs. These issues do not include evaluation or analysis of the resulting model itself. As most methods to analyze models require a steady state, which is not necessarily given or of interest in developmental processes, the analysis of developmental GRNs poses new challenges here as well.

The issues addressed in this study can be summarized as: dealing with sparse experimental data and choosing kinetics for transcriptional regulation. We will briefly discuss the approaches undertaken here to solve these issues before we critically evaluate the model constructed here.

As shown in Sec. 2.1, the available embryonic data needs recalculation to cellular data. We perform such a recalculation by using WISH data to determine spatial expression and different data containing counts of cell numbers for different embryonic territories and time points. From this data, we can infer the number of cells expressing a given gene and thus determine the cellular expression from embryonic expression data. This recalculation crucially depends on data that is very sparse as of today and definitely needs refinement.

The point we want to stress here is that this conversion to cellular data is vital for most modeling approaches concerned with development. Thus, existing experimental data is not necessarily as extensive as it seems at first sight.

The transcriptional kinetics chosen here represent a versatile and intuitive way to connect the regulatory input to the gene to the resulting output in terms of transcription.

The parameters used in these kinetics do not necessarily resemble biochemical constants measurable in experiments.

Using the estimated parameters, the model can be used to simulate the time course and validate the results by comparison to experimental data. The simulated time courses of each gene generally resemble the experimentally determined time course for the cell type expressing the gene.

Nevertheless, the genes are not efficiently shut off in the cell types that are not supposed to express the gene in question. This is most probably due to a lack in inhibitory interactions in the model. Since these interactions are not to be found in the underlying Endomesoderm Network, the need for the refinement of both networks, the Endomesoderm Network and the model presented here, becomes obvious. For the refinement of the networks, new experimental data as well as a reassessment of the existing data is inevitable.

As the Endomesoderm Network contains many features which are believed to be conserved throughout evolution, experimental findings from other species can be an important source of additional information. One example could be the recent findings on non-canonical Wnt-signaling in Xenopus [19].

5. Conclusion

We have successfully created an ODE model on the basis of the Endomesoderm Network. Although this model is not capable of correctly reproducing specification of cells, the present analysis highlights some issues arising in modeling of large developmental GRNs. We hope to thereby improve alertness of computational as well as experimental researchers for the requirements for modeling large developmental GRNs.

The Endomesoderm Network itself must be understood as a snapshot of ongoing research which is constantly refined. Thus, we hope to aid in this refinement by presenting this detailed model of the key components of the network and highlighting the lack of inhibitory interactions.

Supporting Data

Supporting data includes SBML-version of the model

Acknowledgments

We would like to thank Christoph Wierling and Alexander Kühn for fruitful discussion. Clemens Kühn is funded by German Research Foundation via the International Research Training Group "Genomics and Systems Biology of Molecular Networks".

References

[1] Dan, K., Tanaka, S., Yamazaki, K., Kato, Y., Cell cycle study up to the time of hatching in the embryos of the sea urchin, hemicentrotus pulcherrimus, *Development Growth and Differentiation*, 22(3):589–598, 1980.

[2] Davidson E.H., et al, A Provisional regulatory gene network for specification of endomesoderm in the sea urchin embryo, *Dev. Biol.*, 246(1):162–190, 2002.

[3] Davidson, E.H., Erwin, D.H., Gene regulatory networks and the evolution of animal body plans, *Science*, 311(5762):796–800, 2006.

[4] Hertwig, O., Beiträge zur Kenntnis der Bildung, Befruchtung und Teilung des tierischen Eies, *Gegenbaurs Morphologisches Jahrbuch*, 1:374–452, 1876.

[5] Howard-Ashby, M., Materna, S. C., Brown, C. T., Chen, L., Cameron, R.A., Davidson E. H, Identification and characterization of homeobox transcription factor genes in Strongylocentrotus purpuratus, and their expression in embryonic development, *Dev. Biol.*, 300(1):74–89, 2006.

[6] Hucka, M., et al., The Systems Biology Markup Language (SBML): A Medium for Representation and Exchange of Biochemical Network Models, *Bioinformatics*, 19(4):524–531, 2003.

[7] Krüger, R., Heinrich R., Model reduction and analysis of robustness for the Wnt/β-Catenin signal transduction pathway, *Genome Inform.*, 15(1):138–148, 2004.

[8] Lee, E., Salic, A., Krüger, R., Heinrich, R., Kirschner M.W., The roles of APC and Axin derived from experimental and theoretical analysis of the Wnt pathway, *PLoS Biology*, 1(1):116, 2003.

[9] Lee, P.Y., Davidson, E.H., Expression of Spgatae, the Strongylocentrotus purpuratus ortholog of vertebrate GATA4/5/6 factors, *Gene Expr. Patterns*, 5(2):161–165, 2004.

[10] Li, X., Chuang, C.-K., Mao, C.-A., Angerer, L.M., Klein, W.H., Two Otx Proteins Generated from Multiple Transcripts of a Single Gene in Strongylocentrotus purpuratus, *Dev. Biol.*, 187(2):253–266, 1997.

[11] Livi, C.B., Davidson E.H, Expression and function of blimp1/krox, an alternatively transcribed regulatory gene of the sea urchin endomesoderm network, *Dev. Biol.*, 293(2):513–525, 2006.

[12] Longabaugh, W.J.R., Davidson, E.H., Bolouri, H., Computational representation of developmental genetic regulatory networks, *Dev. Biol.*, 283(1):1–16, 2005.

[13] Martins, G.G., Summers, R.G., Morill, J.B., Cells are added to the archenteron during and following secondary invagination in the sea urchin lytechinus variegatus, *Dev. Biol.*, 198:330–342, 1998.

[14] Masuda, M., Sato, H., Asynchronization of cell division is concurrently related with ciliogenesis in sea urchin blastulae. *Development Growth and Differentiation*, 26(33):281–294, 1984.

[15] Oliveri, P., Carrick, D.M., Davidson, E.H., A Regulatory Gene Network That Directs Micromere Specification in the Sea Urchin Embryo, *Developmental Biology*, 246(1):209–228, 2002.

[16] Oliveri, P. , Walton, K.D., Davidson, E.H., McClay, D.R., Repression of mesodermal fate by foxa, a key endoderm regulator of the sea urchin embryo, *Development*, 133(21):4173–4181, 2006.

[17] Poustka, A.J., Kühn, A., unpublished results

[18] Poustka, A.J., Kühn, A., Groth, D., Weise, V., Yaguchi, S., Burke, R.D., Herwig, R., Lehrach, H., Panopoulou, G., A global view of gene expression in lithium and zinc treated sea urchin embryos: new components of gene regulatory networks, *Genome Biol.*, 8(5):R85, 2007

[19] Schambony,A., Wedlich, D., Wnt-5A/Ror2 Regulate Expression of XPAPC through an Alternative Noncanonical Signaling Pathway, *Developmental Cell*, 12:779–792, 2007.

[20] Sea Urchin Genome Sequencing Consortium, The Genome of the sea urchin Strongylocentrotus Purpuratus, *Science*, 314:941–953, 2006

[21] Summers, R.G., Morill, J.B., Leith, J., Marko, M., Piston, D.W., Stonebraker, A.T., A stereometric analysis of karyokinesis, cytokinesis and cell arrangements during and following fourth cleavage period in the sea urchin, Lytechinus Variegatus, *Development Growth and Differentiation*, 35(11):41–57, 1993.

[22] Tanaka, S., Dan, K., Study of the lineage and cell cycle of small micromeres in the embryos of the sea urchin, Hemicentrotus pulcherrimus, *Development Growth and Differentiation* 32(2):145–156, 1990.

[23] Wikramanayake, A.H., Peterson, R., Chen, J., Huang, L., Bince, J.M., McClay,D.R., Klein, W.H., Nuclear β-Catenin-Dependent Wnt8 Signaling in Vegetal Cells of the Early Sea Urchin Embryo Regulates Gastrulation and Differentiation of Endoderm and Mesodermal Cell Lineages, *Genesis*, 39(3):194–205, 2004.

[24] Yuh, C.-H., Dorman, E.R., Davidson, E.H., Brn1/2/4, the predicted midgut regulator of the endo16 gene of the sea urchin embryo, *Dev. Biol.* 281(2):286–298, 2005.

[25] Zi, Z., Klipp, E., SBML-PET: a systems biology markup language based parameter estimation tool, *Bioinformatics*, 22(21):2704–2705, 2006.

[26] `http://sugp.caltech.edu/endomes/`

[27] `http://goblet.molgen.mpg.de/eugene/cgi/eugene.pl`

INSIGHTS INTO THE NETWORK CONTROLLING THE G_1/S TRANSITION IN BUDDING YEAST

MATTEO BARBERIS[1]
barberis@molgen.mpg.de

EDDA KLIPP[1,2]
klipp@molgen.mpg.de

[1] *Max-Planck-Institute for Molecular Genetics, Ihnestraße 73, 14195 Berlin, Germany*
[2] *Humboldt University Berlin, Institute for Biology, Invalidenstr. 42, 10115 Berlin, Germany*

The understanding of complex biological processes whose function requires the interaction of a large number of components is strongly improved by the construction of mathematical models able to capture the underlying regulatory wirings and to predict the dynamics of the process in a variety of conditions. Iterative rounds of simulations and experimental analysis generate models of increasing accuracy, what is called the systems biology approach. The cell cycle is one of the complex biological processes that benefit from this approach, and in particular budding yeast is an established model organism for these studies. The recent publication about the modeling of the G_1/S transition of the budding yeast cell cycle under a systems biology analysis has highlighted in particular the implications of the cell size determination that impinge the events driving DNA replication. During the life cycle of eukaryotic cells, DNA replication is restricted to a specific time window, called the S phase, and several control mechanisms ensure that each DNA sequence is replicated once, and only once, in the period from one cell division to the next. Here we extend the analysis of the G_1/S transition model by including additional aspects concerning the DNA replication process, in order to give a reasonable explanation to the experimental dynamics, as well as of specific cell cycle mutants. Moreover, we show the mathematical description of the critical cell mass (P_S) that cells have to reach to start DNA replication, which value is modulated depending on the different activation of the replication origins. The sensitivity analysis of the influence that the kinetic parameters of the G_1/S transition model have on the setting of the P_S value is also reported.

Keywords: Budding yeast; G_1/S transition; probabilistic model; DNA replication; critical cell size.

1. Introduction

The machinery of the cell cycle is one of the most relevant and fine-tuned processes in the cell, and regulates important cellular processes, e.g., DNA synthesis, budding, and cell proliferation. In many cases, defects in the cell cycle events are known to be a cause of cancer, and more precise information on the regulation of these processes is useful to plan strategies of drug discovery.

The cell cycle regulation has been deeply analyzed by using *in silico* modeling from yeast to mammalian cells [2-9]. The *in vivo* dynamics of several proteins introduced in these models were obtained from results of biochemical experiments, mainly by western blotting analysis (i.e. total protein level, degradation rates). These protein data were used for parameter estimation to refine the computational models. Both data acquisition and parameter fitting are the basic steps for creating a model, and improvements of these processes are crucial for achieving precise simulation of the models that try to represent the experimental dynamics. The quantitative results from these experiments are very useful for creating more complex and precise models. A few kinetic parameters in these models were tuned manually based on biological knowledge available in the literature, and for the major part they were fixed arbitrarily to reproduce the phenotype of a wide range of cell cycle mutants [2,3]. The differential equations that describe the reaction of a model depend on kinetic constants, which are often not accessible to experimental determination, and must therefore be estimated by fitting the model to experimental data. However, with the increase in knowledge about the single reactions involved in a biological pathway, it becomes extremely useful to estimate the kinetic parameters and to manually introduce them into the models.

Recently, we have compiled the huge amount of experimental data available in the literature and modeled the nucleo/cytoplasmic G_1/S transition of the budding yeast *Saccharomyces cerevisiae* [1]. The model was implemented by ordinary differential equations and tested by computer simulation [1]. This map reveals the main known regulatory events that impinge the functionality of this window of the cell cycle. By using time-series quantitative data on proteins reported by Alberghina *et al.* [10] and by Rossi *et al.* [11], we performed the estimation of some critical parameters. Our model [1] is compatible with time-series data measured by western blotting, and it matches the *in vivo* data from among several cell cycle mutants reported in literature. In addition, the network of the G_1/S transition highlighted the feasible approach to determine the critical cell mass, called P_S, that cells have to reach in order to enter S phase and replicate their DNA.

On the basis of the physiological significance of the G_1/S transition network, this work presents an insight on DNA replication. The activation of the replication machinery has still to be highlighted in many of its regulatory events, but a relevant step is the phosphorylation of different substrates by the Cdk1-Clb5,6 kinase complex that induces the firing of the DNA replication origins [12,13]. In [1] we described the steps which lead to the DNA replication with a simple probabilistic model that considers the availability of the Cdk1-Clb5,6 nuclear concentration as the main input. DNA replication, the main event that drives the cell cycle after the G_1/S transition, is analyzed here from the mathematical point of view, providing an explanation for the phenotype of wild type cells grown on different media, and for selected mutants of the network. Moreover, we show the mathematical description of the P_S determination, which value is modulated depending from the different activation of the replication origins. The sensitivity analysis of the influence that the kinetic parameters of the G_1/S transition model have on the setting of the P_S value is also reported.

2. DNA replication at the G_1/S transition

2.1. *Hybrid model for DNA replication events*

The model presented by Barberis *et al.* [1] considers the analysis of the G_1/S transition events for a single cell, in a way to represent the experimental dynamics derived from single elutriated cells [10,11]. However, the real representation of the dynamics of the yeast cell cycle has to consider the behaviour of a cell population. The critical cell size P_S is a quantitative parameter known to characterize each exponentially growing population. Its value can be estimated based on the average protein content (a measure of cell size). DNA replication starts only when the cells reach the P_S value, the value of which changes depending from the nutritional medium where the cells are growing [14,15]. Thus, to estimate the P_S value, it is necessary to find a way to model the DNA replication process. To this purpose, we constructed a hybrid model of the firing of DNA replication origins, where the probabilistic model uses as input the output of the network of the G_1/S transition, the nuclear concentration of the Cdk1-Clb5,6 complex [1].

The DNA replication machinery is a highly complex process [12,13], and many details have still to be highlighted. Thus, the representation of the process cannot be absolutely defined. However, it is possible to make some acceptable assumptions - compatible with the reported literature - to simulate the effect of the G_1/S cascade on the late events of the cell cycle. We consider as relevant step of the DNA replication the phosphorylation of different substrates by Cdk1-Clb5,6 that induces the firing of the replication origins [12,13]. In addition, the probabilistic description of the process involves some approximations on the basis of the scheme reported in Fig. 1 that shows the consecutive steps of the DNA replication initiation. The proteins reported in the scheme (Cdt1 and Cdc6) represent only examples of known actors involved in the process [13], but such molecular details are outside of the scope of the model.

The firing of the DNA replication origins is modeled as a three-step process. As reported in Fig. 1, the first step includes the events that occur from the free replication origins to the formation of the pre-replicative complex (pre-RC) (see [13] for molecular details). The distance between the DNA replication origins is fixed. The time for the formation of the pre-RC complex at each of the replication origins is taken from a normal distribution with mean of 15 minutes, and a standard deviation of 2 minutes. The number of the replication origins is fixed to 440, as reported from [13,16,17]. The second step is dependent on the nuclear Cdk1-Clb5,6 complex, output of the model of the G_1/S transition [1]. In fact, we correlate the concentration of this complex to the onset of DNA replication.

The probability of the activation of the replication origin by Cdk1-Clb5,6 at a certain time is determined by the concentration of this complex at that time. In this case, we consider the period of this step that is necessary for Cdk1-Clb5,6 to exceed a value taken from a normal distribution with a mean of 0.03 μM, and standard deviation of 0.01 μM. Moreover, we consider an additional time due to the fact that as soon as Cdk1-Clb5,6 is available, specific substrates are phosphorylated by the complex for their release from each replication origin. Finally, the third step shows the activation (firing) of the replication origins. The time for each replication origin to reach the fired state is taken from a normal distribution with a mean of 1 minute, and a standard deviation of 0.01 minute. When a replication origin has fired, then DNA replication proceeds bidirectionally from multiple replication origins, as experimentally reported [18,19].

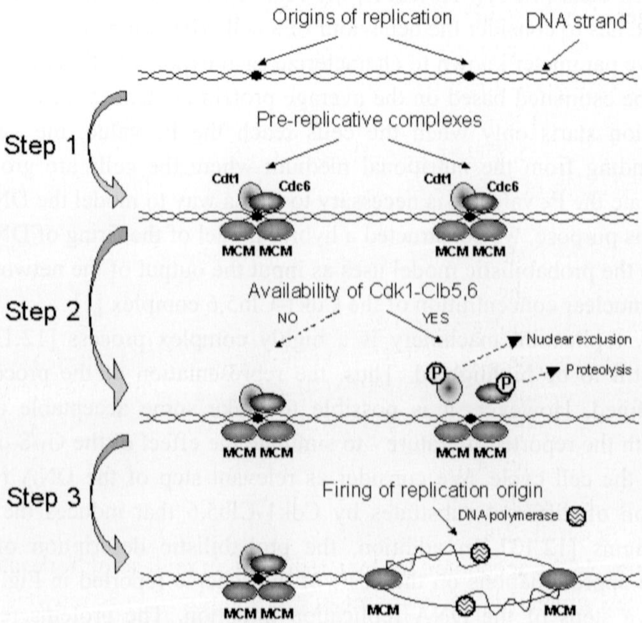

Fig. 1. Schematic representation of relevant events in the firing of the origins of DNA replication. After pre-replication complexes have assembled on the replication origins, phosphorylation of specific targets takes place as a function of the availability of Cdk1-Clb5,6. Origin firing and DNA replication then start bidirectionally.

In the case that the replication reaches the neighboring origin before it fires on its own, that origin is considered as fired. To this purpose, we introduce this correction in an additional part of the code. The code includes the generation of tables with the times of the replication origins activation and the distribution of the activated replication origins in the time when each one reaches the fired state, and the corresponding graphs. Using the Mathematica software, for each origin, the state is updated in three steps (compare Fig. 1) with following duration:

$$t_{step1} = \text{Random [NormalDistribution[15,2]]}$$
$$v_{step2} = \text{Random [NormalDistribution[0.03,0.01]]}$$
$$t_{step2} = \text{Time, when Cdk1-Clb5,6 concentration overcome } v_{step2}$$
$$t_{step3} = \text{Random[NormalDistribution[1,0.01]]}$$

Each origin is activated, when either steps 1,2,3 are passed or 10 minutes after its neighboring origin was activated.

2.2. *Nutritional conditions and DNA replication*

Cell viability requires the coordination between cell growth and cell division, which in budding yeast is achieved by the attainment of a nutritionally modulated critical cell size (P_S) to trigger DNA replication [14,15]. Nutrients are the main environmental determinants that affect cell cycle progression in budding yeast (20,21,22), and it is known that higher growth rates and larger P_S are observed in rich media (23,20,21,10). Data reported in literature show that a very poor carbon source such as ethanol, or a nitrogen source limitation, yield elongation of the S phase [24,25].

In [1] simulations of the onset of DNA replication were reported for cells grown in glucose and in ethanol media. In glucose, the activation of DNA replication origins took place in a coordinated fashion roughly within a period of 70-90 min. On the contrary, in ethanol the dynamics of the G_1/S transition resulted in a longer DNA replication [1]. We considered that in ethanol-growing cells - with a growth rate about 2-fold lower compared with the glucose-growing cells - the fork rate is about one-half (and the time of origins activation is doubling) than the glucose ones. To implement this assumption, we changed in the Mathematica code the value of the time for the formation of the pre-RC complex at each of the replication origins. The value taken from a normal distribution with mean of 33.4 minutes, and a standard deviation of 2 minutes. This assumption agrees with the reported data, in which the longer S phase in yeast cells growing in poor nitrogen medium can be accounted for by a reduction in replication fork rate [25].

In Fig. 2 the simulation of the onset of DNA replication is reported for wild type cells grown either on glucose or on ethanol media. The code was run for five times to observe the variability of the probabilistic model. The effect in ethanol is a dramatic decrease of available concentrations of kinase complexes as compared to the glucose growth. In more detail, a reduced formation of the Cdk1-Cln3 (nuclear) complex as well as of the Cdk1-Cln1,2 (cytoplasmic) and Cdk1-Clb5,6 (nuclear) complexes is observed (upper panels). For wild type cells grown on glucose medium, origin firing occurs at about 80 min, when the Cdk1-Clb5,6 levels overcome the Sic1 levels [1]. As expected, by allowing efficient Cdk1-Clb5,6 nuclear import and following switch-like degradation of Sic1, a sharp spike of Cdk1-Clb5,6 activity is obtained (Fig. 2A). This sharp spike in turn allows sharp and efficient firing of replication origins (Fig. 2B). Conversely, in wild type cells grown on ethanol medium, the Cdk1-Clb5,6 complex is inefficiently imported into the nucleus over a longer period, thus resulting in sparse origin firing (Fig. 2D). In this condition, Cdk1-Clb5,6 can not activate the DNA replication process in the physiological time, thus resulting in a longer S phase (Fig. 2C) as experimentally reported [26].

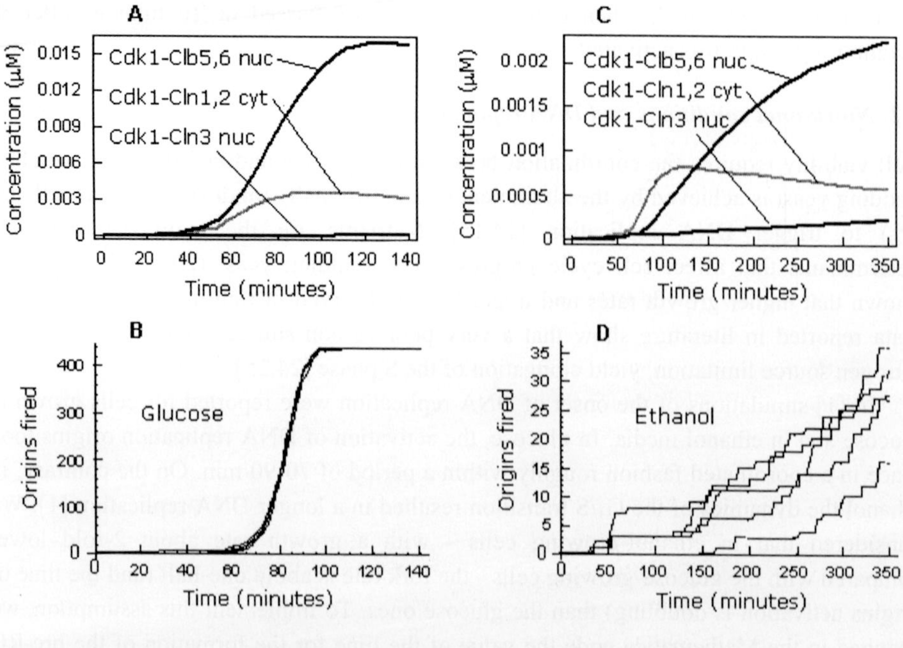

Fig. 2. Distribution of the cyclin-dependent kinase complexes in the firing of DNA replication origins. The cumulative number of fired origins was calculated basing on the probabilistic model for firing of origins in wild type cells grown on glucose (A,B) and on ethanol (C,D) media. Note different scales on the y-axis.

2.3. *Nuclear availability of Cdk1-Clb5,6 and firing of the DNA replication origins*

In the model of the DNA replication events, we related the free concentration of the nuclear Cdk1-Clb5,6 complex to the initiation of DNA replication. Now, we want to focus the attention on the event that controls the availability of Cdk1-Clb5,6, i.e. the phosphorylation of Sic1, an inhibitor of this kinase complex [27,28]. According to the current models reported in literature, Sic1 is involved in the control of DNA replication as a negative regulator of the Cdk activity, and the mathematical models of the cell cycle had of course taken into account only this function [3]. This leaves unanswered a major phenotype of the *sic1Δ* mutant, namely sparse origin firing [29].

We consider that Sic1 is also acting by promoting the Cdk1-Clb5,6 entry into the nucleus [1], based on supporting experimental data [11]. Thus, it would be expected that the Cdk1-Clb5,6 complex enters in the nucleus less efficiently in *sic1Δ* cells. Fig. 3 shows the simulation of the *sic1Δ* mutant grown on glucose media. The amount of nuclear Cdk1-Clb5,6 in wild type cells is about seven/eight times higher than in *sic1Δ* cells (see Fig. 2A for comparison). This fact induces a simulated sparse DNA firing in the mutant cell that starts earlier than in wild type, since no Sic1 degradation is required, and proceeds slowly (Fig. 3B), as experimentally observed [29]. Thus, our model gives a possible rationale explanation for the sparse origin firing observed in the *sic1Δ* mutant.

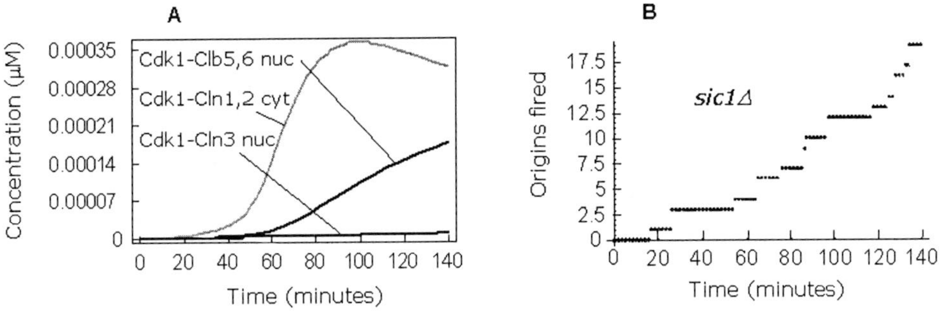

Fig. 3. Distribution of the cyclin-dependent kinase complexes in the firing of DNA replication origins for the *sic1Δ* mutant. grown on glucose medium.

In order to evaluate if the model can explain the starting of DNA replication of different yeast background, we report also the results for deletion mutants or overexpression (indicated as OE- followed by the gene name). Results relating to dosage of *CLN3*, *FAR1*, and *WHI5* genes, which are central to the logic of the G_1/S network [1], are analyzed. The effects of mutations will be always compared with wild type cells. Deletion of *CLN3* gene (*cln3Δ*) prevents the nuclear Cdk1-Cln3 complex formation at the beginning of the G_1/S network, therefore preventing the activation of the SBF/MBF transcription factor. The outcome is a remarkable decrease in *CLN1,2* and *CLB5,6* transcription, and thus in formation of cytoplasmic Cdk1-Cln1,2 and nuclear Cdk1-Clb5,6 that, ultimately, cause a reduction in the number of the activated replication origins (Fig. 4A). This is in agreement with the experimental observations [30,31]. On the other side, *CLN3* overexpression (OE-*CLN3*) leads to the effect that Cln3 overcomes earlier the inhibitory activity of Far1, the force that balance the formation of the nuclear Cdk1-Cln3 complex [10]. This accelerates the formation of cytoplasmic Cdk1-Cln1,2 and nuclear Cdk1-Clb5,6, which results in an anticipated onset of DNA replication (Fig. 4B), as experimentally observed [30,31]. In *far1Δ* cells, there is no balance between the inhibitor Far1 and Cln3, and the nuclear Cdk1-Cln3 complex appears earlier, as well as the cytoplasmic Cdk1-Cln1,2 and nuclear Cdk1-Clb5,6 complexes. The result is a slightly earlier entrance into S phase (Fig. 4C) compared to wild type cells (see Fig. 2B and Ref. [10]). In contrast, in an overexpression of *FAR1* (OE-*FAR1*) the formation of nuclear Cdk1-Cln3 complex occurs to a clearly lower extent. This effect propagates down to the formation of cytoplasmic Cdk1-Cln1,2 and nuclear Cdk1-Clb5,6 and, ultimately, to a reduction in the number of the activated replication origins (Fig. 4D). The *whi5Δ* mutant lacks the initial inhibition of the SBF/MBF transcription factor, the central event of the G_1/S network, and transcription of genes for Cln1,2 and Clb5,6 is turned on immediately. The formation of Cdk/cyclin complexes is brought forward, and the mutant undergoes G_1/S transition about 40 min earlier than wild type. This results in earlier starting of DNA replication (Fig. 4E), as experimentally suggested [32]. Upon overexpression of *WHI5* gene (OE-*WHI5*), there is a stronger inhibition of the SBF/MBF transcription factor and more nuclear Cdk1-Cln3 complex is necessary to phosphorylate Whi5 and release the inhibition. Hence, transcription of Cln1,2 and Clb5,6 is diminished resulting in reduced formation of complexes with Cdk1, and in a strong delay in the DNA replication events (Fig. 4F), as experimentally observed [32]. Overall, the simple probabilistic model connected to the G_1/S network appears to describe the starting of the DNA replication events correctly.

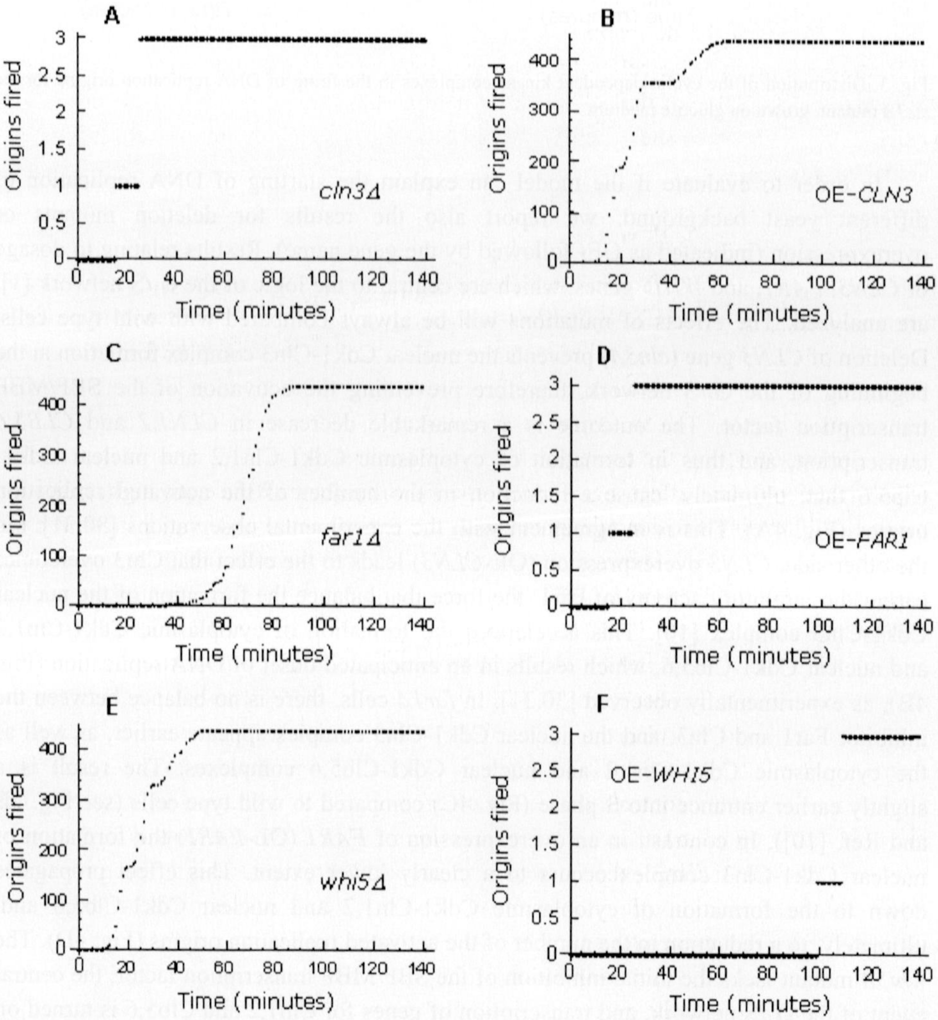

Fig. 4. Simulation of the onset of the DNA replication in glucose-growing cell populations for the deletion mutants *cln3Δ*, *far1Δ*, *whi5Δ* (A,C, and E, respectively), and for their overexpressions OE-*CLN3*, OE-*FAR1*, OE-*WHI5* (B,D, and F, respectively).

3. The critical cell size P_S and the G_1/S transition dynamics

3.1. *Derivation of the P_S value*

To estimate the critical cells size (P_S) that cells have to reach to enter S phase, we need to simulate the onset of DNA replication. In fact, P_S is defined operationally as the protein content of cells that enter S phase [33]. At this point, we estimated P_S as the cell size when 50% of the replication origins were activated in a single cell. In addition, the method permits to calculate the cell size at different % of replication origins activated.

To implement the derivation of the P_S value, we wrote a Mathematica script that calculates the value of the cell volume when a certain % of replication origins is activated. Specifically, the value of the cell volumes is calculated when the first replication origin reaches the state fired, and when 10 %, 50 %, and 90 % of the replication origins are activated, respectively.

3.2. *Fluctuations of the P_S value*

In [1] we analyzed the effects of specific parameters of the model of the G_1/S transition on the setting of P_S. Specifically, we focused on the main parameters that we established to be different from glucose to ethanol. Sensitivity analysis showed that the Ps value is mainly affected by the growth rate. Here we show the influence of each parameter of the G_1/S transition model on the setting of the critical cell mass. In Table 1 are listed the kinetic parameters, and the corresponding reactions (see [1] for the detailed explanation of the reactions), that affect significantly the P_S value (in bold face are highlighted the parameters that play a major role in changing the critical cell mass). Fig. 5 illustrates the sensitivity analysis that shows the P_S changes following the variation of the kinetic parameters from 0.1-fold to 100-fold.

As reported in Table 1, the main kinetic parameters influencing the setting of the critical cell size P_S are related to two regulatory events, which control the entrance into S phase. The first growth-dependent threshold entails the interplay of the activator Cln3 (bound to the kinase Cdk1) and the inhibitor Far1 [10,34]. The nuclear Cdk1-Cln3 activity increases proportionally to Cln3 level, and to the inverse of Far1, thus the threshold mechanism controlling the onset of S phase is operative as long as Cln3 increases faster than Far1. The Cln3/Far1 threshold indicates the reaching of a given size, when the amount of Cln3 - that increases in proportion to cell mass - overcomes Far1. Nuclear Cdk1-Cln3 phosphorylates Whi5, leading to the activation of transcription factors SBF/MBF, so opening up the pathway that leads to the onset of DNA replication. The critical cell size is reached when the cells overcome the second threshold dependent on the balance between the cyclin Clb5 (the activator) and the Cki Sic1 (the inhibitor) [1]. The analysis confirms that the coordination between the two sequential threshold mechanism (*i.e.* the Cln3/Far1 and the Clb5,6/Sic1 thresholds) regulates the critical cell mass necessary to undergo the S phase, thus effectively couples cell growth to the onset of DNA replication.

Table 1. Kinetic parameters and corresponding reactions shown in Fig. 5.

Panel	Reaction	Parameter
A	Protein synthesis	$\mathbf{k_5}$ **(Far1$_{cyt}$)**, k_6 (Cln3$_{cyt}$), $\mathbf{k_7}$ **(Cdk1$_{cyt}$)** k_8 (Whi5$_{cyt}$)
B	Translation mRNA → protein SBF/MBF basal production	$\mathbf{k_3}$ **(*CLN1,2* mRNA$_{cyt}$ → Cln1,2$_{cyt}$)** $\mathbf{k_4}$ **(*CLB5,6* mRNA$_{cyt}$ → Clb5,6$_{cyt}$)** k_{35} (SBF/MBF$_{nuc}$ basal production)
C	Localization (protein nuclear import)	k_{42} (Far1$_{cyt}$), $\mathbf{k_{43}}$ **(Cln3$_{cyt}$)**, $\mathbf{k_{44}}$ **(Cdk1$_{cyt}$)** k_{45} (Whi5$_{cyt}$)
D	Localization (protein complexes nuclear import)	k_{46} (Cdk1-Cln1,2$_{cyt}$) k_{47} (Cdk1-Clb5,6-Sic1$_{cyt}$) k_{48} (Cdk1-Clb5,6$_{cyt}$)
E	Localization (protein export)	k_{49} (Cdk1$_{nuc}$), k_{50} (*CLN1,2* mRNA$_{nuc}$) k_{51} (*CLB5,6* mRNA$_{nuc}$)
F	Association	$\mathbf{k_{24}}$ **(Cln3$_{nuc}$ + Cdk1$_{nuc}$)** $\mathbf{k_{28}}$ **(Cdk1$_{cyt}$ + Clb5,6$_{cyt}$)** k_{30} (Cdk1-Cln3$_{nuc}$ + Far1$_{nuc}$) k_{34} (SBF/MBF$_{nuc}$ + Whi5$_{nuc}$)
G	Dissociation (unphosphorylated complexes)	$\mathbf{k_{25}}$ **(Cdk1-Cln3$_{nuc}$)**, k_{27} (Cdk1-Cln1,2$_{cyt}$) $\mathbf{k_{29}}$ **(Cdk1-Clb5,6$_{cyt}$)** $\mathbf{k_{33}}$ **(Cdk1-Clb5,6-Sic1$_{cyt}$)**
H	Dissociation (phosphorylated complexes)	k_{39} (SBF-MBF-Whi5-P$_{nuc}$) $\mathbf{k_{40}}$ **(Cdk1-Cln3-Far1-P$_{nuc}$)** k_{41} (Cdk1-Clb5,6-Sic1-P$_{nuc}$)
I	Catalytic activity	k_{36} (Cdk1-Cln3$_{nuc}$ on SBF-MBF-Whi5$_{nuc}$) k_{37} (Cdk1-Cln1,2$_{nuc}$ on Cdk1-Cln3-Far1$_{nuc}$)
J	Degradation (in the nucleus)	k_{19} (Far1$_{nuc}$), $\mathbf{k_{20}}$ **(Cln3$_{nuc}$)**
K	Degradation (mRNAs and cyclin-dependent kinase inhibitors in the cytoplasm)	k_{10} (*CLN1,2* mRNA$_{cyt}$) $\mathbf{k_{11}}$ **(*CLB5,6* mRNA$_{cyt}$)** k_{14} (Far1$_{cyt}$), $\mathbf{k_{18}}$ **(Sic1$_{cyt}$)**
L	Degradation (cyclins and Cdk in the cytoplasm)	k_{12} (Cln1,2$_{cyt}$), $\mathbf{k_{13}}$ **(Clb5,6$_{cyt}$)**, $\mathbf{k_{15}}$ **(Cln3$_{cyt}$)** $\mathbf{k_{16}}$ **(Cdk1$_{cyt}$)**

4. Conclusion

A significant part of regulatory events in a living cell are related to the precise functioning of the cell cycle. DNA replication is the main event that drives the cell cycle after the G_1/S transition. Here we show how a probabilistic three-steps model can describe the initiation of the DNA replication process, provides a reasonable explanation for the phenotype of wild type cells grown on different media, and for selected mutants of the network. In addition, with this model we have a valuable tool at hand to estimate the critical cell mass, P_S, which cells need to reach to duplicate the DNA. Due to the relatively simple structure of the model, and considering the nuclear concentration of the Cdk1-Clb5,6 complex as the only input [12,13], the model is very well suited for this task. Moreover, from the sensitivity analysis of the P_S value, specific kinetic parameters are recognized to be important for the fine regulation of the G_1/S transition events.

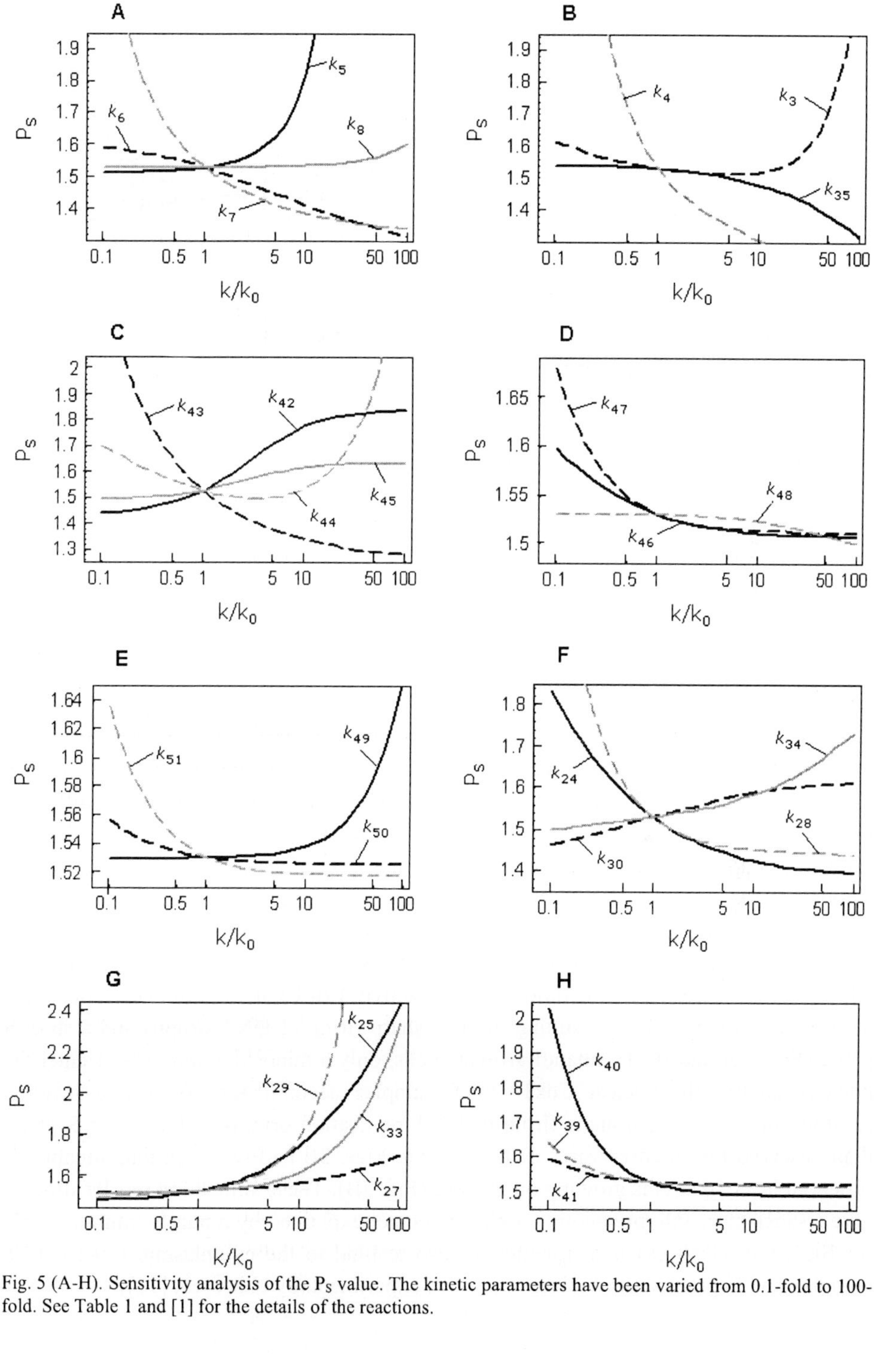

Fig. 5 (A-H). Sensitivity analysis of the P_S value. The kinetic parameters have been varied from 0.1-fold to 100-fold. See Table 1 and [1] for the details of the reactions.

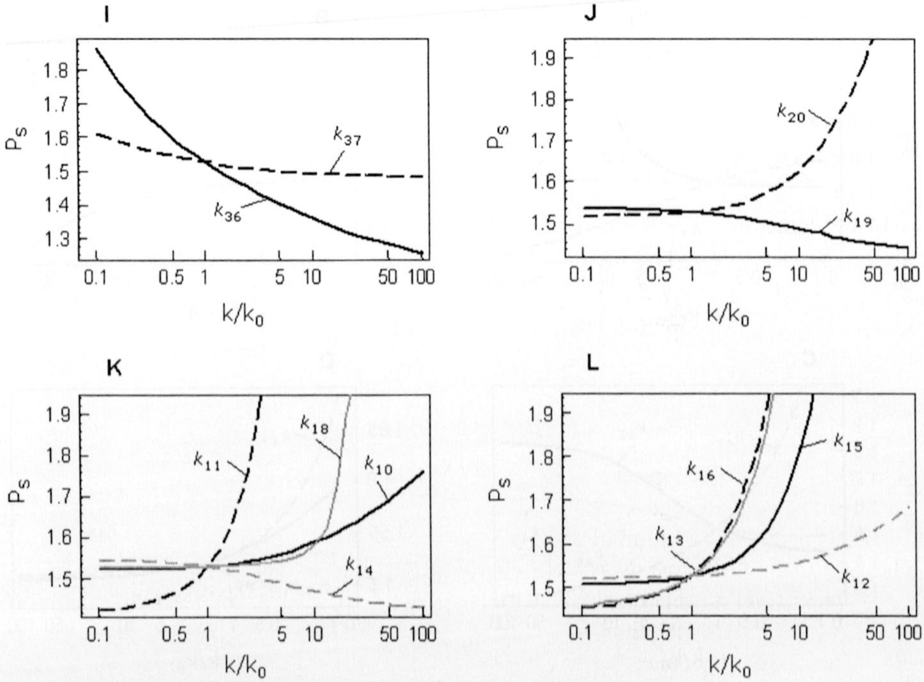

Fig. 5 (I-L). Sensitivity analysis of the P_S value. The kinetic parameters have been varied from 0.1-fold to 100-fold. See Table 1 and [1] for the details of the reactions.

The analysis of the DNA replication initiation focused especially on the critical steps of the G_1/S transition. The involvement of the Cki Sic1 in the control of the replication events represents a good example. The nucleo/cytoplasmic localization of Sic1 differs in glucose- and ethanol-grown cells. It is known that in G_1 cells growing in glucose-supplemented media Sic1 is mostly nuclear, while in G_1 cells grown in ethanol a large amount of Sic1 remains cytoplasmic [11]. This means that in glucose, where Sic1 is almost entirely nuclear, high levels of the nuclear Cdk1-Clb5 complex accumulate, albeit in the inactive form of a ternary complex with Sic1 itself. Such a pre-accumulation allows semi-synchronous liberation of the active Cdk1-Clb5,6 complex after Sic1 degradation in the nucleus, resulting in quite sharp firing of DNA origins and a short S phase (Figs. 2A and B). In ethanol-growing cells, only a minor fraction of Sic1 enter the nucleus, so that the nuclear Cdk1-Clb5,6 complex accumulates slowly and steadily, resulting in a less synchronous firing of DNA replication origins and a longer S phase than observed for glucose-grown cells (compare Figs. 2B and D). A similar situation is observed in *sic1Δ* strains growing on glucose (Fig. 3B). These simulation results allow to give a satisfactory interpretation to a phenotype not explained by a purely inhibitory role for Sic1, but which essential function is also to bind to the cytoplasmic Cdk1-Clb5,6 complex and to promote its nuclear import (as experimentally observed, [11]).

Considering the structure of the network we built as a platform of the essential events that happen in the G_1/S transition, the results obtained now permit us to link this model to the late events of the cell cycle. In particular, we demonstrated that it is crucial to consider the proper network construction - the spatial localization of the regulatory key players - to simulate correctly the onset of DNA replication.

Acknowledgments
This work was supported by the European Commission (Project ENFIN, contract number LSHG-CT-2005-518254), and the German Research Foundation (DFG; the International Research Training Group "Genomics and Systems Biology of Molecular Networks").

References

[1] Barberis, M., Klipp, E., Vanoni, M., and Alberghina, L., Cell size at S phase initiation: an emergent property of the G1/S network, *PloS Computational Biology*, 3:e64.doi:10.1371/journal.pcbi.0030064.

[2] Chen, K.C., Csikasz-Nagy, A., Gyorffy, B., Val, J., Novak, B., and Tyson, J.J., Kinetic analysis of a molecular model of the budding yeast cell cycle, *Mol. Biol. Cell*, 11(1):369-391, 2000.

[3] Chen, K.C., Calzone, L., Csikasz-Nagy, A., Cross, F.R., Novak, B., and Tyson, J.J., Integrative analysis of cell cycle control in budding yeast, *Mol. Biol. Cell*, 15(8):3841-3862, 2004.

[4] Csikasz-Nagy, A., Battogtokh, D., Chen, K.C., Novak, B., and Tyson, J.J., Analysis of a generic model of eukaryotic cell-cycle regulation, *Biophys. J.*, 90(12):4361-4379, 2006.

[5] Obeyesekere, M.N., Tucker, S.L., and Zimmerman, S.O., A model for regulation of the cell cycle incorporating cyclin A, cyclin B and their complexes, *Cell Prolif.*, 27(2):105-113, 1994.

[6] Obeyesekere, M.N., Herbert, J.R., and Zimmerman, S.O., A model of the G1 phase of the cell cycle incorporating cyclin E/cdk2 complex and retinoblastoma protein, *Oncogene*, 11(6):1199-1205, 1995.

[7] Kohn, K.W., Functional capabilities of molecular network components controlling the mammalian G1/S cell cycle phase transition, *Oncogene*, 16(8):1065-1075, 1998.

[8] Aguda, B.D. and Tang, Y., The kinetic origins of the restriction point in the mammalian cell cycle, *Cell Prolif.*, 32(5):321-335, 1999.

[9] Qu, Z., Weiss, J.N., and MacLellan, W.R., Regulation of the mammalian cell cycle: a model of the G1-to-S transition, *Am. J. Physiol. Cell Physiol.*, 284(2):C349-C364, 2003.

[10] Alberghina, L., Rossi, R.L., Querin, L., Wanke, V., and Vanoni, M., A cell sizer network involving Cln3 and Far1 controls entrance into S phase in the mitotic cycle of budding yeast, *J. Cell Biol.*, 167(3):433-443, 2004.

[11] Rossi, R.L., Zinzalla, V., Mastriani, A., Vanoni, M., and Alberghina, L., Subcellular localization of the cyclin dependent kinase inhibitor Sic1 is modulated by the carbon source in budding yeast, *Cell Cycle*, 4(12):1798-1807, 2005.

[12] Bell, S.P. and Dutta, A, DNA replication in eukaryotic cells. *Annu. Rev. Biochem.*, 71:333-374, 2002.

[13] Takeda, D.Y. and Dutta, A, DNA replication and progression through S phase, *Oncogene*, 24(17):2827-2843, 2005.

[14] Wells, W.A., Does size matter?, *J. Cell Biol.*, 158(7):1156-1159, 2002.

[15] Rupes, I., Checking cell size in yeast, *Trends Genet.*, 18(9):479-485, 2002.

[16] Raghuraman, M.K., Winzeler, E.A., Collingwood, D., Hunt, S., Wodicka, L., Conway, A., Lockhart, D.J., Davis, R.W., Brewer, B.J., and Fangman, W.L., Replication dynamics of the yeast genome, *Science*, 294(5540):115-121, 2001.

[17] Wyrick, J.J., Aparicio, J.G., Chen, T., Barnett, J.D., Jennings, E.G., Young, R.A., Bell, S.P., and Aparicio, O.M., Genome-wide distribution of ORC and MCM proteins in *S. cerevisiae*: high-resolution mapping of replication origins, *Science*, 294(5550):2357-2360, 2001.

[18] Newlon, C.S., Petes, T.D., Hereford, L.M., and Fangman, W.L., Replication of yeast chromosomal DNA, *Nature*, 247(5435):32-35, 1974.

[19] Petes, T.D. and Williamson, D.H., Fiber autoradiography of replicating yeast DNA, *Exp. Cell Res.*, 95(1):103-110, 1975.

[20] Lord, P.G. and Wheals, A.E., Variability in individual cell cycles of *Saccharomyces cerevisiae*. *J. Cell Sci.*, 50:361-376, 1981.

[21] Vanoni, M., Vai, M., Popolo, L., and Alberghina, L., Structural heterogeneity in populations of the budding yeast *Saccharomyces cerevisiae*. *J. Bacteriol.*, 156(3):1282-1291, 1983.

[22] Vanoni, M., Rossi, R.L., Querin, L., Zinzalla, V., and Alberghina, L., Glucose modulation of cell size in yeast. *Biochem. Soc. Trans.*, 33(Pt 1):294-296, 2005.

[23] Johnston, G.C., Singer, R.A., Sharrow, S.O., and Slater, M.L., Cell division in the yeast *Saccharomyces cerevisiae* growing at different rates, *J. Gen. Microbiol.*, 118:479-484, 1980.

[24] Rivin, C.J. and Fangman, W.L., Cell cycle phase expansion in nitrogen-limited cultures of *Saccharomyces cerevisiae*, *J. Cell Biol.*, 85(1):96-107, 1980.

[25] Rivin, C.J. and Fangman, W.L. Replication fork rate and origin activation during the S phase of *Saccharomyces cerevisiae*, *J. Cell Biol.*, 85(1):108-115, 1980.

[26] Cipollina, C., Alberghina, L., Porro, D., and Vai, M., SFP1 is involved in cell size modulation in respiro-fermentative growth conditions, *Yeast*, 22(5):385-399, 2005.

[27] Mendenhall, M.D., An inhibitor of p34CDC28 protein kinase activity from *Saccharomyces cerevisiae*, *Science*, 259(5092):216-219, 1993.

[28] Schwob, E., Bohm, T., Mendenhall, M.D., and Nasmyth, K., The B-type cyclin kinase inhibitor p40SIC1 controls the G1 to S transition in *S. cerevisiae*, *Cell*, 79(2):233-244, 1994.

[29] Lengronne, A. and Schwob, E., The yeast CDK inhibitor Sic1 prevents genomic instability by promoting replication origin licensing in late G1, *Mol. Cell*, 9(5):1067-1078, 2002.

[30] Dirick, L., Bohm, T., and Nasmyth, K., Roles and regulation of Cln-Cdc28 kinases at the start of the cell cycle of *Saccharomyces cerevisiae*, *EMBO J.*, 14(19):4803-4813, 1995.

[31] Tyers, M., Tokiwa, G., Nash, R., and Futcher, B., The Cln3-Cdc28 kinase complex of *S. cerevisiae* is regulated by proteolysis and phosphorylation, *EMBO J.*, 11(5):1773-1784, 1992.

[32] Costanzo, M., Nishikawa, J.L., Tang, X., Millman, J.S., Schub, O., Breitkreuz, K., Dewar, D., Rupes, I., Andrews, B., and Tyers, M., CDK activity antagonizes Whi5, an inhibitor of G1/S transcription in yeast, *Cell*, 117(7):899-913, 2004.

[33] Porro, D., Brambilla, L., and Alberghina, L., Glucose metabolism and cell size in continuous cultures of *Saccharomyces cerevisiae*, *FEMS Microbiol. Lett.*, 229(2):165-171, 2003.

[34] Alberghina L., Porro D., Cazzador L., Towards a blueprint of the cell cycle, *Oncogene*, 20(9):1128-1134, 2001.

STEADY STATE ANALYSIS OF SIGNAL RESPONSE IN RECEPTOR TRAFFICKING NETWORKS

ZHIKE ZI[1] zhike_zi@molgen.mpg.de
EDDA KLIPP[1, 2] klipp@molgen.mpg.de

[1]*Computational Systems Biology, Max Planck Institute for Molecular Genetics, Ihnestr. 73, 14195 Berlin, Germany*
[2]*Theoretical Biophysics, Humboldt University Berlin, Institute for Biology. Invalidenstr. 42, 10115 Berlin, Germany*

Receptor trafficking is used to describe the internalization and recycling processes of receptors in the cell. Considerable efforts of quantitative modeling have been made so far in the study of receptor trafficking networks. For the reason of simple mathematical analysis, the canonical receptor trafficking models either ignored the recycling step of receptors or didn't consider the trafficking of empty receptors. Here, we revisit the canonical receptor trafficking models and implement steady state analysis for a general model of receptor trafficking networks, which is composed of the *de novo* appearance of surface receptor, ligand-receptor interaction, internalization, recycling and degradation of both empty and occupied receptors. We present the analytical solution of the two steady states of the receptor trafficking networks before and after the network is exposed to the signal. The results indicate that the distribution of the empty receptor at the cell surface and inside of the cell, before signal is added, is mainly determined by the ratio of internalization rate and recycling rate of empty receptor. Furthermore, the steady state analysis demonstrates that classic Scatchard plot analysis is still valid for the steady state of the complicated receptor trafficking network.

Keywords: signal response; steady state analysis; receptor trafficking network.

1. Introduction

Cells communicate with their extracellular environment by the interaction between the receptors and ligand, which converts the information from the outside environment to inside cell responses such as cell proliferation, apoptosis, differentiation and growth. The receptors at cell surface can be internalized to early endosomes and also be recycled back, which is termed as receptor trafficking [1]. Receptor trafficking is implicated as a potential site for the regulation of signaling pathways by previous experimental data [2,3]. On the other hand, considerable efforts of quantitative modeling have been made in the study of receptor trafficking networks [4-8]. For the reason of simple mathematical analysis, the canonical receptor trafficking models either ignored the recycling step of receptors or didn't consider the trafficking of empty receptors [5,6]. In this work, we revisit the canonical receptor trafficking models and implement steady state analysis for the general model of receptor trafficking networks, which is composed of the *de novo* appearance of surface receptor, ligand-receptor interaction, internalization, recycling and degradation of both empty and occupied receptors. We show the analytical solution of two steady states of the receptor trafficking networks before and after the network is exposed to the signal and perform some qualitative analysis on the two steady states.

2. Mathematical Model of the Receptor Trafficking Network

Referring to the previous canonical models of receptor trafficking networks, we model all the processes with the law of mass action. The structure of the receptor trafficking network is illustrated in Fig. 1. We model the *de novo* synthesis of surface receptor as a constant rate of k_1. Extracellular ligand L can associate with free surface receptors Rs to form a receptor-ligand complex LRs with the forward rate constant k_2 and the ligand-receptor complex disassociates with the backward rate constant k_3. The empty and occupied surface receptors are internalized with the rate constants k_5 and k_7, respectively. Furthermore, the empty and occupied receptors inside the cell can be recycled back to cell surface with recycling rate constants k_4 and k_6, respectively. The degradation rate constants of empty and occupied receptors are set as k_9 and k_{10}, respectively. We also considered the possible dephosphorylation of the activated receptors with the rate constant k_8. The symbols of the parameters involved in the network are summarized in Table 1.

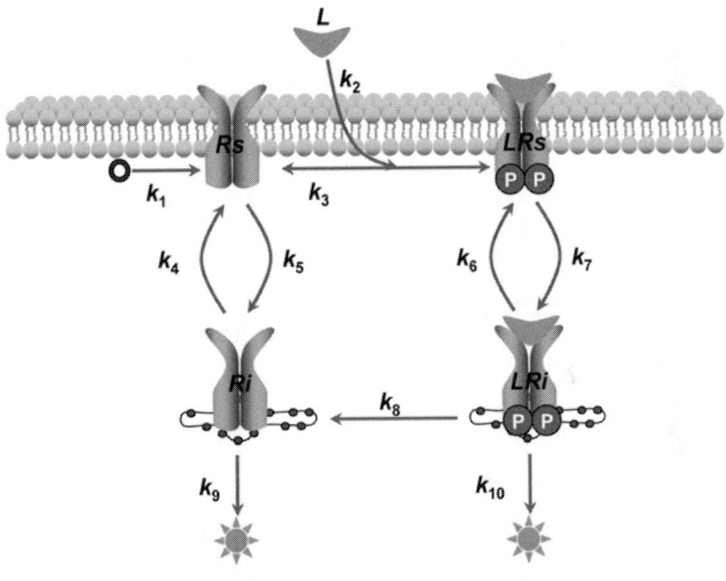

Fig. 1. Schematic description of the receptor trafficking networks. The network includes the de novo appearance of surface receptor, ligand-receptor interaction, internalization, recycling and degradation of both empty and occupied receptors. The symbols *L*, *Rs*, *LRs*, *Ri*, *LRi* represent ligand, cell surface empty receptor, cell surface ligand-receptor complex, internalized empty receptor and internalized ligand-receptor complex, respectively.

The ordinary differential equations for the components in the receptor trafficking networks can be written as

$$\frac{d[Rs]}{dt} = -(k_2[L]+k_5)[Rs]+k_4[Ri]+k_3[LRs]+k_1 \qquad (1)$$

$$\frac{d[Ri]}{dt} = k_5[Rs] - (k_4 + k_9)[Ri] + k_8[LRi] \quad (2)$$

$$\frac{d[LRs]}{dt} = k_2[L][Rs] - (k_3 + k_7)[LRs] + k_6[LRi] \quad (3)$$

$$\frac{d[LRi]}{dt} = k_7[LRs] - (k_6 + k_8 + k_{10})[LRi] \quad (4)$$

Table 1. Parameters for the model of the receptor trafficking network.

Symbols of Parameters	Corresponding Biological Processes	Typical Values	Reference
k_1	de novo synthesis of surface receptor	0.5 nM/min	[9]
k_2	formation of ligand-receptor complex	0.072 nM^{-1}min^{-1}	[8]
k_3	dissociation of ligand-receptor complex	0.34 min^{-1}	[8]
k_4	recycling of internalized empty receptor	0.2 min^{-1}	[10]
k_5	internalization of surface empty receptor	0.12 min^{-1}	[10]
k_6	recycling of internalized ligand-receptor complex	0.2 min^{-1}	[10]
k_7	internalization of surface ligand-receptor complex	0.15 min^{-1}	[10]
k_8	dephosphorylation of ligand-receptor complex	0.1 min^{-1}	
k_9	degradation of empty receptor	0.001 min^{-1}	[10]
k_{10}	degradation of ligand-receptor complex	0.01 min^{-1}	[10]

3. Steady State Analysis of the Receptor Trafficking Network

The concept of steady state is a mathematical idealization, which plays an important role in kinetic modeling [11]. A network is in steady state if the concentrations of the components do not change, which means the corresponding ordinary differential equations are zero. We can do a general steady state analysis of the receptor trafficking network by using only the network structure, without knowing the rate constants for a particular reaction [12].

3.1. Steady State of the Receptor Trafficking Network before the Signal is Added

If there is no ligand added to the receptor trafficking network, the concentration of ligand-receptor complexes *LRs* and *LRi* are zero. Therefore, the receptor trafficking network is composed of cell surface empty receptor and internalized empty receptor before the network is exposed to the external ligand. When the production, internalization, recycling and degradation of the empty receptors arrive at a steady state, we can derive the following system of algebraic equations for the empty receptors.

$$\frac{d[Rs]_1^{ss}}{dt} = -k_5[Rs]_1^{ss} + k_4[Ri]_1^{ss} + k_1 = 0 \quad (5)$$

$$\frac{d[Ri]_1^{ss}}{dt} = k_5[Rs]_1^{ss} - (k_4 + k_9)[Ri]_1^{ss} = 0 \quad (6)$$

The steady state concentrations of cell surface empty receptor *Rs* and internalized receptor *Ri*, obtained by solving the systems of algebraic equations (5-6), are

$$[Rs]_1^{ss} = \frac{k_1(k_4 + k_9)}{k_5 k_9} \tag{7}$$

$$[Ri]_1^{ss} = \frac{k_1}{k_9} \tag{8}$$

The steady state concentration of internalized empty receptor is determined by the synthesis rate and degradation rate of the receptor. The internalization and recycling steps of empty receptor affect the steady state concentration of surface receptor, but it has no effect on the steady state concentration of the empty receptor inside of the cell before the network is exposed to the ligand.

We are also interested in the distribution of the receptor at cell surface and inside of the cell. The distribution of the empty receptor can be evaluated by the ratio of internalized receptor to the surface receptor at steady state, which is

$$\frac{[Ri]_1^{ss}}{[Rs]_1^{ss}} = \frac{k_5}{k_4 + k_9} \approx \frac{k_5}{k_4} = \frac{internalization\ rate}{recycling\ rate} \tag{9}$$

Traditionally, the canonical receptor trafficking models usually ignored the trafficking of the empty receptor [4,7]. Such models assume that most of empty receptors exist at cell surface. From the expression of equation (9), we can see that the internalized receptors can be ignored only when the internalization rate of surface empty receptor is much smaller than the sum of the recycling and degradation rate of internalized empty receptor (ie. $k_5 << k_4+k_9$). The degradation rate of empty receptor is usually much smaller than the internalization rate ($k_9 << k_5$) [13]. Therefore, the distribution of empty receptors is mainly dependent on the ratio of internalization rate and recycling rate.

The assumption that most empty receptors distribute at cell surface will be valid when the recycling rate of empty receptor is much larger than the internalization rate. However, the internalization and recycling rate of receptor are various in different types of cells. For example, Burke *et. al* [10] experimentally measured the internalization rate of empty EGF receptor in HB2 and 184A1 cells, which is 0.03 min^{-1} and 0.15 min^{-1}, respectively. The corresponding recycling rate of empty receptor in HB2 and 184A1 cells is about 0.2 min^{-1} and 0.3 min^{-1}, respectively. From the result of steady state analysis, we can predict that the ratio of internalized empty EGF receptor to surface empty EGF receptor in HB2 cell is smaller than that in 184A1 cell. Therefore, the amount of internalized empty EGF receptors in 184A1 cells should not be ignored compared to the amount of the surface empty receptor. This prediction can be confirmed by the later experimental data by Burke *et. al* [14]. For a quantitative study of receptor trafficking networks, it is necessary to check the validity of the assumption of ignoring the role of internalized empty receptors.

3.2. Steady State of the Receptor Trafficking Network after the Signal is Added

We next investigated the steady state of the receptor trafficking network after it is exposed to the external signal. When the ligand is added, the receptor trafficking network will reach another different steady state. According to the definition of steady state, we can derive the following system of algebraic equations for the new steady state:

$$\frac{d[Rs]_2^{ss}}{dt} = -(k_2[L]_2^{ss} + k_5)[Rs]_2^{ss} + k_4[Ri]_2^{ss} + k_3[LRs]_2^{ss} + k_1 = 0 \tag{10}$$

$$\frac{d[Ri]_2^{ss}}{dt} = k_5[Rs]_2^{ss} - (k_4 + k_9)[Ri]_2^{ss} + k_8[LRi]_2^{ss} = 0 \tag{11}$$

$$\frac{d[LRs]_2^{ss}}{dt} = k_2[L]_2^{ss}[Rs]_2^{ss} - (k_3 + k_7)[LRs]_2^{ss} + k_6[LRi]_2^{ss} = 0 \tag{12}$$

$$\frac{d[LRi]_2^{ss}}{dt} = k_7[LRs]_2^{ss} - (k_6 + k_8 + k_{10})[LRi]_2^{ss} = 0 \tag{13}$$

We can get the new steady state of the network by solving the system of algebraic equations (10-13), which leads to:

$$[Rs]_2^{ss} = \frac{k_1(k_4+k_9)(k_3(k_6+k_8+k_{10})+k_7(k_8+k_{10}))}{D} \tag{14}$$

$$[Ri]_2^{ss} = \frac{k_1(k_3k_5(k_6+k_8+k_{10})+k_2k_7k_8[L]_2^{ss}+k_5k_7(k_8+k_{10}))}{D} \tag{15}$$

$$[LRs]_2^{ss} = \frac{k_1k_2(k_4+k_9)(k_6+k_8+k_{10})[L]_2^{ss}}{D} \tag{16}$$

$$[LRi]_2^{ss} = \frac{k_1k_2k_7(k_4+k_9)[L]_2^{ss}}{D} \tag{17}$$

where

$$D = k_3 k_5 k_9 (k_6 + k_8 + k_{10}) + \\ k_7 k_9 (k_2 [L]_2^{ss} + k_5)(k_8 + k_{10}) + \\ k_2 k_4 k_7 k_{10} [L]_2^{ss} \tag{18}$$

The expression of the new steady state of the receptor trafficking network is too complicated to be handled by mind. Here, we analyze the steady state concentrations of the components in the network under some special conditions and illustrate the effect of different parameters on the steady state of the network after it is exposed to the signal.

- Effect of Ligand: For the ligand concentration, it is usually assumed as a constant. Therefore, the steady state for the ligand (final concentration, $[L]_2^{ss}$) is a certain constant. On the other hand, if the amount of ligand is smaller than the amount of receptors and the ligand is degraded by the cell, the steady state of ligand ($[L]_2^{ss}$) will be zero. In this case, the activated ligand-receptor complex $[LRs]_2^{ss}$ and $[LRs]_2^{ss}$ will come to zero according to the equations (16-17). If we monitor the time course of the activated ligand-receptor complex, we will observe a transient signal response.

- Ratio of steady state concentration of internalized ligand-receptor complex to surface ligand-receptor complex: We can derive this ratio from equations (16-17) as

$$\frac{[LRi]_2^{ss}}{[LRs]_2^{ss}} = \frac{k_7}{k_6 + k_8 + k_{10}} \tag{19}$$

  This ratio is independent of ligand and the trafficking behavior of empty receptors.

- Effect of receptor synthesis rate (k_1): The receptor synthesis rate k_1 appears as a multiplier in the numerator, but not the denumerator, of the equations (14-17). Therefore, it is obviously that the faster the receptor synthesis rate, the higher value of the steady state of various forms of receptor in the receptor trafficking network.

- The total amount of occupied receptors is the sum of the internalized ligand-receptor complex and surface ligand-receptor complex, which is defined as bound receptors (Bound). From equations (16-17), we can derive

$$[Bound]_2^{ss} = [LRs]_2^{ss} + [LRi]_2^{ss} = \frac{\alpha [L]_2^{ss}}{\beta [L]_2^{ss} + \gamma} \tag{20}$$

  where

$$\alpha = k_1 k_2 (k_4 + k_9)(k_6 + k_7 + k_8 + k_{10}) \tag{21}$$

$$\beta = k_2 k_7 (k_4 k_{10} + k_9 (k_8 + k_{10})) \tag{22}$$

$$\gamma = k_5 k_9 (k_3 k_6 + (k_3 + k_7)(k_8 + k_{10})) \tag{23}$$

By this way, we can derive the Bound/Free and Bound relationship at steady state from equation (20):

$$\frac{[Bound]_2^{ss}}{[L]_2^{ss}} = -\frac{\beta}{\gamma}[Bound]_2^{ss} + \frac{\alpha}{\gamma} \tag{24}$$

The corresponding X-intercept for equation (24) is

$$\text{X} - \text{intercept} = \frac{\alpha}{\beta} \tag{25}$$

Equation (24) is similar to the classical Scatchard plot (Rosenthal Plot). This result indicates that the classic Scatchard plot analysis is still valid for the steady state of complicated receptor trafficking network which includes ligand receptor interaction, internalization, recycling, dephosphorylation and degradation of receptors.

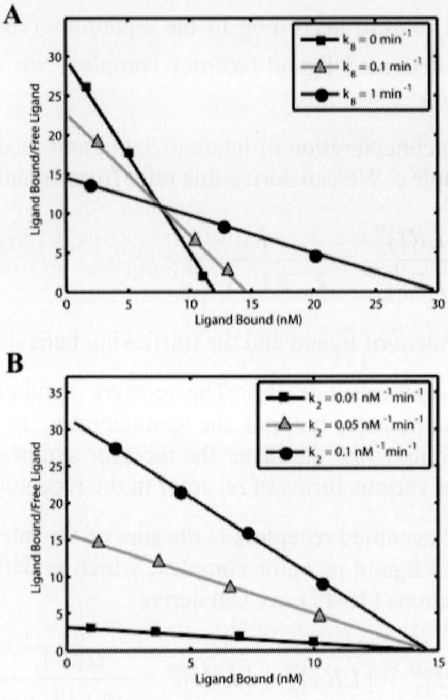

Fig. 2. Steady state Scatchard plot of the relationship between ligand bound and free ligand. The rate constants and corresponding typical values are list in Table 1. (A) Effect of different dephosphorylation rate of activated receptor complex (k_8). (B) Effect of different binding affinities (k_2).

- For the typical values of the parameters in the EGF receptor trafficking network (Table 1), the corresponding steady state binding plots with different dephosphorylation rates of activated receptor complex (k_8) are shown in Fig. 2A. The slope of steady state Scatchard plot decreases with the increase of k_8. On the other hand, different binding affinities of ligand and receptor interaction are found in EGF receptor trafficking networks [15]. According to the equation (21-24), we can conclude that the higher the binding affinity of ligand and receptor interaction (ie. the higher value of k_2), the larger is the absolute value of slope in steady state Scatchard plot, which is confirmed by the computer simulation shown in Fig. 2B.

- When recycling steps of internalized receptors and deactivation of internalized receptor complex are ignored (k_4, k_6 and k_8 are zero) which is assumed in the early work of quantitative modeling of receptor trafficking network [4], we get the following expression

$$\frac{[Bound]_2^{ss}}{[L]_2^{ss}} = -\frac{k_2 k_7}{k_5(k_3 + k_7)}[Bound]_2^{ss} + \frac{k_1 k_2 (k_7 + k_{10})}{k_5 k_{10}(k_3 + k_7)} \qquad (26)$$

Equation (26) is exactly the same expression as that derived by Wiley and Cunningham [4].

4. Conclusion

In this work, we implemented steady state analysis for the general receptor trafficking network before and after it is exposed to the external signal. We can draw the following conclusions based on the analysis results:

(1) Before ligand is added, the steady state concentration of internalized empty receptor is determined by the synthesis rate and degradation rate of the receptor. This is independent of the internalization and recycling rate of empty receptors.

(2) The distribution of the empty receptor at cell surface and inside of the cell, before signal is added, is mainly determined by the ratio of internalization rate and recycling rate of empty receptor.

(3) For quantitatively modeling of receptor trafficking networks, it is necessary to check the validation of the assumption of ignoring the role of internalized empty receptors because this assumption is only valid under a certain special condition.

(4) The classical Scatchard plot is still valid for the steady state of the complicated receptor trafficking network which includes ligand receptor interaction, internalization, recycling, dephosphorylation and degradation of receptors.

5. Acknowledgement

Z. Zi is supported by PhD program of the IMPRS for Computational Biology and Scientific Computing. This work was supported by a grant to E. Klipp., ENFIN, a Network of Excellence funded by the European Commission within its FP6 Programme, under the thematic area "Life sciences, genomics and biotechnology for health", contract number LSHG-CT-2005-518254.

References

[1] Moore, C.A., Milano, S.K., and Benovic, J.L., Regulation of receptor trafficking by GRKs and arrestins, *Annu. Rev. Physiol.*, 69451-69482, 2007.

[2] Schoeberl, B., Eichler-Jonsson, C., Gilles, E.D., and Muller, G., Computational modeling of the dynamics of the MAP kinase cascade activated by surface and internalized EGF receptors, *Nat. Biotechnol.*, 20(4): 370-375, 2002.

[3] Di Guglielmo, G.M., Le Roy, C., Goodfellow, A.F., and Wrana, J.L., Distinct endocytic pathways regulate TGF-beta receptor signalling and turnover, *Nat. Cell Biol.*, 5(5): 410-421, 2003.

[4] Wiley, H.S. and Cunningham, D.D., A steady state model for analyzing the cellular binding, internalization and degradation of polypeptide ligands, *Cell*, 25(2): 433-440, 1981.

[5] Knauer, D.J., Wiley, H.S., and Cunningham, D.D., Relationship between epidermal growth factor receptor occupancy and mitogenic response. Quantitative analysis using a steady state model system, *J. Biol. Chem.*, 259(9): 5623-5631, 1984.

[6] Myers, A.C., Kovach, J.S., and Vuk-Pavlovic, S., Binding, internalization, and intracellular processing of protein ligands. Derivation of rate constants by computer modeling, *J. Biol. Chem.*, 262(14): 6494-6499, 1987.

[7] Lund, K.A., Opresko, L.K., Starbuck, C., Walsh, B.J., and Wiley, H.S., Quantitative analysis of the endocytic system involved in hormone-induced receptor internalization, *J. Biol. Chem.*, 265(26): 15713-15723, 1990.

[8] Starbuck, C. and Lauffenburger, D.A., Mathematical model for the effects of epidermal growth factor receptor trafficking dynamics on fibroblast proliferation responses, *Biotechnol. Prog.*, 8(2): 132-143, 1992.

[9] DeWitt, A.E., Dong, J.Y., Wiley, H.S., and Lauffenburger, D.A., Quantitative analysis of the EGF receptor autocrine system reveals cryptic regulation of cell response by ligand capture, *J. Cell Sci.*, 114(Pt 12): 2301-2313, 2001.

[10] Burke, P.M. and Wiley, H.S., Human mammary epithelial cells rapidly exchange empty EGFR between surface and intracellular pools, *J. Cell Physiol.*, 180(3): 448-460, 1999.

[11] Heinrich, R. and Schuster, S., *The Regulation of Cellular Systems*, Chapman and Hall, 1996.

[12] Kitano, H., Systems biology: a brief overview, *Science*, 295(5560): 1662-1664, 2002.

[13] Sorkin, A. and Waters, C.M., Endocytosis of growth factor receptors, *Bioessays*, 15(6): 375-382, 1993.

[14] Burke, P., Schooler, K., and Wiley, H.S., Regulation of epidermal growth factor receptor signaling by endocytosis and intracellular trafficking, *Mol. Biol. Cell*, 12(6): 1897-1910, 2001.

[15] Felder, S., LaVin, J., Ullrich, A., and Schlessinger, J., Kinetics of binding, endocytosis, and recycling of EGF receptor mutants, *J. Cell Biol.*, 117(1): 203-212, 1992.

USING TRANSCRIPTION FACTOR BINDING SITE CO-OCCURRENCE TO PREDICT REGULATORY REGIONS

HOLGER KLEIN
holger.klein@molgen.mpg.de

MARTIN VINGRON
martin.vingron@molgen.mpg.de

Max Planck Institute for Molecular Genetics, Ihnestr. 63-73, D-14195 Berlin, Germany

Transcription factors (TFs) bind to the regulatory regions of genes in a cooperative manner. This article describes a method to detect pairs of transcription factor binding sites which co-occur in known regulatory regions more often than expected by mere combination of the individual binding sites. We determine frequently co-occurring TF pairs and evaluate the method using known TF interactions.
Furthermore we use co-occurrence scores to assess the regulatory potential of a sequence region by calculating a graph-based score. We show results for the score on known regulatory regions.

Keywords: transcription factor binding sites; co-occurrence; transcription factor interactions; regulatory potential; promoter/enhancer prediction.

1. Introduction

The regulation of transcription is controlled by transcription factors (TFs) binding to specific motifs in the DNA around and upstream of the transcriptional start site (TSS) of transcripts. These transcription factors form complexes with other TFs and cofactors which do not bind the DNA themselves. Hence the binding sites for transcription factors (TFBSs) taking part in these protein complexes usually occur in sterical proximity to each other. The TFBSs can be found organized in cis-regulatory modules or clusters. For metazoans such a cis-regulatory module typically consists of up to ten binding sites for at least three different sequence-specific transcription factors stretched over roughly 500bp [14]. These modules can function to direct complex spatial or temporal expression patterns. For some transcription factors potential interaction partners are well known. The type of interaction can be homotypic (e.g., *GATA-1* with a second protein *GATA-1* [6]), heterotypic (e.g., *NFAT* with *AP-1* [4]), or mediated via co-factors. Usually a transcription factor can interact with several different other factors.

The prediction of individual transcription factor binding sites (TFBSs) is error-prone. The search for putative hits is carried out using position weight matrices (PWMs), and the procedure entails plenty false positive hits, since typical binding sites are short and sometimes degenerate. One remedy for this problem it the application of phylogenetic footprinting, where only TFBSs in evolutionary conserved regions are taken into account. For reviews on TFBS prediction see [2] and [21].

The motivation for this study is the assumption that the predicted binding sites of interacting factors co-localize more often than expected by chance despite the high error rate for the TFBS prediction. The idea is that the signal of co-occurrence patterns is large enough to identify interacting transcription factors and to improve the prediction of regulatory regions.

There are different approaches to identify putative interaction transcription factors. Rateitschak *et al.* [19] annotate conserved regions from the CORG database [7] with predicted binding sites. A log-odds score is used to identify pairs of TFBSs which occur together in upstream regions of genes in a maximum distance. The numbers of expected pairs are computed using the marginals of the co-occurrence count matrix. Hannenhalli *et al.* [10] define a co-localization index based on permutation of TFBSs. A recent study applied the detection of co-occurring TFBSs to identify tissue-specific pairs of TFs [23] in human. Here the number of observed co-localizations in tissue-specific sets of genes is compared to the number observed genome-wide. Previous works on the identification of regulatory regions using clusters of predicted TFBSs are for example COMET, ClusterBuster [8, 9] or MSCAN [11], which use limited sets of motifs for the prediction of regulatory regions. For a review of various other promoter prediction tools see Bajic *et al.* [1].

We had two major goals for this study. First is the prediction of putative synergistic transcription factor pairs, based on the analysis of co-occurrence of their respective binding sites in known regulatory regions. The second goal is the calculation of a regulatory potential score for previously not characterized sequence regions. To achieve these objectives we annotate sets of upstream regions with TFBSs and slide a window over the sequences. We count the number of times each combination of TFBSs shows up in the data set. The expected number of pairs is calculated by recounting the number pairs on annotation sets whose TFBS labels have been permuted before. Using a log-odds score of observed counts over expected counts pairs of binding sites are identified which are present more often than expected. Furthermore we build a graph based on the previously introduced co-occurrence scores and TFBSs predicted in unknown sequence regions and subsequently calculate a matching score. Hence regions with pairs of TFBSs which are common in known regulatory regions achieve higher scores than other regions. This way we hope to add a method to the field of prediction of regulatory regions, which does not rely as much on the presence of CpG-islands as many of the other tools. This would improve the dection of CpG-less promoters and enhancers.

This article is structured in the following way: first we describe how we prepare the sequence data and the transcription factor binding site predictions. In the Methods section the co-occurrence score and the matching score for the calculation of the regulatory potential of a sequence are introduced. In the Results section we show putative interacting transcription factors and test the performance of the score on known and yet unknown interactions. Moreover we present results for the regulatory potential on known regulatory regions.

2. Data

2.1. *Sequence Data*

The regulatory sequence data set we use for the calculation of the co-occurrence scores was prepared as follows. We extracted the upstream regions of all known human genes in the EnsEMBL database (v. 34, based on NCBI 35) [12] in a region of -1500 bp 5'-upstream and +100 bp 3'-downstream relative to the most 5' transcription start site annotated. Overlapping regions in the resulting sequences were merged and the regions which are conserved to the respective upstream regions of orthologous mouse genes were marekd. We masked repeats [20] and predicted the TFBSs on conserved regions.

2.2. *Position Weight Matrices*

In the transcription factor binding site data sets which are available one can observe a certain degree of redundancy. It happens that different TFs have similar binding site specificity, moreover for some TFs multiple binding site descriptions of different quality are available. For that reason we chose a subset of TFBS descriptions for our work. We annotated the set of regulatory sequences with the non-redundant set of vertebrate position weight matrices from TRANSFAC. The TFs were grouped based on their similarity and biological relatedness. From each group the profile with the smallest number of false-positive hits was selected as representative. As of TRANSFAC version 10.3 this set consists of 151 out of 586 total vertebrate profiles [16]. We used the methods of Rahmann *et al.* [18] to annotate putative transcription factor binding sites. The scanning cutoff was chosen in such a way that the probability to get a false positive prediction in a sequence of length 500bp is at 5% (fixed type I error).
We prepared the sequence sets and predicted transcription factor binding sites using the *BioMinerva* set of perl libraries [13].

2.3. *Known Interactions of Transcription Factors*

We built a reference set of PWM-combinations representing combinations of transcription factors known to interact. For the set of PWMs used we flagged all combinations belonging to factors known to interact in TRANSFAC. As of TRANSFAC 10.3 we find 176 pairs of PWMs.

3. Methods

3.1. *Co-occurrence Score*

We define a co-occurrence score as the log-odds score of the observed over the expected number of annotated TFBS pairs in the set of known regulatory sequences. The number of TF pairs was counted using a sliding window over each sequence in the data set. The strand of the hit and the orientation of the TFBS pair was disregarded. Multiple occurrences of a specific combination of TFBSs in the same window were only counted once to reduce the influence of transcription factors that usually bind in homotypic clusters. Since binding motifs may resemble each other we only count non-overlapping hits. Furthermore we count the same pairs in overlapping windows only once.

The co-occurrence score for a pair of PWMs i and j is then defined as the log-odds score of the observed $c_{ij,obs}$ over the expected number of pairs $c_{ij,exp}$.

$$S_{ij} := \log \frac{c_{ij,obs}}{c_{ij,exp}}$$

This was acquired by a repeated permutation of the TFBS labels in the original data set followed by recounting the pairs of TFBSs. The expected number of pairs then equals the average number of pairs from all recounts after permutations.

$$c_{ij,exp} = \frac{\sum_{p=1}^{nperm} c_{ij,p}}{nperm}$$

The score gets high values for pairs that occur in the data set more often than expected by chance. It has a value of 0 if the observed counts equal the expected counts, and it gets a negative value for pairs which are less common than expected by chance. We used pseudocounts to avoid problems on data sets with small numbers of predicted TFBS.

3.2. *Calculation of Regulatory Potential*

We calculated a score that reflects the regulatory potential of a given stretch of sequence by building a bipartite graph. As input data to build the graph the predicted transcription factor binding sites and a reference co-occurrence score matrix were used. The score assigned to the stretch of sequence is then the sum of edge weights for the maximum weighted bipartite matching (MBPM) of the respective graph.

Each TFBS corresponds to a vertex in both partitions of the graph. Subsequently we connect the vertices from the two partitions by edges. Edge weights are set to the co-occurrence score for the given pair of TFBSs.

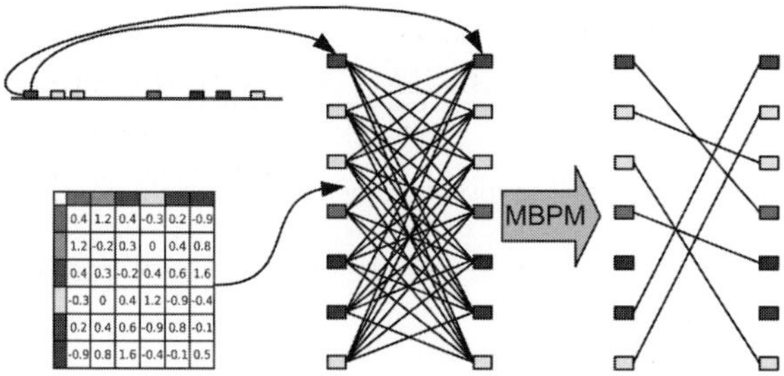

Fig. 1. A bipartite graph is built out of the predicted TFBSs and the co-occurrence scores. Subsequently a maximum weighted bipartite matching is carried out. The score for the respective sequence window is the sum of edge weights of the maximum weighted bipartite matching.

The usage of a bipartite graph permits to take into account the two most important interactions for each factor as represented by the co-occurrence score. The maximum weighted bipartite matching on the graph allows for one edge connecting to a vertex in the opposite partition for each vertex. The sum of edge weights in the maximum weighted bipartite matching is maximal. There is no other combination of edges possible whose sum of edge weights is higher [5]. We define the aforementioned sum as the regulatory potential of the given stretch of sequence.

4. Results and Discussion

4.1. *Co-occurrence of Predicted TFBSs*

In the results shown here we use a sliding window size of 100bp. This choice is justified by experimental data from the database TRANSCompel (v. 10.3) [16]: Of 375 known composite elements from vertebrates 98.16% have a distance between the experimentally determined transcription factor binding sites < 100bp (see Fig. 2). Moreover we tested the influence of the window size on the dissimilarity of the score distributions for combinations of TFs known and not known to interact on another vertebrate sequence set with a smaller set of PWMs. The results suggest that the dissimilarity of the two distributions was largest for the window size of 100bp (data not shown).

In Fig. 3 we show the cumulative histogram for the scores for pairs known to interact in TRANSFAC and for pairs not known to interact. The distributions of scores overlap, but the scores related to known interactions are shifted to larger values compared to the scores for the unknown interactions. The p-value for a comparison

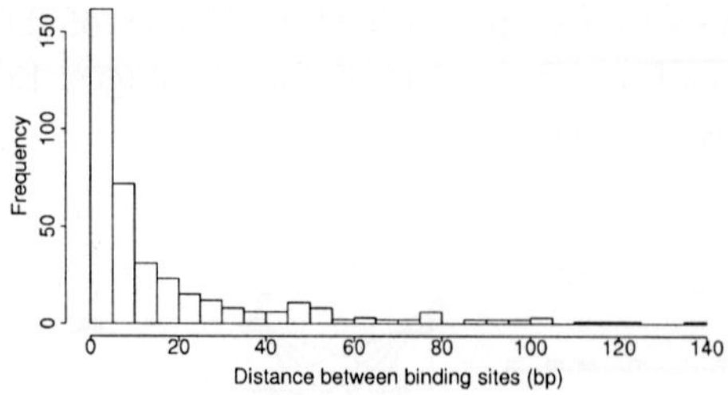

Fig. 2. Distance distribution of transcription factor binding sites from composite elements from TRANSCompel 10.3.

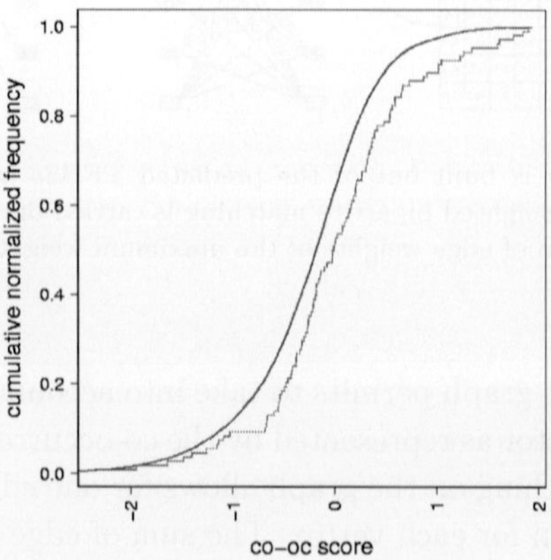

Fig. 3. Cumulative histogram for scores of known TF pairs (red) and TF pairs not known to interact (blue).

of the two score distributions using a Wilcoxon two sample test is 4.34×10^{-4}. Table 1 shows the twenty top scoring pairs of PWMs. Whereas only four are annotated as interacting in TRANSFAC, on manual inspection interaction information for three other pairs can be found in the literature and in TRANSFAC itself. For the factor *Stra13* (PWM: V$STRA13_01) the formation of homodimers is mentioned in TRANSFAC full text, but an interaction entry is missing. The same is true for *CDP*. In protein-protein interaction databases like UniHI [3] additional hints for direct and indirect interactions can be found.

Despite the precautions taken we observe a tendency for transcription factor binding sites to occur in homotypic clusters. Nine of the twenty top scoring pairs are combinations of two different factors. A comparison of the score distributions of homotypic and heterotypic pairs reveals a shift to higher values for the homotypic combinations. For simulations carried out on randomized TFBS data the score distributions were identical (data not shown). The observed bias that hits of individual motifs to cluster in the proximity of each other is in agreement with other studies e.g. in *Drosophila* [15] and human [24].

Table 1. Top scoring PWM pairs.

PWM 1	PWM 2	Score
V$CHX10_01	V$CHX10_01	1.93
V$STRA13_01	**V$STRA13_01**	**1.9**
V$POU3F2_02	**V$POU3F2_02**	**1.9**
V$CAAT_01	**V$CAAT_01**	**1.89**
V$CDC5_01	V$POU3F2_02	1.88
V$TEF_Q6	**V$TEF_Q6**	**1.84**
V$IPF1_Q4	V$S8_01	1.75
V$NFY_C	**V$NFY_C**	**1.75**
V$CART1_01	V$CART1_01	1.75
V$CDP_01	**V$CDP_01**	**1.73**
V$STRA13_01	V$USF_01	1.73
V$CAAT_01	V$NFY_C	1.72
V$LHX3_01	V$LHX3_01	1.68
V$CDC5_01	V$TEF_Q6	1.67
V$CDP_01	V$HNF6_Q6	1.66
V$CDP_01	V$LHX3_01	1.63
V$CDP_01	V$E4BP4_01	1.62
V$HP1SITEFACTOR_Q6	V$HP1SITEFACTOR_Q6	1.61
V$USF_01	**V$USF_01**	**1.61**
V$HMGIY_Q3	V$HNF6_Q6	1.61

Note: Known interactions in bold type.

Clearly not all interacting pairs of transcription factors obtain a high score for their respective PWMs. One reason might be a too stringent cutoff while scanning for the individual binding sites, so that some functional TFBSs are missed. Looking at a few examples it seems that the known interactions, whose TFBS combinations seem to be underrepresented have at least one partner with many known interactions. It is imaginable that the interaction in question is a specific one while at least one partner is involved in many other interactions. Examples for PWM combinations with a described interaction and co-occurrence scores < 0 are V$P53_01 (*P53*, seven known interactions in TRANSFAC) and V$YY1_02 (*YY1*, 13 known interactions), or V$DR4_Q2 (bound by a plethora of hormone receptors, e.g., *RAR*, which on its own already has more than 14 interaction entries in TRANSFAC) and V$MEF2_01 (*MEF-2*, three known interactions).

4.2. *Regulatory Potential on Known Regulatory Regions*

4.2.1. *Pax6*

To test the regulatory potential score described in section 3.2 we annotate known regulatory regions. The size of the sliding window for which the scores of the examples in this section are calculated is 200bp.

Our first example are the well understood regulatory regions of *Pax6*. The region around the *Pax6* locus in *Mus musculus* on chromosome 2 is annotated with the

MBPM score (Fig. 4). Furthermore the areas which are known to have a regulatory function [17] are marked. In all known regulatory regions but the ones upstream of exon 1 the MBPM score shows peaks that are higher than scores from the surrounding region. Some peaks are located in regions not known to influence the regulation of *Pax6*, examples are the one between exons 8 and 9 and the one directly upstream of exon 12. It remains to be seen if these regions fulfill a regulatory function.

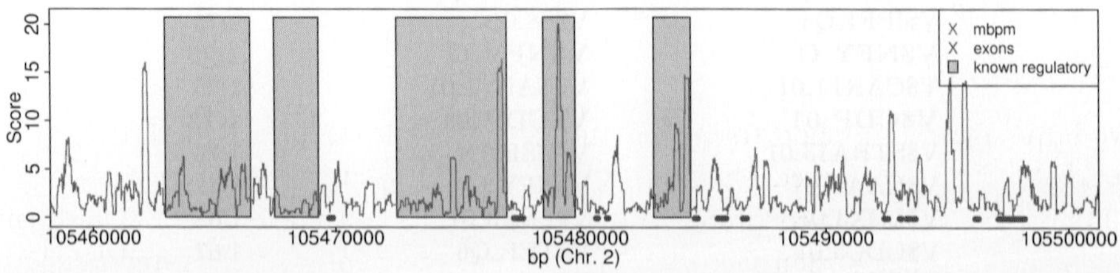

Fig. 4. *Pax6* regulatory regions and regulatory potential score. The exons of murine *Pax6* are marked in blue. The known enhancers and promoter regions of *Pax6* are marked in yellow [17].

4.2.2. *VISTA Enhancer Data Set*

We annotate the VISTA enhancer data set [22] with the MBPM score. Visel *et al.* generated this set using comparative genomics followed by experimental verification of the predictions. The experimentally verified enhancer regions are extended by 5000 bp upstream and downstream and the complete region was annotated with the described MBPM score.

In the example from Fig. 5 we plot the score around the experimentally verified enhancer element 174. The enhancer region itself is marked in yellow. This enhancer, located on chromosome 1, has been shown to be specific for forebrain and partly for limb *in vivo*. The genes surrounding this enhancer are *LMO4* and *PKN2*.

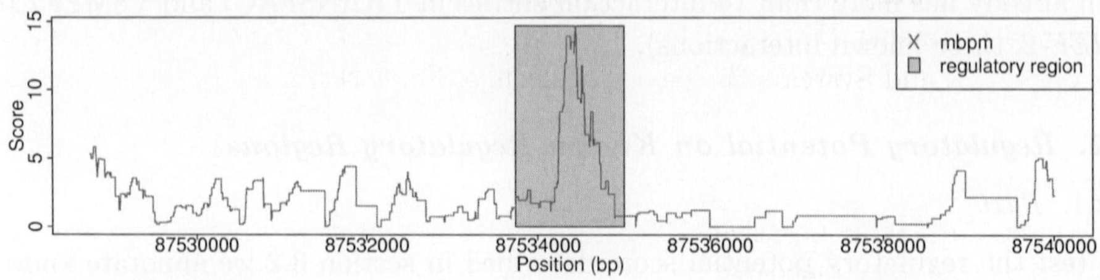

Fig. 5. Enhancer element 174 from the VISTA enhancer data set annotated with the MBPM score. The enhancer element is marked in yellow.

We take the whole set of enhancers that were tested positive for enhancer activity *in vivo*, extend it in a similar way to the procedure described for the example above and annotate it with the MBPM score. For roughly 21% of the enhancers in the set the highest score for the complete region is achieved within the known enhancer region, while the fraction of sequence covered by the known enhancers is ca. 12%. During the extension of the known enhancer regions we did not check for the presence of promoters or other enhancers, which could explain high MBPM scores even outside the annotated regions from the VISTA data set.

4.3. *Discussion*

We have shown a method to identify co-occurring pairs of predicted transcription factor binding sites based on a log-odds score for observed pair counts of PWM hits over expected pair counts of PWM hits. The score distribution for PWM pairs belonging to transcription factors which are known to interact is shifted to higher values compared to the distribution of scores for pairs not known to interact yet. The two score distributions overlap largely though, which makes it hard to define a threshold above which one can assume two transcription factors to interact. In accordance with other studies we find a bias towards higher scores for homotypic pairs of transcription factor binding sites.

Moreover we have described a method to assess the regulatory potential using the described co-occurrence score and TFBS predictions in uncharacterized sequence regions. We can identify known regulatory regions, while a systematic examination of the influence of different parameters (reference score matrix, threshold for the TFBS prediction, performance on different data sets) is still in progress. Also several other ways for the calculation of a graph-based score describing the regulatory potential of a window of unknown sequence are researched.

Acknowledgments

We would like to thank Szymon M. Kiełbasa for discussions and providing the *BioMinerva* library and Utz J. Pape, Hugues Richard and Hannes Luz for discussions and comments. HK is supported by the International Research Training Group for Genomics and Systems Biology of Molecular Networks.

References

[1] Bajic,V.B., Tan,S.L., Suzuki,Y., and Sugano,S., Promoter prediction analysis on the whole human genome, *Nat. Biotechnol*, 22(11):1467–1473, 2004.

[2] Bulyk,M.L., Computational prediction of transcription-factor binding site locations, *Genome Biol.*, 5(1):201, 2003.

[3] Chaurasia, G., Iqbal, Y., Hanig, C., Herzel, H., Wanker, E.E., and Futschik, M.E., UniHI: an entry gate to the human protein interactome, *Nucleic Acids Res.*, 35(Database issue):D590–594, 2007.

[4] Chen, L., Glover, J.N., Hogan, P.G., Rao, A., and Harrison, S.C., Structure of the

DNA-binding domains from NFAT, Fos and Jun bound specifically to DNA, *Nature*, 392(6671):42–8, 1998.

[5] Cormen, T.H., Leiserson, C.E., Rivest, R., and Stein, C., *Introduction to Algorithms, Second Edition*, MIT Press, 2001.

[6] Crossley, M., Merika, M., and Orkin, S.H., Self-association of the erythroid transcription factor GATA-1 mediated by its zinc finger domains, *Mol. Cell Biol.*, 15(5):2448–2456, 1995.

[7] Dieterich, C., Grossmann, S., Tanzer, A., Ropcke, S., Arndt, P., Stadler, P., and Vingron M., Comparative promoter region analysis powered by CORG, *BMC Genomics*, 6(1):24, 2005.

[8] Frith, M.C., Li, M.C., and Weng, Z., Cluster-Buster: Finding dense clusters of motifs in DNA sequences, *Nucleic Acids Res.*, 31(13):3666–3668, 2003.

[9] Frith, M.C., Spouge, J.L., Hansen, U., and Weng, Z., Statistical significance of clusters of motifs represented by position specific scoring matrices in nucleotide sequences, *Nucleic Acids Res.*, 30(14):3214–3224, 2002.

[10] Hannenhalli, S. and Levy, S., Predicting transcription factor synergism, *Nucleic Acids Res.*, 30(19):4278–84, 2002.

[11] Johansson, O., Alkema, W., Wasserman, W.W., and Lagergren, J., Identification of functional clusters of transcription factor binding motifs in genome sequences: the MSCAN algorithm, *Bioinformatics*, 19(1):i169–176, 2003.

[12] Hubbard, T.J.P., *et al.*, Ensembl 2007, *Nucleic Acids Res.*, 35(Database issue):D610–617, 2007.

[13] Kielbasa, S., The biominerva framework (in preparation), 2007.

[14] Levine, M. and Tjian, R., Transcription regulation and animal diversity, *Nature*, 424(6945):147–151, 2003.

[15] Lifanov, A.P., Makeev, V.J., Nazina, A.G., and Papatsenko, D.A., Homotypic regulatory clusters in Drosophila, *Genome Res.*, 13(4):579–588, 2003.

[16] Matys, V., *et al.*, TRANSFAC and its module TRANSCompel: transcriptional gene regulation in eukaryotes, *Nucleic Acids Res.*, 34(Database issue):D108–110, 2006.

[17] Morgan R., Conservation of sequence and function in the Pax6 regulatory elements, *Trends Genet.*, 20(7):283–287, 2004.

[18] Rahmann, S., Muller, T., and Vingron, M., On the power of profiles for transcription factor binding site detection, *Statistical Applications in Genetics and Molecular Biology*, 2(1):7, 2003.

[19] Rateitschak, K., Muller, T., and Vingron, M., Annotating significant pairs of transcription factor binding sites in regulatory DNA, *In Silico Biol.*, 4(4):479–487, 2004.

[20] Smit, A.F.A., Hubley, R., and Green, P., Repeatmasker open-3.0, `http://www.repeatmasker.org`, 1996-2004.

[21] Stormo, G.D., DNA binding sites: representation and discovery, *Bioinformatics*, 16(1):16–23, 2000.

[22] Visel, A., Minovitsky, S., Dubchak, I., and Pennacchio, L.A., Axel Visel, Simon Minovitsky, Inna Dubchak, and Len A Pennacchio. VISTA Enhancer Browser–a database of tissue-specific human enhancers, *Nucleic Acids Res.*, 35(Database issue):D88–92, 2007.

[23] Yu, X., Lin, J., Zack, D.J., and Qian, J., Computational analysis of tissue-specific combinatorial gene regulation: predicting interaction between transcription factors in human tissues, *Nucleic Acids Res.*, 34(17):4925–4936, 2006.

[24] Zhu, Z., Shendure, J., and Church, G.M., Discovering functional transcription-factor combinations in the human cell cycle, *Genome Res.*, 15(6):848–855, 2005.

IDENTIFICATION OF ACTIVATED TRANSCRIPTION FACTORS FROM MICROARRAY GENE EXPRESSION DATA OF KAMPO MEDICINE-TREATED MICE

RUI YAMAGUCHI[1]
ruiy@ims.u-tokyo.ac.jp

MASAHIRO YAMAMOTO[2,3]
yamamoto_masahiro@mail.tsumura.co.jp

SEIYA IMOTO[1]
imoto@ims.u-tokyo.ac.jp

MASAO NAGASAKI[1]
masao@ims.u-tokyo.ac.jp

RYO YOSHIDA[4]
yoshidar@ism.ac.jp

KENJI TSUIJI[2]
di055036@sc.itc.keio.ac.jp

ATSUSHI ISHIGE[2]
ishige@sc.itc.keio.ac.jp

HIROAKI ASOU[3]
asou@tmig.or.jp

KENJI WATANABE[2]
toyokeio@sc.itc.keio.ac.jp

SATORU MIYANO[1]
miyano@ims.u-tokyo.ac.jp

[1] *Human Genome Center, Institute of Medical Science, University of Tokyo, 4-6-1 Shirokanedai, Minato-ku, Tokyo 108-8639, Japan*
[2] *Department of Kampo medicine, Keio University School of Medicine 35 Shinanomachi, Shinjuku-ku, Tokyo 160-8582, Japan*
[3] *Department of Neuro-glia cell biology, Tokyo Metropolitan Institute of Gerontology, 35-2 Sakaecho, Itabashi-ku, Tokyo 173-0015, Japan*
[4] *The Institute of Statistical Mathematics 4-6-7 Minami-Azabu, Minato-ku, Tokyo 106-8569, Japan*

We propose an approach to identify activated transcription factors from gene expression data using a statistical test. Applying the method, we can obtain a synoptic map of transcription factor activities which helps us to easily grasp the system's behavior. As a real data analysis, we use a case-control experiment data of mice treated by a drug of Kampo medicine remedying degraded myelin sheath of nerves in central nervous system. Kampo medicine is Japanese traditional herbal medicine. Since the drug is not a single chemical compound but extracts of multiple medicinal herb, the effector sites are possibly multiple. Thus it is hard to understand the action mechanism and the system's behavior by investigating only few highly expressed individual genes. Our method gives summary for the system's behavior with various functional annotations, e.g. TFAs and gene ontology, and thus offer clues to understand it in more holistic manner.

Keywords: transcription factor activity; gene expression; Kampo medicine; multiple sclerosis; statistical absolute evaluation; MetaGP; myelination; remyelination.

1. Introduction

Transcription factors (TFs) have a central role in gene regulatory systems. The transcription factor activities (TFAs) affect the transcription of genes which are regulated by the TFs. To investigate when and under what condition they become effective is important for understanding the system's behavior. However, to observe TFAs is difficult, since the gene expression levels of TFs are often low. Moreover,

TFAs are activities of proteins, and thus, it would be hard to interpret the TFs' gene expressions directly into the TFAs. Several methods have been proposed to estimate the TFAs from the gene expression data, e.g. kinetic models [14], network component analysis [4, 7, 8, 12], state-space models [11].

As typical biological studies, experiments are often conducted in case-control manner. The main interest is to find differences between the cases and the control. To this purpose, a simple way is to find genes that are differentially expressed between the cases and the controls. For more precise understanding of such genes, the activities of TFs that regulate them is essential. Therefore, we can extend the problem from the finding differentially expressed genes to the discovery of TFs that show different activities between the cases and the controls. Although we may not observe difference in TFAs directly, we can observe that of gene expressions. Thus it is beneficial to develop a method to identify TFs having significantly different activities from the gene expression data. Here we introduce an approach for that purpose based on a statistical test.

To test differentially activated TFs, we use a statistical test for a set of genes which was proposed by Gupta *et al.* [6]. We name the method the Meta Gene Profiler (MetaGP) method. The MetaGP method accumulates statistical evidences from a set of genes in order to build a test with higher power than those for individual genes. There exist many other testing methods for a set of genes, which is often annotated by a term in Gene Ontology (GO) [3], have been proposed for the similar purpose. Unlike the most of existing methods, the main characteristic of MetaGP is that it evaluates the statistical evidences of a set of genes independently from information of the other set of genes. In this regard, we refer it as statistical absolute evaluation. Due to the characteristic, it makes sense to compare the results of tests for the same set of genes but observed in different conditions, e.g. case-control experiments with multi cases, time-course experiments, because the statistical evidences are evaluated with the same standard. The MetaGP method will be explained in detail in the following section.

Although the method was used with GO terms in [6], as they mentioned in the paper, it is applicable for any predefined set of genes like gene set enrichment analysis (GSEA) [13, 17]. To test TFAs, we define a set of genes annotated by a TF as the genes which have the binding site of the TF in the promoter regions. We compile this information from a TFs' data base, TRANSFAC [10]. In the analysis, 306 unique TFs were found for the genes mapped on the microarrays that we used. Applying MetaGP with TFs exhaustively, we can obtain a synoptic map of the activities and expect to achieve deeper knowledge of the gene regulations characterizing the systems of the subject.

As a real data analysis, we apply the method to a microarray data obtained from a set of case-control experiments for mice [16]. In one case, mice were dosed with cuprizone, a well-known neuro toxin, inducing demyelination for nerves in the central nervous system, thus, mimicking demyelinating diseases, e.g. multiple sclerosis (MS). In another case, a drug, Ninjin'yoeito (NYT), was additionally dosed to mice with the cuprizone during the same period. NYT is a drug of Kampo medicine which is Japanese traditional herbal medicine originating in ancient Chinese traditional medicine. It is the only drug known to remedy demyelination induced by aging or dosage of cuprizone up to the present date [15, 16]. However, little is known regarding the action mechanism since the drug is not a single molecule but a mixture of extracts from medicinal plants. Hence, obtaining the synoptic map of TFAs by MetaGP and comparing the activities across different cases would help to understanding the action mechanism of the drag and the system's behavior. We also apply the method with GO terms for the same purpose but to obtain information regarding the system from another aspect.

The organization of this paper is as follows. In Section 2, we explain the method to test the group of genes, e.g. those which correspond to TFAs or GO terms. In Section 3, we apply the method to the expression data obtained from the case-control experiment, in which demyelination-induced mice were dosed with a herbal medicine leading to demyelination and remyelination, and compared the significance of TFAs and GO terms in the several different experimental conditions. Section 5 is the concluding remarks.

2. Method

2.1. *MetaGP*

In this paper, to evaluate significance of differentially activated TFs from gene-expression data, we employ a testing procedure for a set of genes proposed by [6]. We call the method MetaGP. To this end, we define the gene sets corresponding to TFAs and apply MetaGP with them to data set. As a result, we can obtain p-values for TFAs.

The MetaGP method accumulates statistical evidences from a set of genes in order to build a test with higher power than those for individual genes and to help the biological interpretation with functional annotation terms. The tests for individual genes, e.g. t-test for differentially expressed genes, tend to have low statistical powers, due to the small sample size. For the similar purpose, many other testing methods for a set of genes have been proposed, e.g. GSEA [13, 17], FatiGO [1, 2], and so on. See for a review [9].

Most of those methods use a test based on a 2×2 contingency table for a list of genes e.g. [1]. Each of the four cells in the table is characterized by two properties of genes, that is, (i) annotated by the term or not and (ii) differentially expressed or not. For the table, each of the cells is filled by the number of genes categorized by the two properties. Then the p-value for over-representation of the set of genes annotated by the term is calculated by using an independence test of the 2×2 contingency table, e.g. Fisher's exact test. The categorization of the differential expression of each genes is determined by setting a threshold value for a ranked list of genes, e.g. a list ordered by t-statistic. Therefore the test for the gene set with the annotation depends on the rank information relative to the gene set without the annotation.

On the other hand, the main characteristic of MetaGP is that it evaluates the statistical evidences of a set of genes independently from information of the other set of genes. Due to the characteristic, it makes sense to compare the results of tests for the same set of genes but observed in different conditions, e.g. case-control experiments with multi cases, time-course experiments, since the statistical evidences are evaluated in an absolute manner. On the contrary, it is not appropriate to compare the results of 2×2 tests for the same set of genes observed in different conditions, since the result of each test depends on the rank statistics relative to the complementary set and the rank information can not be compared in an absolute manner across different conditions.

The formulation of MetaGP is explained according to [6] in the following section.

2.2. *Formulation of MetaGP*

Suppose that we test the null hypothesis $H_0^{(i)}$ for the ith gene where the total number of genes is denoted by d $(i = 1, \cdots, d)$. For example, to identify differentially expressed genes between case and control samples, one may apply t-test under the null hypothesis $H_0^{(i)} : \mu_0^{(i)} = \mu_1^{(i)}$ for $i = 1, \cdots, d$. Here $\mu_0^{(i)}$ and $\mu_1^{(i)}$ denote group means of the control and the case samples for the ith gene.

The objective of MetaGP is to retrieve the annotations corresponding to the set of genes which are relevant to the underlying gene regulation mechanism on the basis of the statistical evidences. As remarked in [6], the method can be applicable not only for GOs but also for retrieving generic biological knowledge by defining the set of genes, e.g. TFAs, biological pathways.

Let us denote a set of genes, which are classified in a term of a generic biological knowledge by $\mathcal{F}$. For instance, $\mathcal{F}$ represents a set of genes annotated by a term in GO. In this study, $\mathcal{F}$ denotes a set of genes annotated by a TF's name. The annotation procedure is described in Section 3.3. The problem of evaluating the significance of $\mathcal{F}$ is stated by the statistical test with the null hypothesis H_0 and the alternative H_1 as follows:

$$H_0 : H_0^{(i)} \text{ is true for all } i \in \mathcal{F},$$
$$H_1 : H_0^{(i)} \text{ is false for one or more } i \in \mathcal{F}.$$

For instance, to understand functional gene regulations, one aims to retrieve GO terms in which more genes indicating the false $H_0^{(i)}$ are involved.

In order to evaluate this test, they present a testing procedure which exploits a technique of the statistical meta-analysis, known as the normal inversion method. Let p_i denote the p-values of the ith gene. The testing procedure for a $\mathcal{F}$ is then described as follows:

(1) Transform p_i to the random deviate z_i according to

$$z_i = \Phi^{-1}(1 - p_i), \quad i \in \mathcal{F},$$

where Φ^{-1} stands for the inverse function of the standard normal cumulative distribution function.

(2) Compute z-score by

$$z = \sum_{i \in \mathcal{F}} z_i \Big/ \sqrt{|\mathcal{F}|},$$

where $|\mathcal{F}|$ denotes the number of genes labeled by the term $\mathcal{F}$.

(3) Compute an integrated p-value of the annotation $\mathcal{F}$, denoted by $p_{\mathcal{F}}$, by the reverse transformation of z-score as

$$p_{\mathcal{F}} = 1 - \Phi(z).$$

The basic theory of statistics indicates that each z_i is independently and identically distributed according to the standard normal distribution if and only if the individual null hypothesis $H_0^{(i)}$ is true. Moreover, under the assumption that the null hypotheses for all of the genes in $\mathcal{F}$ are true (H_0 is true), and independent to each other, the integrated z-score also follows the standard normal distribution. Based on these statistical properties, the integrated p-value captures the enrichment of the term in the following way: For a part of genes in $\mathcal{F}$, of which the null hypotheses are true, the p-values are uniformly distributed over $[0, 1]$, while p-values for the other genes, of which the null hypotheses are not true, in $\mathcal{F}$ are clustered around the region close to zero. This property can be formally stated if a set of tests is statistically unbiased. Hence, as the proportion of the genes in $\mathcal{F}$, for which the null hypothesis are not true, becomes larger, the computed z-score is shifted towards a higher value. Correspondingly, the integrated p-value becomes smaller.

3. Data and Analysis

To investigate the efficiency of MetaGP with TFAs, we applied the method to a real data set obtained by case-control experiments with mice [16]. As mentioned in Section 1, the experiment were conducted to see the dose effect of NYT on the gene expressions, which is a herbal drug and known to be the only drug to remedy demyelination of myelin sheath of the central nervous system up to the present date. However, the action mechanism has not been clarified. One of the difficulty comes from that the target of NYT might be multiple, since it consists of extract from a mixture of 12 raw medicinal plants. Thus obtaining a synoptic map of multiple TFs' activities for different dosage and time conditions by MetaGP, we can overview the system's behavior and expect to achieve clues to reveal the action mechanism.

3.1. *Data: Demyelination-/Remyelination-Induced Mice Treated with Kampo Medicine*

The summary of the experiments is as follows. There are one control-experiment and three case-experiments. The experiments will be denoted by E1, E2, E3, and E4, respectively (see Table 1). The experimental period was 7 weeks. Male C57BL/6J

Table 1. The types and conditions of experiments.

Experiment ID	E1	E2	E3	E4
Case/Control	Control	Case	Case	Case
NYT		√		√
Cuprizone			√	√

mice were used. At the beginning of the period, they are 8-week-old.
In the control experiment E1, mice were fed with a normal diet during the 7 weeks.
In the case experiment E2, mice were fed with a diet containing 1% NYT during the 7 weeks, which is a spray-dried extract of a mixture of 12 raw medicinal plants.
In the case experiment E3, to model the demyelination and remyelination process, mice were fed with a diet containing 0.2% cuprizone, a well-known neuro toxin, for the first 5 week. The demyelination started at the 3rd week and the degree of demyelination had a peak at the 5th week. During the following two weeks, the normal diet were fed. In the following two weeks, the remyelination process occurred. The myelin sheath looks almost recovered at the 7th week.
In the case experiment E4, NYT was additionally dosed with cuprizone to mice during the first 5 weeks. As a result the demyelination was suppressed. During the following two weeks, the dosage of cuprizone was quitted and that of NYT was continued. At the 7th week, the shape of the regenerated myelin sheath were better than those for the above cuprizone-dosed experiment (E3).
To obtain gene expression data, for each experiment (E1, E2, E3, and E4), three mice were sacrificed at the 3rd and the 7th week. White matter of each mouse's brain was homogenized in Trizol reagent with a Phycontron, and total RNA was isolated, labeled, and prepared for hybridization with Mouse Expression 430A 2.0 Array according to the recommended protocol. As the result, three DNA microarray samples were obtained for each experiment and time point. See [16] for more detail about the data and experiments.

3.2. *Annotation to Gene Groups*

3.2.1. *TFs*

We defined a gene group to be annotated with a TFA as a set of genes in which each gene has the binding site of the TF in the promoter region. To select such genes placed on the microarray, we searched a database of TFs, TRANSFAC.
Each searching region spanned 2500 bps from the transcription starting site (TSS) toward the upstream. The confidence level of the sequences to be accepted as binding sites was 0.98. As a result, the number of the gene sets annotated by unique TFAs became 306. We note that it is highly possible not all genes are really regulated by the TFs, even if they have the particular sequences with the confidence level. The accuracy to select genes regulated by the TFs would be improved by incorporating other information from e.g. ChIP-chip experiments.

3.2.2. *GO terms*

As a set of genes to be annotated with a GO term, we used the predefined group of genes for each GO term [3]. In the analysis, we used all three gene ontology, i.e. Biological Process (BP), Cellular Component (CC), Molecular Function (MF). As a result, the number of the gene sets annotated by GO terms became over 20000.

3.3. *MetaGP for TFs and GO terms*

We applied the MetaGP method to the gene sets annotated by TFAs or GO terms with the gene expression data as follows:

(i) To obtain the p-values for individual genes, Welch's t-test was done for the relative difference between μ^{ijk} and μ^{i1k}, where μ^{ijk} is the mean expression value of the ith gene of the jth case experiment i.e. Ej ($j \in \{2,3,4\}$), at the kth week ($k \in \{3,7\}$) and μ^{i1k} is that for the control experiment, i.e. E1. Under the null hypothesis H_0^{ijk} : $\mu^{ijk} = \mu^{i1k}$, the p-value, p^{ijk} was obtained for each gene, case sample, and time.

(ii) The individual p-values corresponding to an annotation $\mathcal{F}$ were integrated by MetaGP. The integrated p-value for the annotation $\mathcal{F}$ of the jth case experiment and the kth week denoted by $p_{\mathcal{F}}^{jk}$ were calculated by integrating p^{ijk}'s, $i \in \mathcal{F}$.

(iii) Finally, we obtained the integrated p-values for all TFs and GO terms in our lists.

4. Discussion

After obtaining the integrated p-values for 306 TFs, in order to see the characteristics of the significance of all TFs, we did a cluster analysis of the p-values as shown with the heatmap in Fig. 1. That is a synoptic map of the TFAs.

As shown in the upper half, a number of TFs are differentially activated for both the cuprizone-dosed (W3E3, W7E3) and the cuprizone-NYT-dosed (W3E4, W7E4) cases, while NYT-dosed (W3E2, W7E2) cases are not. Thus these TFAs might be induced by the dose effect of cuprizone.

In the lower half, we can find more variety of patterns. Investigating these TFs activated for particular combinations and the corresponding gene expressions systematically, we can expect to obtain newer insight about the action mechanism of the medicines and processes of demyelination and remyelination.

Since the experiments intended to find a way to cure demyelination diseases, e.g. MS, it is interesting to compare our result with TFAs reported in existing literature regarding the disease. Gobin *et al.* [5] reported upregulated TFs in MS lesion of patient's brain, i.e. RFX, NF-κB, IRF1, STAT1, USF, CREB, and CIITA, which control major histocompatibility complex (MHC). In the paper, the existence of the TFs were confirmed by immunohistochemical staining.

Fig. 2 shows the integrated p-values of the TFs in our data; except for CIITA which was not included in our list. Note that, in each panel, a hight of bar represents

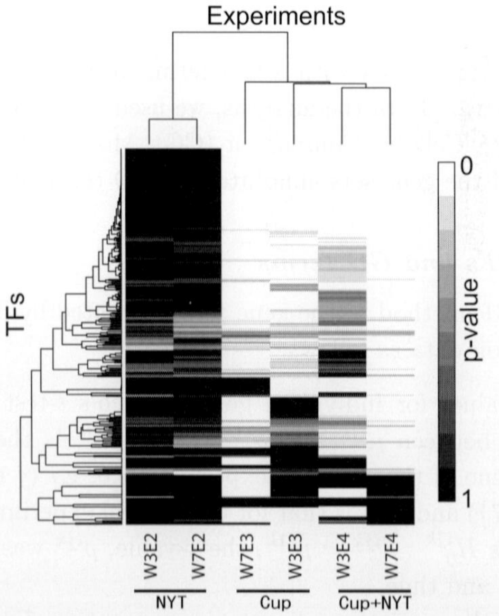

Fig. 1. The clustered profiles of the integrated p-values of the differentially activated TFs, $p_{TF_i}^{jk}$, for each of TFs (rows) and experiments (columns). The number of TFs is 306, i.e. $i = 1, \cdots, 306$, and the names of TFs are not shown. At the bottom, the label "WkEj" suggests the experiment at the kth week by the experiment Ej.

$1 - p_{\mathcal{F}}^{jk}$, thus the higher one corresponds to the smaller $p_{\mathcal{F}}^{jk}$. The mean of the label "WkEj" is the same as in Figure 1. Interestingly, we can see variety of patters in the panels. These observation have helpful information for the system's response to the dosage and the action mechanism of the drugs as discussed below. In the panel (a) RFX, the TF is activated in the experiment E4 (cuprizone + NYT dosage), but not in the experiments E2 (NYT dosage) and E3 (cuprizone dosage). It suggests the existence of interaction effect from both medicines to the pathway of RFX. In the panels (b) NF-κB and (c) IRF-1, we can see the coherent pattern in the experiment E3 and the E4. It may suggest that mainly cuprizone affected on the pathways of these TFs, and thus, NYT could not control the cuprizone's effect. In the panel (d) STAT1, at the former period (the 3rd week), there are no significant activities. At the latter period (the 7th week), the TF is activated in the experiment E3 but not in E4. It suggests the existence of a time effect. In the latter period, NYT seems to repress the effect of cuprizone and to stabilize the system. In the panel (e) USF and (f) CREB, there are also time effect for the effect of NYT. At the former period. NYT was not effective for both TFs. At the latter period, the interaction effect existed for USF. On the other hand, the repressing effect by NYT can be seen for CREB.

We also assessed GO terms by MetaGP. The number of GO terms considered is

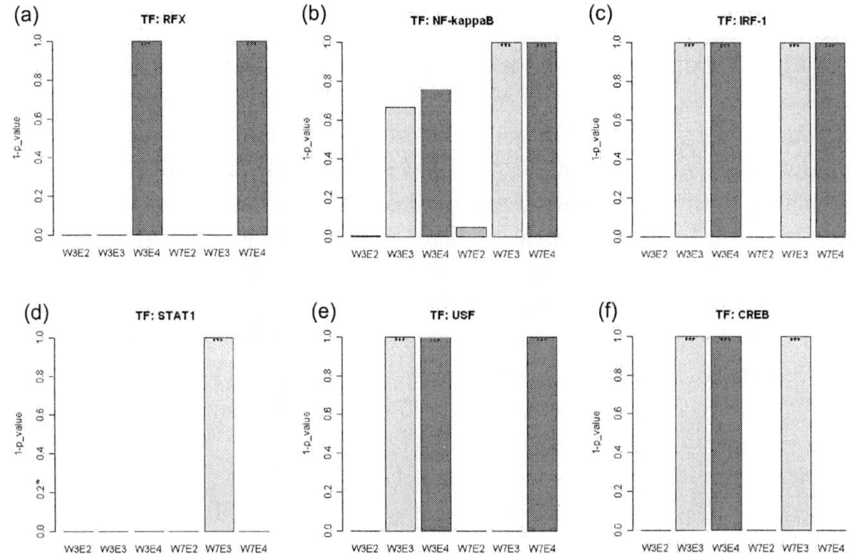

Fig. 2. The integrated p-values of the differentially activated TFs for each experiment relative to the control. The TFs, (a) RFX, (b) NF-κB, (c) IRF1, (d) STAT1, (e) USF, and (f) CREB, are reported as activated TFs in the brain of multiple sclerosis patient [5]. Note: The vertical axis is for $1-p$. The marks "$*$", "$**$", and "$***$" in the bars mean "$p \leq 0.1$", "$p \leq 0.05$", and "$p \leq 0.01$", respectively.

more than 20000. Here we only show two of them, i.e. "myelination" and "structural constituent of myelin sheath" (Fig. 3), because of limitation of space. Since these terms must relate to myelination/remyelination, it becomes a validation of the method. Investigating the results of other terms in more detail, we can acquire information regarding the system's behavior from another point of view.

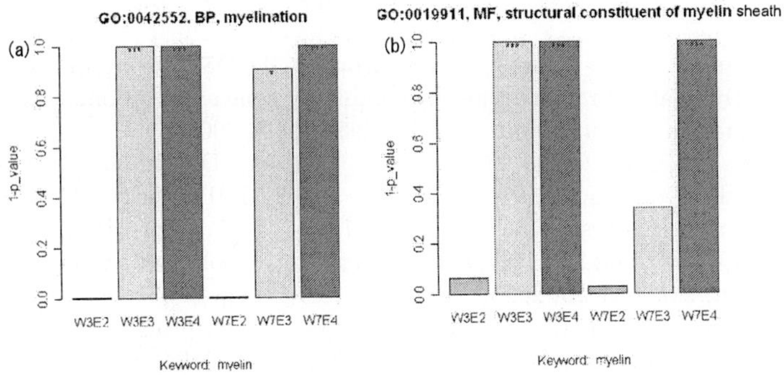

Fig. 3. The integrated p-values of the GO terms including the word "myelin" in its definition. The format is the same as in Fig. 2.

5. Concluding Remarks

In this study, we proposed a method to identify activated TFs, that possibly decide or control cell status, from gene expression data by using the statistical test-based method named MetaGP. MetaGP tests a set of genes with an annotated notion and is an absolute evaluation method [6]. Unlike the rank-based methods, the MetaGP method makes sense to compare the results of the same set of genes obtained in different conditions, e.g. data of case-control experiments with multiple cases, and time course experiments, etc. Applying the method with TFs, we can obtain a synoptic map of the activity of multiple TFs in different conditions.

The main contribution of this study is three folds. i) We applied MetaGP with TFs to a real data. We defined the annotated groups of genes by TFs and applied MetaGP. We showed the results in a synoptic map of TFAs by which we can easily grasp the tendency of those activities in the system. ii) We applied the MetaGP method to the data obtained from case-control experiments including multiple cases and time courses. Because the results of the tests are fairly comparable between different conditions, then we compared the activities and discussed action mechanism of the drugs. The information would be useful for drug discovery and improvement. iii) In our knowledge, this is possibly the first study of Kampo medicine data analyzed with a systems-biological method and view.

Since a drug of Kampo medicine is not a single chemical compound but extracts of multiple medicinal herb, the effector sites are possibly multiple. Thus it is hard to understand the action mechanism and the system's behavior by investigating only few highly expressed individual genes. MetaGP with TFs and other annotations, e.g. GO terms, can give summaries for the systems' behavior from various points of view, and thus offers clues to obtain insights regarding the system in more holistic manner.

References

[1] Al-Shahrour, F., Diaz-Uriarte, R., and Dopazo, J., A web tool for finding significant associations of gene ontology terms with groups of genesy. *Bioinformatics*, 20(4):578–580, 2004.

[2] Al-Shahrour, F., Diaz-Uriarte, R., and Dopazo, J., Discovering molecular functions significantly related to phenotypes by combining gene expresesionon data and biological information. *Bioinformatics*, 21(13):2988–2993, 2005.

[3] Ashburner, M., Ball, C. A., Blake, J. A., Botstein, D., Butler, H., Cherry, J. M., Davis, A. P., Dolinski, K., Dwight, S. S., Eppig, J. T., Harris, M. A., Hill, D. P., Issel-Traver, L., Kasarskis, A., Lewis, S., Matese, J. C., Richardson, J. E., Ringwald, M., Rubin, G. M., and Sherlock, G., Gene ontology: tool for the unification of biology. *Nat. Genet.*, 25(1):25–29, 2000.

[4] Galbraith, S. J., Tran, L. M., and Liao, J. C., Trancriptome nework analysis with limited microarray data. *Bioinformatics*, 22(15):1886–1894, 2006.

[5] Gobin, J. P., Montagne, L., van Zutphen, M., van den Elsen, P. J., van der Valk, P., and de Groot, C. J. A., Upregulation of transcription factors controlling MHC expression in multiple sclerosis lesions. *Glia*, 36(1):68–77, 2001.

[6] Gupta, P. K., Yoshida, R., Imoto, S., Yamaguchi, R., and Miyano, S., Statistical ab-

solute evaluation of gene ontology terms with gene expression data. *LNBI*, 4463:146–157, Springer-Verlag, 2007.

[7] Kao, K. C., Yang, Y. L., Boscolo, R., Sabatti, C., Roychowdhury, V., Liao, J. C., Transcriptome-based determination of multiple transcription regulator activities in *escherichia coli* by using network component analysis. *Proc. Natl. Acad. Sci. USA*, 101:641–646, 2004.

[8] Kao, K. C., Tran, L. M., Liao, J. C., A global regulatory role of gluconeogenic genes in *escherichia coli* revealed by transcriptome network analysis. *J. Biol. Chem.*, 280:36079–36087, 2005.

[9] Khatri, P., Drăghici, S., Ontological analysis of gene expression data: current tools, limitations, and open problems. *Bioinformatics*, 21(18):3587–3595, 2005.

[10] Knäppel, R., Dietze, P., Lehnberg, W., Frech, K., and Wingender, E. Transfac retrieval program: a network model database of eukaryotic transcription regulating sequences and proteins. *J. Comput. Biol.*, 1:191–198, 1994.

[11] Li, Z., Shaw, S. M., Yedwabnick, M. J., and Chan, C., Using a state-space model with hidden variables to infer transcription factor activities. *Bioinformatics*, 22(6):747–754, 2006.

[12] Liao, J. C., Boscolo, R., Yang, Y.-L., Tran, L. M., Sabatti, C., and Roychowdhury, V. P., Network component analysis: reconstruction of regulatrory signals in biological systems. *Proc. Natl. Acad. Sci. USA*, 100(26):15522–15527, 2003.

[13] Mootha, V. K., Lindgren, C. M., Eriksson, K. F., Subramanian, A., Sihag, S., Lehar, J., Puigserver, P., Carlsson, E., Ridderstrale, M., Laurila, E., Houstis, N., Daly, M. J., Patterson, N., Mesirov, J. P., Golub, T. R., Tamayo, P., Spiegelman, B., Lander, E. S., Hirschhorn, J. N., Altshuler, D., and Groop, L. C., PGC-1alpha-responsive genes involved in oxidative phosphorylation are coordinately downregulated in human diabetes. *Nat. Genet.*, 34(3):267–273, 2003.

[14] Nachman, I., Regev, A., and Friedman, N., Inferring quantitative models of regulatory networks from expression data. *Bioinformatics*, 20(Suppl. 1):i248–i256, 2004.

[15] Nakahara, J., Tan-Takeuchi, K., Seiwa, C., Gotoh, M., Kaifu, T., Ujike, A., Inui, M., Yagi, T., Ogawa, M., Aiso, S., Takai, T., and Asou, H., Signaling via immunoglobulin Fc receptors induces oligodendrocyte precursor cell differentiation. *Dev. Cell*, 4(6):841–852, 2003.

[16] Seiwa, C., Yamamoto, M., Tanaka, K., Fukutake, M., Ueki, T., Takeda, S., Sakai, R., Ishige, A., Watanabe, K., Akita, M., Yagi, T., Tanaka, K., and Asou, A., Restoration of FcRγ/Fyn signaling repairs central nervous system demyelination. *J. Neurosci. Res.*, 85(5):954–966, 2007.

[17] Subramanian, A., Tamayo, P., Mootha, V. K., Mukherjee, S., Ebert, B. L., Gillette, M. A., Paulovich, A., Pomeroy, S. L., Golub, T. R., Lander, E. S., and Mesirov, J. P. Gene set enrichment analysis: a knowledge-based approach for interpreting genome-wide expression profiles. *Proc. Natl. Acad. Sci. USA*, 102(43):15278–15279, 2005.

BREAST CANCER STRATIFICATION FROM ANALYSIS OF MICRO-ARRAY DATA OF MICRO-DISSECTED SPECIMENS

GABRIELA ALEXE[1,2,*] GUL S. DALGIN[3,*] DANIEL SCANFELD[1] PABLO TAMAYO[1]
galexe@broad.mit.edu sdalgin@bu.edu scanfeld@broad.mit.edu tamayo@broad.mit.edu

JILL P MESIROV[1] SHRIDAR GANESAN[4] CHARLES DELISI[3]
mesirov@broad.mit.edu ganesash@umdnj.edu delisi@bu.edu

GYAN BHANOT[2,4,5]
gyanbhanot@gmail.com

[1] *The Broad Institute of MIT and Harvard, Cambridge MA, 02142, USA*
[2] *Institute for Advanced Study, Princeton, NJ, 08540, USA*
[3] *Boston University, Boston, MA, 02215, USA*
[4] *Cancer Institute of New Jersey, New Brunswick, NJ, 08903, USA*
[5] *Rutgers University, Piscataway, NJ 08854, USA*
[*] *Joint first authors*

We describe a new method based on principal component analysis and robust consensus ensemble clustering to identify and elucidate the subtypes of breast cancer disease. The method was applied to microarray gene expression data using micro-dissection of samples from 36 breast cancer patients with at least two of three pathological stages of disease. Controls were normal breast epithelial cells from 3 disease free patients. Our method identified an optimum set of genes and strong, stable clusters which correlated well with clinical classification into Luminal, Basal and Her2+ subtypes based on ER, PR and Her2 status. It also revealed a hierarchical portrait of disease progression through various grades and stages and identified genes and functional pathways for each stage, grade and disease subtype. We found that gene expression heterogeneity across subtypes is much greater than the heterogeneity of progression from DCIS to IDC within a subtype, suggesting that the disease subtypes are distinct disease processes. The averaging over data perturbations and clustering methods is critical in the robust identification of subtypes and gene markers for grade and progression.

Keywords: breast cancer; microarray analysis; disease subtypes; progression; principal component analysis; consensus ensemble clustering.

1. Introduction

The probability for women in the US to get breast cancer in their lifetime is ~ 10-13%. Standard treatment includes surgery, radiation, and hormonal, chemo and/or biological therapy. 60-80% of tumors are positive for the estrogen receptor (ER+) and respond to treatment with hormonal agents such as Tamoxifen [10,18]. 20-40% have amplification of the Her2 gene [20], a marker of higher recurrence and poor prognosis. The outcome of Her2+ tumors can be improved using humanized anti-Her2 antibody trastuzumab (Herceptin). 10-15% of tumors neither express the estrogen receptor nor have Her2 amplification [25]. These tumors, called Basal-like [21,23], are high grade with poor prognosis and no known targeted therapy. Overall, therapy for breast cancer is often confounded by the fact that tumors with similar histopathology have divergent course and varied response to therapy [24].

Microarrays should be able to identify the genes and pathways altered in cancer initiation, progression and metastasis. However, their success has been limited by practical considerations, the most important of which are sensitivity to noise and method of analysis [22]. Such limitations have often resulted in publications with ambiguous and contradictory results and biologically non-intuitive genes and pathways for stratification, making it difficult to move analysis results from bench to bedside.

In this paper, we develop a robust method which addresses these issues. We first use Principal Component Analysis (PCA) [15] to identify the overall structure of clusters in the data and to select the subset of genes that distinguish the clusters. We use this gene set and a new consensus ensemble clustering technique, which averages over several clustering methods and many data perturbations, to identify strong, stable clusters. We use simple criteria to find the optimum number of clusters and describe methods to identify robust markers for subtype/grade separation and disease progression.

Applied to a public breast cancer microarray data set [17], our method results in stable lists of genes and pathways that distinguish high and low grade tumors and progression. A hierarchy of clusters paints a portrait of the disease at varying levels of granularity. With two clusters, the normal samples separate from the disease samples. At the next level of clustering we get a separation of low and high grade samples. The optimal number of clusters identified is seven and correspond to two low grade (LG1 and LG2) and four high grade (HG1-HG4) sub-clusters with strong markers which can distinguish them with sensitivity/specificity in the 80-100% range. Using the gene markers and measured ER, PR, Her2+ levels, we match the sub-clusters to standard clinical classification of breast cancer as in [26]. The low grade clusters correspond to one Luminal A subtype and one Luminal B subtype. The high grade samples correspond to two additional Luminal B subtypes, one Her2+ subtype and one Basal subtype. A major overall observation is that *each sub-cluster contains samples from non-invasive and invasive tumors from the same patient,* which suggests that the sub-clusters are distinct diseases.

2. Results

Data provided in [17] (www.geneexpression_ma.org) consisted of microarrays from micro-dissected samples from 36 breast cancer patients. 31 patients had at least two out of three stages of disease: atypical ductal hyperplasia or ADH, ductal carcinoma *in situ* or DCIS and invasive ductal carcinoma or IDC respectively. Five patients had pre-invasive disease (ADH) only. Samples were also collected from normal breast epithelial tissue from three healthy women during mammoplasty. Multiple samples were analyzed for each stage using a 12,000 gene cDNA microarray. The samples were pathologically classified into grades I, II and III. The expression levels of "normal cells" from cancer patients matched those of normal epithelial cells from disease free patients; which meant that normal samples from cancer patients could serve as controls. The data provided in [17] was limited to 1940 genes across 93 microarrays; 32 from disease free patients, 8 ADH, 30 DCIS and 23 IDC samples.

Fig 1. shows the flow chart for our method. The data was normalized and missing entries imputed robustly. PCA showed that 85% of the data variation was due to the first 32 PCs and 207 genes, which were identified from high values of their coefficients in the corresponding eigenvectors. The optimal number of clusters k_{opt} was estimated using gap statistics [29] and silhouette scores [16]. A variety of clustering techniques and data perturbations were then averaged to identify $k = 2,3,\ldots k_{opt}$ clusters using an agreement matrix.

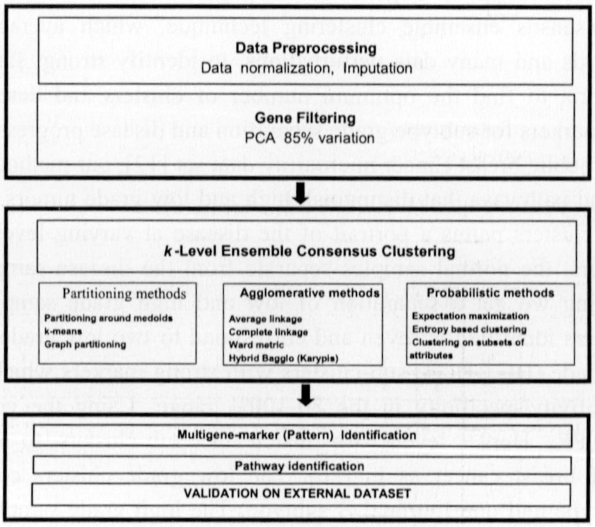

Fig 1: Flow chart of the analysis method.

The clustering results are shown in Fig 2. At k=2, the samples separated into the Normal (N) group with all normal samples and one ADH sample, and a Cancer (BCA) group, with only cancer samples. At k=3, the normal group was unaltered and the BCA group split into Low grade (LG: 18 samples labeled grade I and 9 samples labeled grade II) and High grade (HG: 13 samples labeled grade II and 19 samples labeled grade III). As k increased from 4 through 7, the LG group split into 2 subgroups (LG1, LG2) and the HG group into 4 subgroups (HG1-HG4). Fig 2. suggests that disease progression is a hierarchical process which is readily and robustly identified by our clustering procedure. Table 1 shows the number of samples in each subtype broken down into the clinical characteristics of stage, ER, PR, Her2, node and grade status in the ADH, DCIS and IDC category for k = 2,3 and 7. Based on this and gene signatures (see below), we identify LG1 as Luminal A; LG2, HG3, HG4 as Luminal B; HG1 as Basal and HG2 as Her2+.

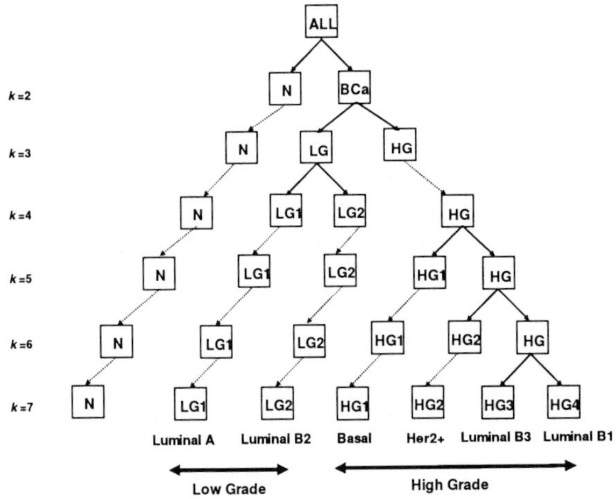

Fig 2: Consensus ensemble clustering tree showing the recursive splitting of samples into six cancer subtypes. The normal cluster separates at the k=2 split and remains distinct.

Table 1: The number of samples in various subtypes as a function of their clinical characteristics.

Cluster level	Group	Size	Stage				ER			PR			Her2			Node		Grade		
k			ADH	DCIS	IDC	N	Pos	Neg	ND	Pos	Neg	ND	Pos	Neg	ND	Pos	Neg	1	2	3
2	N	33	1			32														
	BCA	60	7	30	23		47	10	3	42	15	3	10	37	9	44	14	18	22	19
3	LG	28	7	13	8		26		2	21	5	2	4	18	6	20	8	18	9	
	HG	32		17	15		21	10	1	21	10	1	6	19	3	24	6		13	19
7	LG1	11	4	5	2		11			8	3		1	10		7	4	9	2	
	LG2	17	3	8	6		15		2	13	2	2	3	8	6	13	4	9	7	
	HG1	5		2	3			5			5		1	4		3				5
	HG2	10		7	3		7	3		5	5		3	4	1	9	1		2	8
	HG3	13		6	7		10	2	1	12		1	2	7	2	10	3		7	6
	HG4	4		2	2		4			4				4		2	2		4	

The ER and Her2 signatures suggest that both LG1 and LG2 are Luminals in the standard nomenclature [21], with LG2 presenting more aggressive features than LG1. In the nomenclature of [27] we identify LG1 as the Luminal A subtype and LG2 as one Luminal B subtype. All samples in the HG1 subgroup were ER and PR negative while those in the HG3 and HG4 subgroups were mostly ER and PR positive. The HG2 samples have mixed ER and PR signatures. The HG1 subgroup had the worst prognosis based on clinical markers. HG3 markers included a group of down-regulated genes in chromosomal region 17q23-25 which harbors the ERBB2 amplicon 17q 22.24. This identifies the clinical signature of HG3 as HER2-. Based on these observations, we identify HG1 as Basal [21,23], HG2 as Her2+, and HG3 and HG4 as additional subtypes of Luminal B [21].

Table 2: Characteristic and progression markers for grade and subtype.

Characteristic markers							
LG				HG			
marker	score	pval	FDR	marker	score	pval	FDR
EYA2	-0.89	0.01	0.06	X123	-2.02	0.01	0.03
ANXA1	-0.73	0.01	0.06	GNG7	-1.90	0.00	0.03
RUNX3	-0.73	0.01	0.06	SH3BGRL2	-1.89	0.01	0.03
DKFZp762A22	-0.61	0.02	0.06	LOH11CR2A	-1.86	0.01	0.03
GPRC5B	-0.61	0.02	0.06	IMAGE3917549	-1.85	0.02	0.03
RBSK	2.83	0.02	0.02	UBE2C	1.28	0.22	0.02
FLJ12924 fis, c	2.85	0.02	0.02	CDKN3	1.28	0.21	0.02
CRIP1	2.86	0.01	0.02	TACC3	1.28	0.21	0.02
sim RIKEN cD	3.07	0.01	0.02	HSPC150	1.30	0.21	0.02
503671	3.12	0.01	0.02	TRAM	1.31	0.19	0.02

Progression markers							
LG				HG			
marker	score	pval	FDR	marker	score	pval	FDR
AIP-1	0.89	0.01	0.06	MMP11	0.80	0.00	0.02
RBSK	0.73	0.01	0.06	TACC3	0.58	0.00	0.02
STLC2	0.67	0.01	0.06	COL5A2	0.56	0.01	0.02
COL6A1	0.65	0.02	0.06	AEBP1	0.55	0.01	0.02
"clone137308"	0.63	0.02	0.21	ADAM12	0.50	0.01	0.04
AI123717	0.61	0.02	0.21	CDC25C	0.44	0.02	0.05
N22687	0.57	0.02	0.45	ANKT	0.39	0.05	0.07
MGC15737	0.56	0.02	0.57	COL15A1	0.37	0.05	0.07
TSTA3	0.53	0.04	0.57	AI003775	0.35	0.07	0.20
FBLN2	0.42	0.09	0.57	TEM1	0.33	0.07	0.20

Characteristic markers																							
LG1				LG2				HG1				HG2				HG3				HG4			
marker	score	pval	FDR	marker	score	pval	FDR	marker	score	pval	FDR	marker	score	pval	FDR	marker	score	pval	FDR	marker	score	pval	FDR
PRC1	-0.20	0.02	0.04	CGI-57	-2.30	0.00	0.05	MLPH	-7.45	0.00	0.04	KIAA0210	-1.34	0.01	0.00	GSTP1	-1.37	0.00	0.09	MGC4248	-0.84	0.00	0.03
EYA2	-0.20	0.02	0.04	NP25	-1.63	0.00	0.05	IGBP1	-5.12	0.00	0.04	SATB1	-1.31	0.00	0.00	ZD52F10	-1.17	0.00	0.09	EFS2	-0.57	0.00	0.03
ANXA1	-0.20	0.02	0.04	COL4A2	-1.33	0.00	0.05	DKFZp586H0	-2.80	0.00	0.04	FLJ13384 fis, clon	-1.11	0.02	0.01	AQP5	-1.12	0.00	0.09	CSDA	-0.43	0.00	0.03
ENO1	-0.20	0.02	0.04	FLJ20392	-1.27	0.01	0.05	ZNF175	-2.63	0.00	0.04	HDAC5	-1.03	0.02	0.01	KIAA0015	-1.04	0.01	0.09	PRO2032	-0.42	0.00	0.03
IL1RN	-0.21	0.02	0.04	BE222894	-1.11	0.00	0.05	T68510	-2.12	0.00	0.04	BCL2	-1.01	0.03	0.02	NCALD	-1.03	0.00	0.09	TPM2	-0.40	0.00	0.03
ACAA1	2.81	0.06	0.03	KIAA1361	2.39	0.02	0.02	RIKEN cDNA	2.78	0.20	0.01	MGC9753	0.98	0.09	0.05	H11	0.87	0.01	0.09	SNRPB	0.43	0.01	0.03
VEGFB	2.80	0.01	0.03	CACNA1D	2.47	0.03	0.02	AI284295	2.85	0.20	0.01	MLN64	1.02	0.03	0.02	PSMD12	0.89	0.01	0.08	NDUFA2	0.55	0.01	0.03
ITGA7	2.67	0.03	0.03	FLJ12924 fis, clon	2.55	0.01	0.02	CENPA	3.34	0.20	0.01	RRM2	1.04	0.02	0.01	UBE2C	0.91	0.01	0.08	AA151092	0.63	0.01	0.03
DKFZP761I212	2.63	0.02	0.03	MGC:10083 IMAGI	2.80	0.01	0.02	JC5314 CDC2	3.49	0.20	0.01	FLJ10074	1.11	0.02	0.01	FLJ22087	0.93	0.01	0.08	FLJ23602 fis,	0.86	0.01	0.03
DGKD	2.59	0.02	0.03	MGC4643	3.13	0.01	0.02	T47163 hypoth	5.45	0.20	0.01	HEC	1.12	0.02	0.01	LOC51321	1.15	0.00	0.08	MGC4692	1.28	0.01	0.03

Progression markers																							
LG1				LG2				HG1				HG2				HG3				HG4			
marker	score	pval	FDR	marker	score	pval	FDR	marker	score	pval	FDR	marker	score	pval	FDR	marker	score	pval	FDR	marker	score	pval	FDR
MIG-6	2.88	0.00	0.02	NPPB	0.88	0.00	0.14	H2BFQ	1.83	0.00	0.02	AA992253	1.31	0.00	0.14	MMP11	1.37	0.00	0.05	283748	3.29	0.00	0.01
FLJ21213	2.86	0.00	0.02	ST5	0.92	0.01	0.14	SCAMP3	2.00	0.00	0.02	MGC4825	1.34	0.01	0.21	WDR5	1.17	0.00	0.05	IDH2	3.36	0.00	0.01
FLJ23293	2.84	0.00	0.02	FLJ14987	0.95	0.01	0.14	KIAA0943	5.12	0.00	0.02	GHRH	1.11	0.02	0.35	KIAA0682	1.12	0.00	0.05	NS1-BP	3.19	0.00	0.01
FLJ10483	2.40	0.00	0.02	RAD50	0.96	0.01	0.14	HFL1	2.80	0.00	0.02	PRELP	0.98	0.02	0.35	TACC3	1.15	0.01	0.05	KLK6	2.99	0.00	0.01
RAE1	2.24	0.00	0.02	APBB1	0.79	0.02	0.16	NUDE1	2.12	0.00	0.02	MGC12936	1.02	0.03	0.48	COL5A2	0.89	0.01	0.05	CA5B	2.90	0.00	0.01
LOXL1	2.20	0.11	0.44	"cl 137308"	0.86	0.02	0.17	CEACAM6	5.45	0.20	0.78	CCNE1	0.88	0.08	0.61	TEM1	1.04	0.01	0.05	SLC35A3	2.64	0.37	0.02
ZMPSTE24	2.07	0.11	0.44	AKAP12	0.72	0.03	0.17	KIAA0182	2.85	0.20	0.78	RRAS2	1.11	0.09	0.61	AA628867	0.72	0.02	0.06	KIAA0440	3.75	0.37	0.02
DHCR24	1.96	0.11	0.44	COL6A1	0.63	0.04	0.19	PRG5	3.49	0.20	0.78	SAA1	0.92	0.09	0.61	MAPK1	0.80	0.03	0.06	KIAA0375	3.99	0.37	0.02
STAT5A	1.69	0.11	0.44	TSTA3	0.64	0.04	0.19	FLJ20667 fis,	2.78	0.20	0.78	R09201	0.90	0.10	0.61	ANKT1	0.58	0.07	0.06	IKBKB	2.94	0.37	0.02
KIAA0523	1.45	0.11	0.44	AIP-1	0.65	0.05	0.21	ABCA8	2.17	0.20	0.78	IKBKB	0.75	0.18	0.68	AI123617	0.58	0.08	0.06	N55063	2.40	0.37	0.02

Using the Signal-to-Noise-Ratio (SNR) test and leave-one-out (LOO) experiments for Weighted Voting (WV) and kNN models, we identified the top 10 markers which distinguish the LG samples from the rest (HG and N) with 90% accuracy. HG samples could be distinguished from the rest (LG and N) with an accuracy of 97%. Table 2 presents the top ten genes for classification of grade and subtype. Each marker set distinguished a subtype from all others with accuracy exceeding 90% in LOO experiments for WV and kNNs. Progression markers were identified for LG/HG and within subtypes as the top up-regulated genes by SNR which distinguish DCIS from IDCs. In LOO experiments for WV and kNN, the average accuracy for predicting progression was 76% in HG/LG and 68-73% within subtypes. We believe that these low accuracies derive from the limited sample sizes and genes in this data. The p values were obtained using permutation experiments and the FDR rates inferred from these. Fig 3. maps the genes identified for progression in different grades into pathways for disease progression using the classification of Hanahan and Weinberg [11,12].

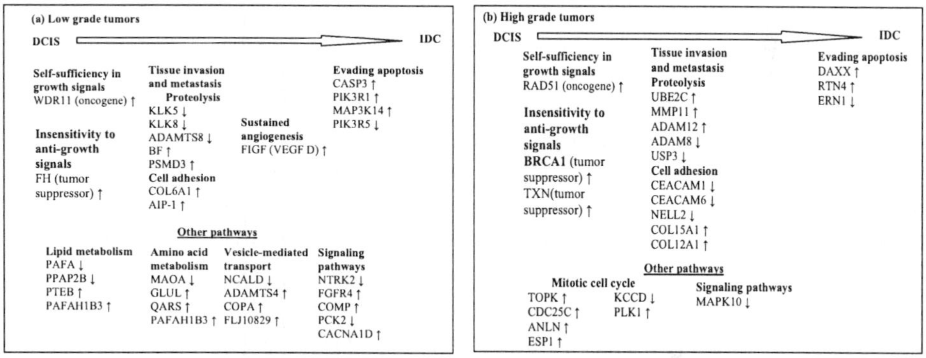

Fig 3: Pathways involved in the progression of low and high grade tumors.

3. Discussion

The use of ensemble consensus clustering was critical to identifying the subtypes. PCA by itself could identify useful markers, but could not find the rich stratification discovered by consensus ensemble *k*-clustering. Hierarchical clustering by itself was sensitive to bootstrap, indicating that its clusters are unstable to data perturbation. Robustness of clustering was obtained only after the averaging over clustering techniques and data perturbations.

The development of cancers is accompanied by alterations in cell physiology [12] involving disturbances in many regulatory mechanisms: environment independent growth, insensitivity to antigrowth factors, evasion of apoptosis, limitless replicative potential, sustained angiogenesis and tissue invasion and metastasis. We created progression model for each group based on the pathways (KEGG and GO biological process) and disease associations (OMIM, Genetic Association Database) of genes. The analysis included genes known to be involved in cancer (tumor suppressors and oncogenes) as well as genes not directly associated with cancer but which have a function critical to tumor development (proteolysis and cell adhesion markers linked to tissue invasion and metastasis). We found that progression from DCIS to IDC occurred along different pathways in low and high grade. In low grade tumors, progression correlated with alteration in lipid metabolism, transcriptional regulation, vesicle-mediated transport, amino-acid and derivative metabolism. High grade tumor progression correlated with alterations in genes in the cell cycle, ATM signaling pathway, BRCA1, BRCA2. Table 3 presents a summary of the enriched pathways in low-and high grade subtypes.

Table 3: Pathways enriched in various grades and subtypes.

Group	Enriched pathway
LG	lipid metabolism, transcriptional regulation, vesicle-mediated transport, amino-acid and derivative metabolism
LG1	small GTPase, mediated signal transduction, intracellular trafficking and vesicular transport
LG2	proteolysis collagens mRNA processing
HG	mitotic cell cycle, ATM signaling pathway, role of BRCA1, BRCA2 and ATR in cancer susceptibility, cell cycle: G2/M checkpoint
HG1	ion transport
HG2	cell cycle proteolysis
HG3	collagens proteolysis
HG4	Proteolysis

The progression models inferred for each subgroup suggest group-specific genes that are activated/repressed which contribute to the six key steps towards tumor progression as well as specific pathways altered in each subgroup. The number of available genes (1940) limits the identification of progression markers. Nevertheless, the evaluation of these markers in the context of key steps necessary for tumor progression would be valuable in the analysis of the subtypes as distinct diseases.

The main observation of the original paper of Ma et al [17] was that the molecular signature of breast cancer is already present in the early (ADH) stage of the disease. The genes that distinguish ADH from Normal and high from low grade progressively change their levels away from Normal as the disease progresses to DCIS and IDC. Our results, particularly the hierarchy we see when the data is grouped into k=2,3,...7 clusters (Fig 2.) agree with this observation.

Our methods identified six different subtypes of breast cancer with distinct patterns of progression. From histopathology, four subtypes (LG1, LG2, HG3, HG4) had a strong Luminal signature (ER+, PR+, Her2–); one subtype (HG1) had the triple negative (ER-, PR-, Her2–) characteristic of the Basal subtype, and one subtype (HG2) had a predominantly Her2+ signature (mixed ER, mostly Her2+). The validation of these subtypes on a larger dataset with more genes is currently underway.

At k=7, each of the six BCA clusters always contained samples in both DCIS and IDC stages *from the same patient.* This strong heterogeneity in the genetic signature of subtypes (progression within a subtype is less distinct than the subtypes themselves) suggests that breast cancer decides its progression path early and progression happens along different pathways in each subtype.

4. Methods

After normalization [17] and imputation of genes with < 5% missing entries using a dynamical *k*NN approach [1] we were left with 1927 genes for 93 samples. PCA [7,16,30] was done using singular value decomposition on this matrix and the eigenvectors of the largest eigenvalues that accounted for 85% of the variation in the data were used to find the subset of genes with coefficients in the top 25% in absolute value in these eigenvectors. This collection of genes was then used to find robust clusters in the data.

The optimal number of clusters was identified using gap statistics [29] and silhouette scores [16]. Next, ensemble consensus clustering [19,28] was used to divide the data into k=2, 3,..., k_{opt} clusters. The technique has two parts: [1] a method to generate clustering solutions using different methods applied to many perturbations of the data, and [2] a consensus function to combines the clusters into a single output clustering.

150 datasets were created by bootstrapping samples and genes and each was partitioned into k=2,..., k_{opt} clusters using representative methods from the three major clustering methods: *Partitioning*: partition around medoids (PAM) [16], *k*-means [13] and graph partitioning [31]; *Agglomerative*: hierarchical clustering based on average linkage, complete linkage and Ward metric [16], including bagglo [31]; *Probabilistic*: expectation maximization (EM) method [5], entropy-based-clustering (ENCLUST) [4], clustering on subsets of attributes (COSA) [8]. Each clustering method was applied 50 times with different parameter initialization on the full dataset, and once on each of the 150 datasets. The 200 resulting clusters were combined into an agreement matrix of size N_{sample} x N_{sample} for each method, whose entries m_{ij} represented the fraction of times a pair of samples (i,j) occurred in the same cluster out of the number of times the pair was selected in the 200 datasets. For each k, the agreement matrices were averaged across the clustering techniques. The samples were then sorted using simulated annealing to create a *k*-block diagonal structure in the combined agreement matrix.

For optimum sensitivity, we used the full collection of genes to identify a large pool of markers to distinguish the group of interest from its complement. This was done using SNR [9] with a permutation p-value of 0.1 and a False Discovery Rate (FDR) [2]of 0.5. We then identified a smaller subset of these genes by using stringent criteria which combined (a) a permutation p-value of 0.05 (b) stability to sample perturbation through bootstrapping (c) stability to leave-one-out experiments in top 25% genes selected by WV and kNN classifiers which distinguish the two classes with specificity and sensitivity above 0.75. The analysis used GenePattern from the Broad Institute: (http://www.broad.mit.edu/cancer/software/genepattern/).

We used the bioinformatics public resources DAVID [6], iHOP [14], and MatchMiner [3] for functional and pathway annotation. We also used 14 functional annotation sources including KEGG and GO annotations, Biocarta pathways, linked to DAVID as well as the Functional Classification Tool implemented in DAVID using Kappa Statistics.

Acknowledgments:
We thank Dr. Xiao-Jun Ma (Arcturus, CA) for providing the gene expression breast cancer data and Dr. Stefano Monti (The Broad Institute of MIT and Harvard) for discussions on estimating the number of clusters.

References

[1] Alexe, G., Dalgin, G.S., Ramaswamy, R., DeLisi, C., and Bhanot, G., Data Perturbation Independent Diagnosis and Validation of Breast Cancer Subtypes Using Clustering and Patterns, *Cancer Informatics* 2 (online): 2006.

[2] Benjamini, Y. and Hochberg, Y., Controlling the false discovery rate: a practical and powerful approach to multiple testing, *Journal of the Royal Statistics Society (Series B),* 57:289-300, 1995.

[3] Bussey, K.J., Kane, D., Sunshine, M., Narasimhan, S., Nishizuka, S., Reinhold, W.C., Zeeberg, B., Ajay, W., and Weinstein, J.N., MatchMiner: a tool for batch navigation among gene and gene product identifiers, *Genome Biol.,* 4(4):R27, 2003.

[4] Cheng, C-H., Fu, A.W., and Zhang, Y., Entropy-based subspace clustering for mining numerical data In: *Proceedings of the fifth ACM SIGKDD international conference on Knowledge discovery and data mining* San Diego, California, United States ACM Press, 1999.

[5] Dempster, A., Laird, N., and Rubin, D., Maximum likelihood from incomplete data via the EM algorithm, *Journal of the Royal Statistical Society (Series B),* 39:11-38, 1977.

[6] Dennis, G., Sherman, B.T., Hosack, D.A,, Yang, J., Gao, W., Lane,H.C., and Lempicki, R.A., DAVID: Database for Annotation, Visualization, and Integrated Discovery. *Genome Biol.,* 4:R60, 2003.

[7] Everitt, B.S. and Dunn, G., *Applied Multivariate Data Analysis*, London: Arnold, 2001.

[8] Friedman, J.H. and Meulman, J.J., Clustering objects on subsets of attributes, *Journal of the Royal Statistical Society (Series B),* 66:815-850, 2004.

[9] Golub, T.R., Slonim, D.K,, Tamayo, P., Huard, C., Gaasenbeek, M., Mesirov, J.P., Coller, H., Loh, M.L., Downing, J.R., Caligiuri, M.A., Bloomfield, C.D., and Lander, E.S., Molecular classification of cancer: class discovery and class prediction by gene expression monitoring, *Science,* 286(5439):531-537, 1999.

[10] Gruvberger, S., Ringner, M., Chen, Y., Panavally, S., Saal, L.H., Borg, A., Ferno, M., Peterson, C., and Meltzer, P.S., Estrogen receptor status in breast cancer is associated with remarkably distinct gene expression patterns, *Cancer Res.,* 61(16):5979-5984, 2001.

[11] Hanahan, D. and Folkman, J., Patterns and emerging mechanisms of the angiogenic switch during tumorigenesis, *Cell,* 86(3):353-364, 1996.

[12] Hanahan, D. and Weinberg, R.A., The hallmarks of cancer, *Cell,* 100(1):57-70, 2000.

[13] Hartigan, J.A., *Clustering algorithms*, New York: John Wiley & Sons, 1975.

[14] Hoffmann, R. and Valencia, A., A gene network for navigating the literature, *Nat. Genet.,* 36(7):664, 2004.

[15] Jolliffe, I.T., *Principal Component Analysis*, Springer, 2002.

[16] Kaufmann, L. and Rousseeuw, P.J., *Finding Groups in Data: An Introduction to Cluster Analysis*, John Wiley & Sons, 1990.

[17] Ma, X.J,, Salunga, R., Tuggle, J.T,, Gaudet, J., Enright, E., McQuary, P., Payette, T., Pistone, M., Stecker, K., Zhang, B.M,, Zhou, Y.X,, Varnholt, H., Smith, B., Gadd, M., Chatfield, E., Kessler, J., Baer, T.M,, Erlander, M.G., and Sgroi, D.C., Gene expression profiles of human breast cancer progression, *Proc. Natl. Acad. Sci. USA,* 100(1):5974-5979, 2003.

[18] Mauriac, L., Aromatase inhibitors: effective endocrine therapy in the early adjuvant setting for postmenopausal women with hormone-responsive breast cancer, *Best Pract Res. Clin. Endocrinol Metab,* 20(Suppl 1):S15-29, 2006.

[19] Monti, S., Tamayo, P., Mesirov, J., and Golub, T., Consensus Clustering: A resampling-based method for class discovery and visualization of gene expression microarray data, *Machine Learning Journal,* 52:91-118, 2003.

[20] Morris, S.R. and Carey, L.A., Molecular profiling in breast cancer, *Rev. Endocr Metab Disord,* 2007.

[21] Perou, C.M., Sorlie, T., Eisen, M.B., van de Rijn, M., Jeffrey, S.S,, Rees, C.A., Pollack, J.R., Ross, D.T,, Johnsen, H., Akslen, L.A., Fluge, O., Pergamenschikov, A., Williams, C., Zhu, SX., Lonning, P.E., Borresen-Dale, AL., Brown, P.O., and Botstein, D., Molecular portraits of human breast tumours, *Nature,* 406(6797):747-752, 2000.

[22] Quackenbush, J., Microarray analysis and tumor classification, *N. Engl. J. Med.,*354(23): 2463-2472, 2006.

[23] Rakha, E.A., El-Rehim, D.A., Paish, C., Green, A.R., Lee, A.H., Robertson, J.F., Blamey, R.W., Macmillan, D., and Ellis, I.O., Basal phenotype identifies a poor prognostic subgroup of breast cancer of clinical importance, *Eur. J. Cancer,* 42(18): 3149-3156, 2006.

[24] Sorlie, T., Perou, C.M., Fan, C., Geisler, S., Aas, T., Nobel, A., Anker, G., Akslen, L.A., Botstein, D., Borresen-Dale, A.L., and Lonning, P.E., Gene expression profiles do not consistently predict the clinical treatment response in locally advanced breast cancer, *Mol. Cancer Ther.,* 5(11):2914-2918, 2006.

[25] Sorlie, T., Wang, Y., Xiao, C., Johnsen, H., Naume, B., Samaha, R.R., and Borresen-Dale, A.L., Distinct molecular mechanisms underlying clinically relevant subtypes of breast cancer: gene expression analyses across three different platforms, *BMC Genomics,* 7:127, 2006.

[26] Sorlie, T., Wang, Y., Xiao, C., Johnsen, H., Naume, B., Samaha, R.R., and Borresen-Dale, A.L., Distinct molecular mechanisms underlying clinically relevant subtypes of breast cancer: gene expression analyses across three different platforms, *BMC Genomics,* 7:127, 2006.

[27] Sorlie, T., Wang, Y., Xiao, C., Johnsen, H., Naume, B., Samaha, R.R., and Borresen-Dale, A.L., Distinct molecular mechanisms underlying clinically relevant subtypes of breast cancer: gene expression analyses across three different platforms, *BMC Genomics,* 7:127, 2006.

[28] Strehl, A. and Ghosh, J., Cluster ensembles: a knowledge reuse framework for combining partitionings, In: *Eighteenth national conference on Artificial intelligence*, Edmonton, Alberta, Canada , 2002.

[29] Tibshirani, R., Walther, G., and Hastie, T., Estimating the number of clusters in a dataset via the Gap statistic, *Journal of the Royal Statistics Society (Series B),* 411-423, 2001.

[30] Wall, M.E., Rechtsteiner, A., and Rocha, L.M., *A Practical Approach to Microarray Data Analysis*, Norwell, MA: Kluwer, 2003.

[31] Zhao, Y. and Karypis, G., *Clustering in Life Sciences*, Humana Press, 2003.

GRAPH-THEORETICAL COMPARISON REVEALS STRUCTURAL DIVERGENCE OF HUMAN PROTEIN INTERACTION NETWORKS

MATTHIAS E. FUTSCHIK[1]
m.futschik@staff.hu-berlin.de

ANNA TSCHAUT[2]
tschaut@zedat.fu-berlin.de

GAUTAM CHAURASIA[1,3]
g.chaurasia@biologie.hu-berlin.de

HANSPETER HERZEL[1]
h.herzel@biologie.hu-berlin.de

[1]*Institute for Theoretical Biology, Charité, Humboldt-Universität, Berlin, Germany*
[2]*Department of Educational Science and Psychology, Freie Universität, Berlin, Germany*
[3] *Max-Delbrück-Center for Molecular Medicine, Berlin-Buch, Berlin, Germany*

Protein interactions constitute the backbone of the cellular machinery in living systems. Their biological importance has led to systematic assemblies of large-scale protein-protein interaction maps for various organisms. Recently, the focus of such interactome projects has shifted towards the elucidation of the human interaction network. Several strategies have been employed to gain comprehensive maps of protein interactions occurring in the human body. For their efficient analysis, graph theory has become a favourite tool. It can identify characteristic features of interaction networks which can give us important insights into the general structure of the underlying molecular networks. Although such graph-theoretical analyses have delivered us a variety of interesting results, their general validity remains to be demonstrated. We therefore examined whether independently assembled human interaction networks show common structural features. Remarkably, while some general graph-theoretical features were found, we detected a strong dependency of network structures on the method used to generate the network. Our study strongly indicates that graph-theoretical analysis can be severely compromised by the observed structural divergence and reassessment of earlier results might be warranted.

Keywords: Protein interaction; graph theory; human interactome.

1. Introduction

As protein interactions are essential for cellular processes, their systematic identification has become an important target in molecular biology. Initial efforts to assemble comprehensive lists of interactions have been undertaken for model organisms such as *S. cerevisiae*, *D. melanogaster* and *C. elegans* [1-3]. Recently, the elucidation of the human protein interactome (i.e. the complete set of protein interactions occurring in the human body) has become the major focus for many research groups [4-10]. A variety of experimental and computational strategies to map the human interactome have been pursued. All of these approaches have their unique strengths and weaknesses. Currently, the major three strategies (with their advantages and disadvantages) are as follows [11]:

- ***Literature-based interaction maps***. Protein interactions are derived from literature searches performed either by human experts or computational text-mining approaches. *Advantages*: i) This approach is not biased towards a specific experimental technique, ii) interactions are measured under a variety of conditions, and iii) maps include interactions that require post-transcriptional modifications specific to humans. *Disadvantages:* i) The false positive rates are difficult to estimate, and ii) the approach is highly biased towards proteins which are currently popular research targets.

- ***Orthology-based interaction maps.*** This approach is based on the assumption that protein interactions are evolutionarily conserved. Thus, interactions between proteins detected in other organisms are extrapolated to their human orthologs. *Advantages:* i) The method is entirely computational, enabling rapid and cost-effective construction of human interaction networks, and ii) it gains power through the abundance of interaction data for model organisms such as *S. cerevisiae*, *C. elegans*, *D. melanogaster* and *M. musculus*. *Disadvantages*: i) It is purely predictive, and ii) a considerable rate of false positives can arise through two types of errors: 1) the mapping to wrong orthologs, and 2) interactions are not conserved.

- ***Yeast two-hybrid (Y2H)-based maps.*** The Y2H method comprises on a screening approach using a set of modified proteins. Proteins are fused either to DNA-binding or transcriptional activator domains. Both types of fused proteins are subsequently co-expressed in yeast. If interaction occurs, a functional transcription factor (such as GAL4) is formed and a reporter gene is transcribed. *Advantages*: i) The Y2H assay enables systematic and rapid large-scale screening for interaction, ii) it is not biased towards known interactions, and iii) it can detect transient interactions. *Disadvantages*: i) Interactions are measured outside native surrounding (except for yeast proteins) and thus are unrelated to any physiological function, and ii) assayed proteins need to be located in the nucleus.

All of these mapping efforts have generated huge networks. For their analysis, graph theory has become an important tool. It has been applied to a variety of complex networks in various fields ranging from the World Wide Web to social networks. The aim of graph-theoretical analysis is the identification of characteristic network features and properties. In the field of systems biology, graph-theory has become a method of choice for the study of large interaction networks [12]. Although the notation of molecular networks as simple graphs is clearly an oversimplification, the use of graph-theoretical tools has been demonstrated to be very useful for the general understanding of disease processes [13], internal network structures [14] and evolutionary processes [15].

Although the application of graph-theory to interaction networks has produced many intriguing findings, it is not clear if these results are of general validity or specific to the analysed network. We therefore critically examined whether independently assembled human interaction networks show common structural features.

2. Methods and Materials

2.1. *Construction of Protein-Protein Interaction Maps*

To assess whether current human interaction networks display common structures, we have selected eight interaction maps representing the three described approaches. These networks were subsequently scrutinized for common as well as differing structural features.

Two literature-based networks were assembled based on data from the Human Protein Reference Database (HPRD) and Biomolecular Interaction Network Database (BIND) [4,8]. For construction of a third literature-based network (termed here as COCIT), we used a published list of interacting proteins derived by text-mining [16]. Orthology-based networks were generated using data from the Online Predicted Human Interaction Database (OPHID) and HOMOMINT database [5,7]. In addition, we selected an alternative collection of inferred interactions to build a third orthology-based interaction network (termed here as ORTHO) [6]. Finally, two Y2H-based networks were derived from results of recently published Y2H screens for human protein interactions [9,10].

Note that all networks were independently assembled. For comparison, proteins were mapped to their corresponding EntrezGene IDs. The sizes of the generated networks are displayed in table 1. Further details can be found in references [11,17].

2.2. *Graph-theoretical measures*

For analysis, protein interaction maps were converted to graphs with proteins as nodes and interactions as links or edges. The resulting graphs can be characterized using a variety of graph-theoretical measures:

The most fundamental characteristic of a node in a graph is its number of links to other nodes. It is referred as the *degree* of a node. The *degree distribution* $P(k)$ gives the fraction of proteins with k interactions in the total network. It can be used to distinguish different network classes. For example, the degree distribution follows a Poisson distribution for random networks of Erdös-Rényi type. Such networks have a typical node degree. In contrast, the power-law distribution ($P(k) \sim k^{-\gamma}$) is characteristic for the class of scale-free networks. The scale-free network architecture has been associated with robustness against failure of single components [12]. A hallmark of scale-free topology is the appearance of so called network *hubs* i.e. highly connected nodes. The exponent γ determines the role of the hubs in the network. The smaller γ is, the larger the fraction of nodes connected to hubs is in the network.

The *shortest path* length between two nodes is defined as the minimum number of links included in the path between the nodes. For calculation, we used the shortest path algorithm by Dijkstra [18].The mean average path length of a network is the average

Table 1: Human protein interaction networks compared in this study. The following abbreviations were used: **P**- number of mapped proteins, **I** - number of mapped interactions, **AD** – average degree, **NN**- Number of sub-networks that include more than 1000, between 1000 and 101, between 100 and 10 or between 10 and 1 proteins, **MPL** - mean path length, **D** - network diameter, γ – degree exponent, **CC** - average clustering coefficient.

Network	P	I	AD	NN	MPL	D	γ	CC	Method
MDC-Y2H	1703	3186	3.7	1/0/38/4	4.9	13	1.63	0.01	Y2H-assay
CCSB-H1	1549	2754	3.5	1/0/90/27	4.4	12	1.46	0.05	Y2H-assay
HPRD	5908	15658	5.2	1/0/135/140	5	15	2.44	0.13	Literature
BIND	2677	4233	2.9	1/3/169/256	5.9	16	1.90	0.17	Literature
COCIT	3737	6580	3.5	1/7/545/0	5.9	20	2.18	0.43	Literature
OPHID	2284	8962	7.8	1/3/95/0	4.8	15	1.36	0.23	Orthology
ORTHO	3503	9641	5.4	1/2/183/9	6.5	17	2.14	0.19	Orthology
HOMOMINT	2556	5582	4.2	1/0/85/45	5.1	12	2.76	0.07	Orthology

shortest path lengths between all possible pairs of proteins. The network's diameter is the maximum shortest path between two nodes included. To measure the local tendency of neighbors to be linked, the clustering coefficient can be utilized [12]. It is defined as $C=2n/m(m-1)$ where n is the number of links between m neighbors. A large clustering coefficient indicates that neighbors of a node are likely to interact to each other.

To avoid artifacts, self-interactions were excluded in the graph-theoretical analysis and all calculations were performed based on the largest connected graph for each map. The analysis was carried out in the R language using the Bioconductor packages *graph* and *GraphAT* [19,20].

2.3. *Generation of random graphs*

We assessed the significance of the results by comparison with two background network models: i) Random graphs with the same number of nodes and interactions, but without conservation of the degree distribution. Interactions were assigned to randomly selected pairs of proteins until the random graph had the same number of interactions as the original network. ii) Random graphs with conservation of number of nodes and interactions as well as of the degree distribution. To construct such graphs, we started with the original network and repeatedly exchanged interactions in a random manner: Edges between node A and B (A-B) and between C and D (C-D) will be changed to A-C and B-D, if such edges are not present yet. Thus, the degree of A, B, C and D is conserved, whereas the connections are changed.

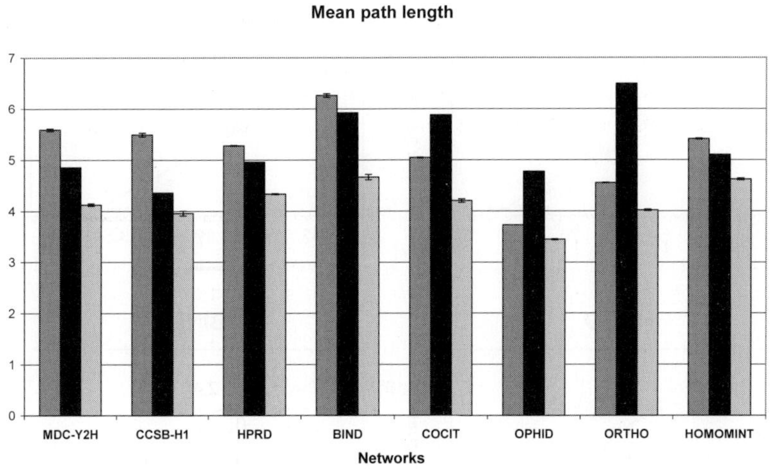

Fig. 1: Mean path lengths of interaction networks: Black bars correspond to original graphs, dark gray bars correspond to random graphs with the same number of proteins and interactions and light gray bars correspond to random networks with conserved degree distribution. Errors bars show the standard deviations derived for three independent randomizations.

3. Results

3.1. *Connectivity of networks*

Using graph-theoretical measures, fundamental topological properties of protein interaction maps can be compared and characterized. After converting all interaction maps to graphs, we analyzed their internal connectivity (table 1). For all graphs, the vast majority of proteins were connected in a main network, which appears to be a general feature of protein-protein interaction networks, being also observed in other species [1-3,21]. The remaining proteins formed predominantly smaller networks of less than 10 proteins. Only for BIND, COCIT, OPHID and ORTHO, medium sized networks (including 100-1000 proteins) emerged. Whether such separated 'islands' are artifacts reflecting the fragmentary state of proteins maps or functionally separated units remains subject for further research.

3.2. *Small worlds*

A main conclusion of previous studies was that protein interaction networks display 'small world' properties having small mean path lengths. This is also the case for the networks compared here (table 1 and Fig. 1). Their mean path length is similar and ranges from 4.4 (CCSB-H1) to 6.5 (ORTHO). For most networks, it is smaller than expected for the corresponding random graphs. For all networks, however, the mean path length is

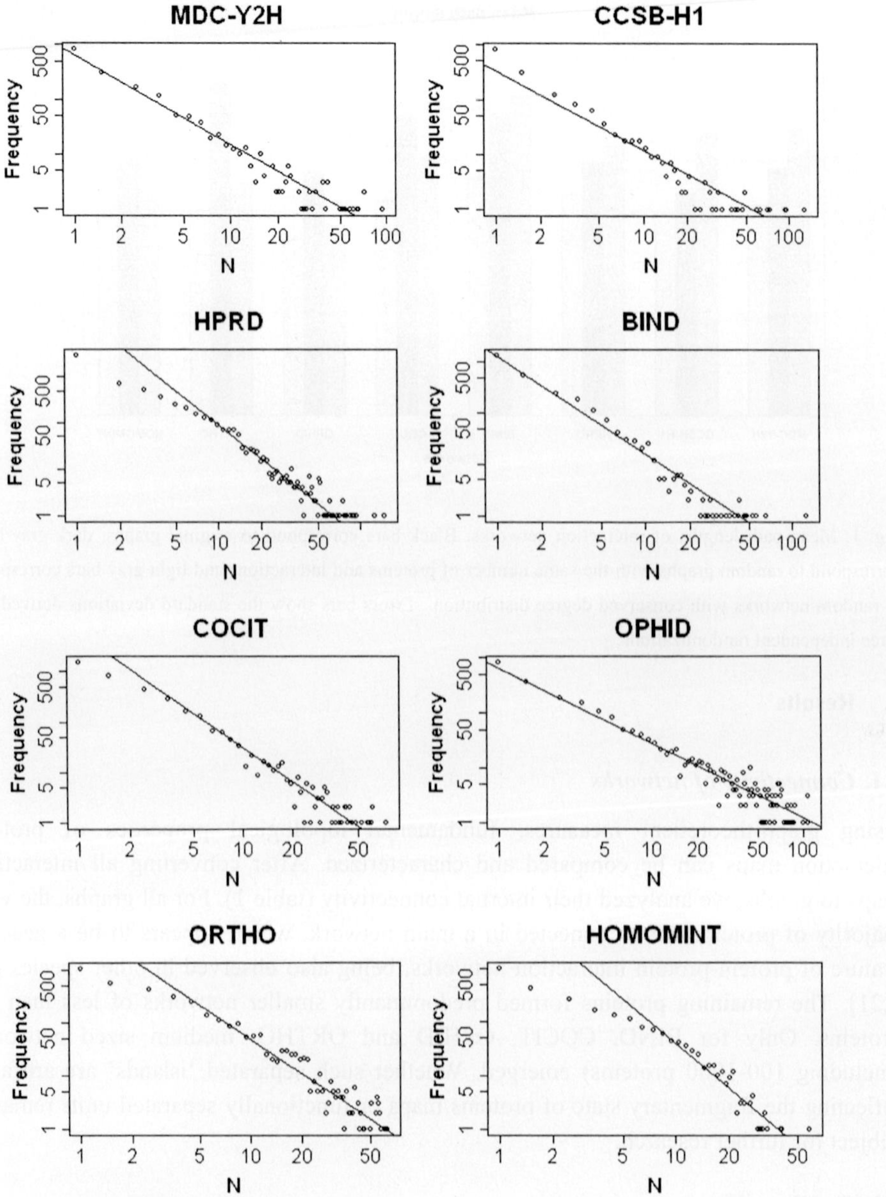

Fig. 2: Degree distributions. The number of proteins was plotted as a function of the number of neighbors that proteins in the interaction maps have. For all maps, the degree frequencies follow a power-law $P(k) \sim k^{-\gamma}$ with some derivations for HPRD, COCIT, ORTHO and HOMOMINT. The exponent γ was derived by linear regression of the logged data.

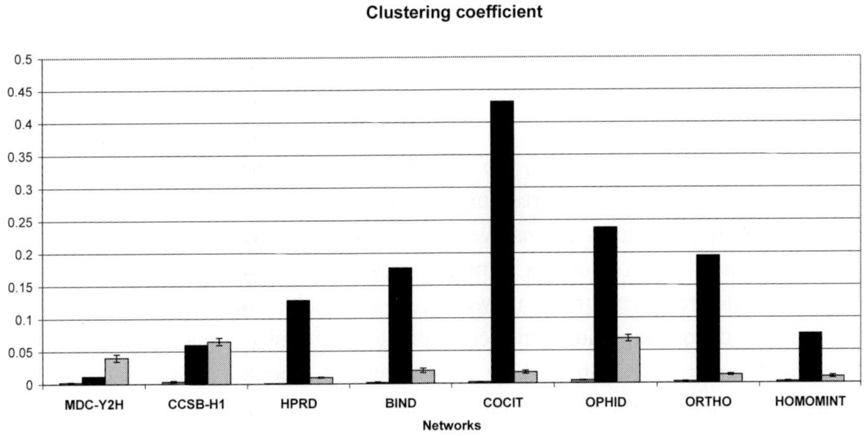

Fig. 3: Mean clustering coefficient of interaction networks. The same representation as in figure 1 was used.

larger than expected for the corresponding random scale-free networks, pointing to the existence of internal structures (see also *Supplementary Materials*). The diameter (i.e. the largest path length within a network) ranges noticeably between 12 (CCSB-H1, HOMOMINT) and 20 (COCIT).

3.3. *Degree distribution*

An important determinant of a network's structure is the degree distribution $P(k)$. We found that all networks display power-law distribution implying a general emergence of hubs (Fig. 2). However, some deviations can be observed. Networks derived from BIND, OPHID or Y2H-assays followed most closely the power-law distribution, in contrast the remaining ones show a relative depletion of interaction-poor proteins. Furthermore, the exponent γ varies by a factor of two between networks indicating that the role of hubs differs considerably (see section 2.2). Notably, networks that obey closely the power-law distribution also tend to have smaller mean path lengths.

3.4. *Modularity*

Cellular networks have been proposed to exhibit modular structure i.e they can be divided into separable highly connected sub-networks [12,14]. A commonly used measure for modularity is the clustering coefficient reflecting the cohesiveness of the neighborhood of network nodes. In our analysis, the average clustering coefficient ranges remarkably by a factor of almost 50 from 0.01 to 0.45 (Fig. 3). The smallest coefficients were found for Y2H-based networks; they were similar to the expected values for random scale-free networks, leading to the conclusion that the Y2H-based maps do not display particularly strong neighborhood cohesiveness. A possible reason could be a large number of undetected interactions (false negatives). In contrast, clustering coefficients for literature- and orthology-based networks were considerably larger than for the corresponding random networks, implying that these networks are highly modular.

3.5. *Hierarchical structure*

Besides the assessment of modularity, the clustering coefficient has been employed to study hierarchical modular structures of networks. The concept of hierarchical modularity implies that modules themselves are made up by smaller modules. It was introduced by Ravasz and co-workers aiming to resolve the apparent contradiction of modularity and scale-free structure of networks. In the analysis of metabolic networks, they associated a decreasing clustering coefficient for highly connected nodes with hierarchical modular organization [14]. In such networks, poorly connected nodes (i.e. the majority of proteins) are situated in modules and thus have a large clustering coefficient. In contrast, hubs connecting these distinct modules display only small clustering coefficients. We observed this dependency of clustering coefficient on degree for most networks compared (Fig. 4). For orthology-based networks, however, this pattern is absent suggesting the lack of a hierarchical structure in these networks. Alternatively, large highly connected complexes could result in proteins having both a large number of interactions and large clustering coefficient.

4 Discussion and Conclusions

Graph theory represents an important and popular approach for the analysis of large-scale interaction networks [12]. It is frequently used to obtain a general characterization of molecular networks. It is also of importance for systems biology in revealing modular structures which can subsequently be modeled more quantitatively. Nevertheless, results of graph-theoretical analyses should be taken with caution, since they are generally based on the assumption that the studied network is error-free and complete. This, however, is hardly the case for current protein interaction networks. Whereas the effects of sparse sampling have been intensively studied, the impact of the method used to generate interaction networks has been neglected so far [22,23].

Here we presented a first graph-theoretical comparison of major human interaction networks. It shows that the method used for the network assembly strongly influence the structure of the network. While all interaction networks showed small world properties and corresponded to scale-free networks, we detected considerable structural divergence regarding their modularity and hierarchical structure. This observation has to be taken into account for an unbiased application of graph theory. General conclusions about the structure of the human protein interaction network should therefore be verified against potential interference by the chosen assembly method. As many previous analyses of network structures are based on single networks, a reassessment of their results might be warranted.

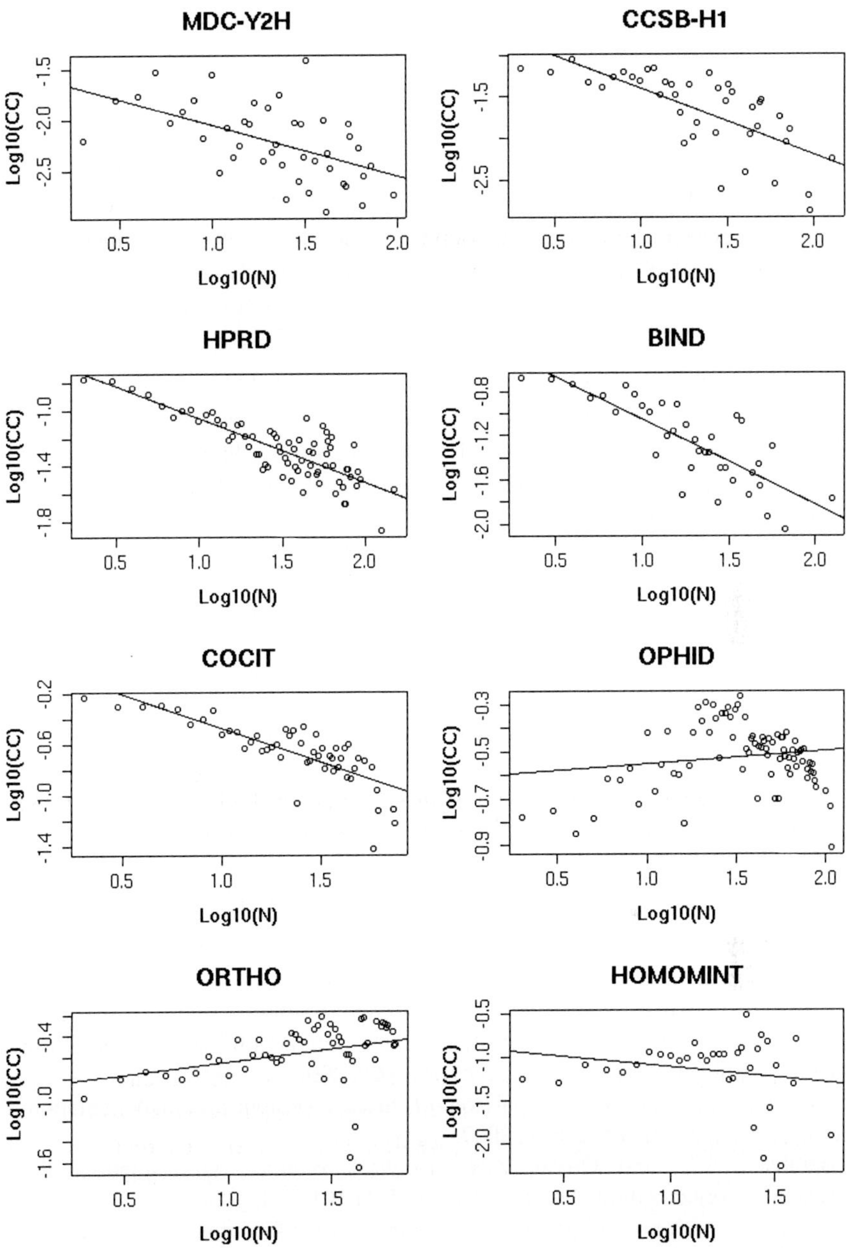

Fig. 4: Clustering coefficient. Plots show the dependence of the clustering coefficient on the degree of proteins. The clustering coefficients shown were derived by averaging over all proteins having the same degree. The solid line shows the linear fit.

Supplementary materials

Supplementary materials can be found at: http://itb.biologie.hu-berlin.de/Members/futschik/ ibsb2007

Acknowledgements

The work presented here was supported by the SFB 618 grant of the *Deutsche* Forschungsgesellschaft (DFG). We thank Bronwyn Carlisle for careful proofreading.

References

[1] Uetz, P., *et al.*, A comprehensive analysis of protein-protein interactions in Saccharomyces cerevisiae, *Nature*, 403(6770):623-627, 2000.
[2] Li, S., *et al.*, A map of the interactome network of the metazoan C.elegans, *Science*, 303(5654):540-543, 2004.
[3] Giot, L., *et al.*, A protein interaction map of Drosophila melanogaster, *Science*, 302(5651):1727-1736, 2003.
[4] Bader, G.D., *et al.*, BIND--The Biomolecular Interaction Network Database, *Nucleic Acids Res.*, 29(1):242-245, 2001.
[5] Brown, K.R. and Jurisica, I., Online predicted human interaction database, *Bioinformatics*, 21(9):2076-2082, 2005.
[6] Lehner, B. and Fraser, A.G., A first-draft human protein-interaction map, *Genome Biol.*, 5(9): R63, 2004.
[7] Persico, M., *et al.*, HomoMINT: an inferred human network based on orthology mapping of protein interactions discovered in model organisms, *BMC Bioinformatics*, 6(Suppl 4):S21, 2005.
[8] Peri, S., *et al.*, Development of human protein reference database as an initial platform for approaching systems biology in humans, *Genome Res.*, 13(10):2363-2371, 2003.
[9] Rual, J.F., *et al.*, Towards a proteome-scale map of the human protein-protein interaction network, *Nature*, 437(7062):1173-1178, 2005.
[10] Stelzl, U., *et al.*, A human protein-protein interaction network: a resource for annotating the proteome, *Cell*, 122(6):957-968, 2005.
[11] Futschik, M.E., *et al.*, Comparison of human protein-protein interaction maps, *Bioinformatics*, 23(5):605-611, 2007.
[12] Barabasi, A.L. and Oltvai, Z.N., Network biology: understanding the cell's functional organization, *Nat. Rev. Genet.*, 5(2):101-113, 2004.
[13] Lim, J., *et al.*, A protein-protein interaction network for human inherited ataxias and disorders of Purkinje cell degeneration, *Cell*, 125(4):801-814, 2006.
[14] Ravasz, E., *et al.*, Hierarchical organization of modularity in metabolic networks, *Science*, 297(5586):1551-1555, 2002.
[15] Wagner, A., The yeast protein interaction network evolves rapidly and contains few redundant duplicate genes, *Mol. Biol. Evol.*, 18(7):1283-1292, 2001.

[16] Ramani, A.K., *et al.*, Consolidating the set of known human protein-protein interactions in preparation for large-scale mapping of the human interactome, *Genome Biol.*, 6(5):R40, 2005.

[17] Chaurasia, G., *et al.*, UniHI: an entry gate to the human protein interactome, *Nucleic Acids Res.*, 35(Database issue):D590-594, 2007.

[18] Dijkstra, E.W., A note on two problems in connection with graphs, *Numerische Mathematik*, 1:269-271, 1959.

[19] Carey, V.J., *et al.*, Network structures and algorithms in Bioconductor, *Bioinformatics*, 21(1):135-136, 2005.

[20] Balasubramanian, R., *et al.*, A graph-theoretic approach to testing associations between disparate sources of functional genomics data, *Bioinformatics*, 20(18): 3353-3362, 2004.

[21] Lee, I., *et al.*, A probabilistic functional network of yeast genes, *Science*, 306(5701):1555-1558, 2004.

[22] Han, J.D., *et al.*, Evidence for dynamically organized modularity in the yeast protein-protein interaction network, *Nature*, 430(6995):88-93, 2004.

[23] de Silva, E., *et al.*, The effects of incomplete protein interaction data on structural and evolutionary inferences, *BMC Biol.*, 4:39, 2006.

NEW AMINO ACID INDICES BASED ON RESIDUE NETWORK TOPOLOGY

JIAN HUANG[1, 2]
hjian@kuicr.kyoto-u.ac.jp

SHUICHI KAWASHIMA[3]
shuichi@hgc.jp

MINORU KANEHISA[1, 3]
kanehisa@kuicr.kyoto-u.ac.jp

[1]*Bioinformatics Center, Institute for Chemical Research, Kyoto University, Gokasho Uji, Kyoto 611-0011, Japan*
[2]*School of Life Science and Technology, University of Electronic Science and Technology of China, Chengdu 610054, China*
[3]*Human Genome Center, Institute of Medical Science, University of Tokyo, Minato-ku, Tokyo 108-8639, Japan*

Amino acid indices are useful tools in bioinformatics. With the appearance of novel theory and technology, and the rapid increase of experimental data, building new indices to cope with new or unsolved old problems is still necessary. In this study, residue networks are constructed from the PDB structures of 640 representative proteins based on the distance between Cα atoms with an 8 Å cutoff. All these networks show typical small world features. New amino acid indices, termed relative connectivity, clustering coefficient, closeness and betweenness, are derived from the corresponding topological parameters of amino acids in the residue networks. The 4 new network based indices are closely clustered together and related to hydrophobicity and β propensity. When compared with related amino acid indices, the new indices show better or comparable performance in protein surface residue prediction. Relative connectivity is the best index and can reach a useful performance with an area under the curve about 0.75. It indicates that the network property based amino acid indices can be useful complements to the existing physicochemical property based amino acid indices.

Keywords: amino acid index; residue network; connectivity; closeness; betweenness; clustering coefficient.

1. Introduction

Any given property of amino acids can be represented by a set of 20 numerical values, usually called a propensity scale or amino acid index [1-3]. As scales of different physicochemical and biochemical properties, amino acid indices have been widely used in various bioinformatics studies, such as predicting protein secondary structures [4], transmembrane sequences [5], surface [6, 7] and linear B cell epitopes [8-10]. Sometimes, however, the existing indices perform poorly [9], indicating that better methods or new amino acid indices are needed. With the appearance of novel theory and technology, and the rapid increase of experimental data, it is necessary to revise old amino acid indices, and build new ones.

Recently, graph and network theory have become a paradigm for research on complex biological systems [11-16]. Proteins have also been studied intensively as networks formed by amino acid residues and their interactions [17-29]. To avoid confusion with protein-protein interaction networks, these networks are usually called residue networks or amino acid networks. In residue networks, nodes stand for amino acids and two nodes are linked together when the distance between the two nodes is shorter than a given threshold (see Fig. 1). Though constructed with various distance cutoffs and based on different residues or atoms, all residue networks studied so far have small world features [17-29]. Nearly all these networks have a normal degree distribution rather than a scale-free power-law degree distribution; though the latter is often seen in other biological networks [12, 15]. It is also very interesting that topological parameters of residue networks have shown relationships to protein folding [17-19], dynamics [23], stability [25], functional sites and residues [21, 22]. Therefore, topological properties of amino acids in residue networks may play an important role in exploring protein structure and function.

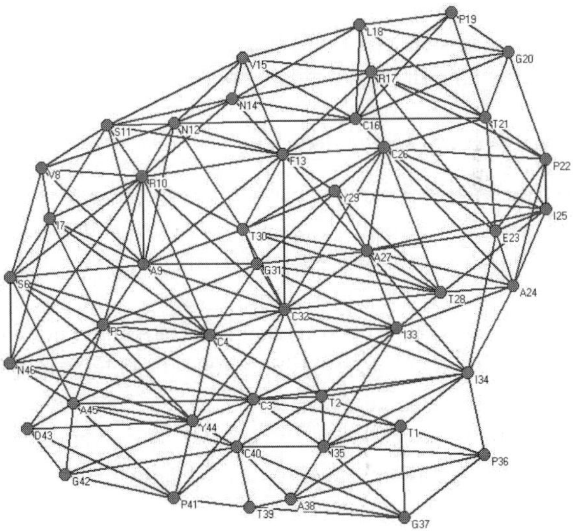

Fig 1. Residue network of crambin. Constructed from the PDB structure 1CRN, based on the distance between Cα atoms with an 8 Å cutoff and visualized with the Pajek program [30].

As reported previously, our group began to construct and maintain a database of amino acid indices almost 20 years ago [1-3]. However, most existing amino acid indices, if not all, are derived from physicochemical properties of amino acids, such as size, charge, polarity and hydrophobicity. The topological properties of amino acids in the residue networks have not been adequately studied and included into the AAindex database [3].

In this study, we built new amino acid indices based on the local and global topology of residue networks. We also studied the relation between these topological properties and the physicochemical properties of amino acids through cluster analysis with existing indices in the AAindex database [3]. The application of these new indices is demonstrated in protein surface residue prediction.

2. Methods and Data Sets

2.1. *Constructing residue networks*

A set of 640 representative proteins were selected from the PDB release #2006_08_20 through the web interface of the PDB-REPRDB database [31]. All the structures are determined with X-ray diffraction at a resolution of less than 2 Å and with a sequence longer than 40 residues. Structures with Cα or backbone coordinates only, or with more than one chain, or with any chain break, fragment, mutant or non-standard residue are excluded. To eliminate sequential or structural homology among the selected structures, their sequence identity and Root Mean Square Deviation (RMSD) are required to be below 30% and above 10 Å respectively.

Each residue in a structure is considered as a node. Two nodes are linked together when the Euclidian distance between their Cα atoms is shorter than 8 Å. The calculation of the Euclidian distance between two atoms has been described in detail elsewhere [32]. For each structure, a corresponding residue network is constructed.

2.2. *Analyzing the topology of residue networks*

The topology of a residue network can be characterized with local and global parameters. Connectivity, which is often termed "degree", is one of the most important local parameters. The connectivity of residue *r* (*Kr*) is the number of neighbors linked to residue *r*.

Clustering coefficient is another local parameter of a residue network. The clustering coefficient of residue *r* (*Cr*) reflects the probability that the neighbors of residue *r* are also neighbors of each other. In residue network, if residue *r* has *Kr* neighbors, the number of all possible links among these neighbors is *Kr (Kr-1)/2*. However, the actual links among these neighbors are counted and represented by "*Ar*". Then *Cr* is given as:

$$Cr = \frac{Ar}{Kr(Kr-1)/2} = \frac{2Ar}{Kr(kr-1)} \quad (1)$$

In network theory, closeness is a global measure for centrality. The closeness of residue *r* (*Or*) to other residues in the residue network is defined as:

$$Or = N - 1 \Big/ \sum_{s \in V, r \neq s} D(r,s) \quad (2)$$

where *N* is the network size. *D(r,s)* is the shortest path between residue *r* and another residue *s*. *V* is the set of all residues in the network.

Betweenness is another global centrality measure of a node within a network. Nodes that lie on many shortest paths between other nodes have higher betweenness than those that do not. The betweenness of residue *r* (*Br*) in the residue network is defined as:

$$Br = \left. \sum_{q \neq r \neq s \in V} \frac{D(q,r,s)}{D(q,s)} \right/ (N-1)(N-2) \tag{3}$$

where *D (q,r,s)* is number of shortest path between residue *q* and *s* pass through *r*, *D(q,s)* is all shortest paths between residue *q* and *s*. *(N-1)(N-2)* equals to the number of ordered pairs of residues not including *r*.

For each residue in all the residue networks, the four topological parameters are computed. For each residue network, the average parameters are calculated and the connectivity distribution is analyzed. The diameter and average path length of each residue network are also computed. These parameters are further analyzed together with protein size and structural class.

2.3. *Deriving new amino acid indices*

For each of the 20 amino acid commonly found in proteins, its topological properties (connectivity, clustering coefficient, closeness and betweenness) are averaged over all the residue networks constructed above. A set of 20 values for each topological parameter makes the raw amino acid index. All the 4 newly derived, raw amino acid indices are then normalized with

$$Rx = Ox / M\ x \tag{4}$$

where *Ox* is the original value of the raw amino acid index and *Mx* is the mean of that index set. The set of normalized results *Rx* makes the new, relative amino acid index based on a topological property.

2.4. *Clustering new indices with existing indices*

Hierarchical cluster analysis is applied to explore the relationships between these network based new amino acid indices and the 494 published indices in the AAindex database [3]. This is done with the program Amino Acid Explorer [33], which is based on a method reported previously by our group [1, 2].

A similar analysis is done on the 4 new indices and 11 highly related indices, which are clustered into the same branch. A minimum spanning tree is also built from the 4 new indices and 67 other indices that contain any of the following strings "hydroph", "polar," "size," "volume," "charge," and "electr" in their description.

2.5. *Predicting surface residues with new and related indices*

Seven amino acid indices are used to predict protein surface residues on 3 data sets. The first data set consists of 640 representative proteins, from which the new indices are derived. The second data set has 25 representative proteins, randomly picked from data set 1. The third data set has 25 representative proteins also. However, they are selected from newly released PDB structures, fulfilling the previously described requirements for data set 1.

Surface residues are assigned based on their solvent accessible area at different cutoff values of 1, 10, 20, 50 and 100 Å^2. The solvent accessible area is computed with the NACCESS program [34] using default parameters.

The amino acid indices tested include the 4 new indices and three related indices. The new indices are Relative Connectivity (*Rk*), Relative Clustering Coefficient (*Rc*), Relative Closeness (*Ro*) and Relative Betweenness (*Rb*). The three related indices are 8 Å contact number (*N8*) [6], Parker's hydrophilicity (*Ph*) and Levitt's index (*Li*) [8]. *N8* was clustered close together with the 4 network based indices and had good performance in surface residue prediction. *Ph* was derived from experiment data and related to surface residues. Both *Ph* and *Li* have been confirmed to be one of the best indices for B cell epitope prediction [10]. If an index correlates negatively to the surface possibility, it is multiplied by -1 when used in predicting surface residues.

The prediction is completed with the classical sliding window method. In brief, a window slides from the N-terminal to C-terminal of the query protein sequence. The mean propensity value of the window is then assigned to the residue in the middle of the window. At the N- and C- termini, we use asymmetric windows to avoid omitting prediction examples. Different window sizes of 1, 3, 5, 7, and 9 are tested.

Receiver Operating Characteristics (ROC) curves are constructed and visualized with the ROCR package [35]. The area under the curve (*Aroc*) is used to evaluate the performance of each prediction [36].

3. Results

3.1. *Residue networks are small worlds*

All the residue networks constructed show typical small world features such as short average path length and high clustering coefficient. The connectivity distribution in residue network is normal rather than power-law (see Fig. 2) and the average path length scales up logarithmically with the network size.

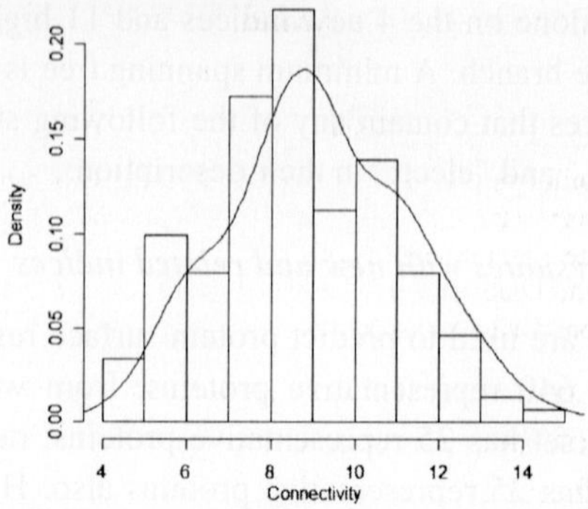

Fig 2. Connectivity distribution of sperm wale myoglobin. Residue network based on PDB structure 1A6M is shown as an example, which is in agreement with normality by Shapiro-Wilk test ($P = 0.008$).

While diameter and connectivity logarithmically scale up with residue network size, closeness, betweenness and clustering coefficient scale down logarithmically. The 640 representative proteins are assigned to "all α", "all β", "α/β", "α+β" and "others" class according to the SCOP database [37]. All the network parameters studied above show no significant differences among different structural classes.

3.2. *Topologically derived new amino acid indices*

Four new amino acid indices termed Relative Connectivity (*Rk*), Relative Clustering Coefficient (*Rc*), Relative Closeness (*Ro*) and Relative Betweenness (*Rb*) are derived from topological parameters of the residue networks and listed in Table 1.

Table 1. Four new amino acid indices based on residue network topology.

	A	C	D	E	F	G	H	I	K	L	M	N	P	Q	R	S	T	V	W	Y
Rk	1.05	1.17	0.88	0.85	1.07	0.99	0.99	1.11	0.88	1.07	1.04	0.93	0.92	0.93	0.94	0.96	0.99	1.12	1.05	1.05
Rc	0.99	0.89	1.11	1.13	0.92	1.08	1.00	0.89	1.10	0.92	0.95	1.07	1.01	1.06	1.04	1.05	1.01	0.90	0.93	0.94
Ro	1.00	1.13	0.95	0.95	1.03	0.99	1.01	1.04	0.96	1.02	1.02	0.96	0.96	0.97	0.98	0.98	0.99	1.04	1.01	1.02
Rb	0.96	1.60	0.63	0.61	1.31	0.77	1.03	1.43	0.61	1.30	1.24	0.72	0.83	0.73	0.82	0.80	0.90	1.35	1.20	1.16

3.3. *New indices are related to hydrophobicity and β propensity*

After hierarchical clustering, all the 4 new indices are closely clustered together and related to the hydrophobicity and β propensity indices. As shown in Fig. 3, connectivity and clustering coefficient are both highly related to β propensity, and the betweenness measure is highly related to hydrophobicity. Closeness directly links to betweenness, through which the 4 new indices are joined together.

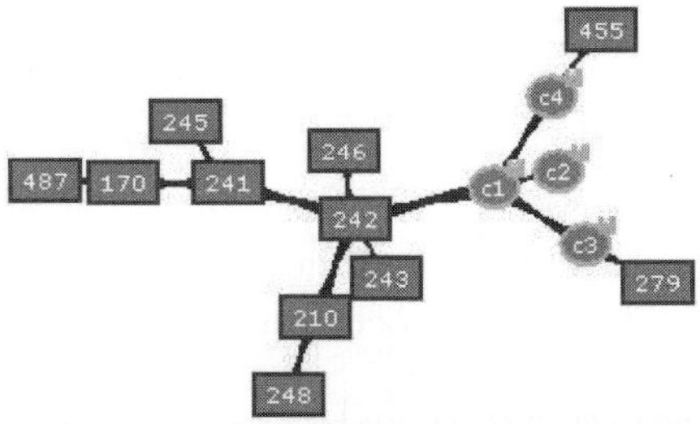

Fig 3. Minimum spanning tree built from the 4 new indices and 11 highly related indices. The new indices are displayed in circles. C1: betweenness; C2: closeness; C3: connectivity; C4: clustering coefficient. The rectangles stand for highly related indices in AAindex [3] labeled with corresponding serial number. 242: Average gain in surrounding hydrophobicity; 279: Weights for beta-sheet at the window position of 2; 455: Beta-sheet propensity derived from designed sequences.

The minimum spanning tree is also built from the 4 new indices and 67 published indices that contain any of the following string "hydroph", "polar," "size," "volume," "charge," and "electr" in their description. The close relation between the network-based

4 new indices and hydrophobicity or hydrophilicity is confirmed; in contrast, their relationships to amino acid size and charge are weak (data not shown).

3.4. *Performance in predicting surface residues with new indices*

Seven indices have been applied to predict surface residues on three data sets with five different sliding window sizes and surface cutoffs. Among them, relative connectivity (*Rk*) always performs best. Relative closeness (*Ro*) and relative betweenness (*Rb*) are comparable to the 8 Å contact number index (*N8*) given by Ooi *et al* [6]. Though relative clustering coefficient (*Rc*) does not perform as well as the other 3 new indices, it is still better than Parker's hydrophilicity (*Ph*) and Levitt's index (*Li*) [8] (see Fig. 4).

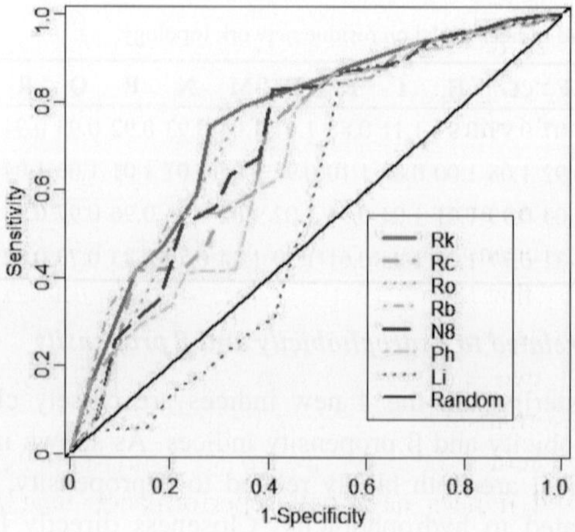

Fig 4. ROC curves for seven indices. The curves above are constructed from predictions on Data set 3 with sliding window size 1 and surface cutoff 100 Å^2. In this condition, the *Aroc* of *Rk* is about 0.75; its accuracy, sensitivity and specificity can reach 72%, 74% and 72% respectively. For a random prediction, *Aroc* is 0.5; for a perfect method, *Aroc* is 1; *Aroc* value higher than 0.7 is usually considered as a useful prediction performance.

When the size of sliding window decreases, all indices except *Li* perform better (result not shown). When the surface cutoff increases, 4 new indices tend to perform better; but the performance of *N8*, *Ph* and *Li* tend to decrease (see Table 2).

Table 2. *Aroc* from predictions on data set 3 with window size 1 and various surface cutoffs (1, 10, 20, 50 and 100 Å^2)

	SC1	SC10	SC20	SC50	SC100
Rk	0.734	0.734	0.735	0.736	0.753
Rc	0.703	0.703	0.700	0.693	0.705
Ro	0.720	0.720	0.722	0.721	0.727
Rb	0.710	0.713	0.713	0.708	0.722
N8	0.721	0.716	0.717	0.715	0.711
Ph	0.683	0.677	0.671	0.664	0.665
Li	0.658	0.635	0.628	0.600	0.557

4. Discussion

Amino acid indices are useful tools in bioinformatics. Our group has been building and maintaining a database of amino acid indices for almost 20 years [1-3]. However, most published amino acid indices, if not all, are based on physicochemical properties of amino acids, such as size, charge, polarity and hydrophobicity. Proteins can be considered as networks of amino acid residues and their interactions [17-29]. In this study, we confirmed the small world properties of such networks and built 4 new indices based on residue network topological parameters.

A very recent paper reported a very good agreement between connectivity and amino acid hydrophobicity [29]. Our results from hierarchical cluster analysis indicated that the 4 new indices do relate to hydrophobicity, but β propensity as well. As several topological parameters of residue networks have shown useful relationships to protein folding [17-19], dynamics [23], stability [25], functional sites and residues [21, 22], network topology based indices might be helpful for exploring protein structure and function.

Compared with related amino acid indices such as *Ph* and *Li*, the new indices show better performance in protein surface residue prediction. The problem of surface residue prediction is related to that of B cell epitope prediction, due to the requirement for epitopes to be surface accessible to interact with an antibody. *Ph* and *Li* have been proved to be the best two indices so far in linear B cell epitope prediction [8-10]. However, even the performance of *Ph* and *Li* are unsatisfactory [9], indicating that better methods or new amino acid indices are needed for B cell epitope prediction. Since the network topology based indices have better performance than *Ph* and *Li* in protein surface residue prediction, they might also perform better in B cell epitope prediction. This will be an area of future study for us.

In conclusion, it indicates that network topology based amino acid indices can be useful complements to the existing physicochemical property based amino acid indices.

Acknowledgments

We thank Dr Alex Gutteridge for copyediting the manuscript and giving help on R. This work was supported by grants from the Ministry of Education, Culture, Sports, Science and Technology and the Japan Science and Technology Agency. The computational resource was provided by the Bioinformatics Center, Institute for Chemical Research, Kyoto University. The support from NSFC project (30600138) is also acknowledged.

References

[1] Nakai, K., Kidera, A., and Kanehisa, M., Cluster analysis of amino acid indices for prediction of protein structure and function, *Protein Eng.*, 2(2):93-100, 1988.

[2] Tomii, K. and Kanehisa, M., Analysis of amino acid indices and mutation matrices for sequence comparison and structure prediction of proteins, *Protein Eng.*, 9(1):27-36, 1996.

[3] Kawashima, S. and Kanehisa, M., AAindex: amino acid index database, *Nucleic Acids Res.*, 28(1):374, 2000.

[4] Kazemian, M., Moshiri, B., Nikbakht, H., and Lucas, C., A new expertness index for assessment of secondary structure prediction engines, *Comput. Biol. Chem.*, 31(1):44-47, 2007.

[5] Zhao, G. and London, E., An amino acid "transmembrane tendency" scale that approaches the theoretical limit to accuracy for prediction of transmembrane helices: relationship to biological hydrophobicity, *Protein Sci.*, 15(8):1987-2001, 2006.

[6] Nishikawa, K. and Ooi, T., Prediction of the surface-interior diagram of globular proteins by an empirical method, *Int. J. Pept. Protein Res.*, 16(1):19-32, 1980.

[7] Nishikawa, K. and Ooi, T., Radial locations of amino acid residues in a globular protein: correlation with the sequence, *J. Biochem*, 100(4):1043-1047, 1986.

[8] Pellequer, J.L., Westhof, E., and Van Regenmortel, M. H., Predicting location of continuous epitopes in proteins from their primary structures, *Methods Enzymol.*, 203:176-201, 1991.

[9] Blythe, M.J. and Flower, D.R., Benchmarking B cell epitope prediction: underperformance of existing methods, *Protein Sci.*, 14(1):246-248, 2005.

[10] Larsen, J. E., Lund, O., and Nielsen, M., Improved method for predicting linear B-cell epitopes, *Immunome Res.*, 2:2, 2006.

[11] Watts, D.J. and Strogatz, S.H., Collective dynamics of 'small-world' networks, *Nature*, 393(6684):440-442, 1998.

[12] Barabasi, A.L. and Albert, R., Emergence of scaling in random networks, *Science*, 286(5439):509-512, 1999.

[13] Milo, R., Shen-Orr, S., Itzkovitz, S., Kashtan, N., Chklovskii, D., and Alon, U., Network motifs: simple building blocks of complex networks, *Science*, 298(5594):824-827, 2002.

[14] Ravasz, E., Somera, A.L., Mongru, D.A., Oltvai, Z.N. and Barabasi, A.L., Hierarchical organization of modularity in metabolic networks, *Science*, 297(5586):1551-1555, 2002.

[15] Barabasi, A.L. and Oltvai, Z.N., Network biology: understanding the cell's functional organization, *Nat. Rev. Genet*, 5(2):101-113, 2004.

[16] Palla, G., Derenyi, I., Farkas, I., and Vicsek, T., Uncovering the overlapping community structure of complex networks in nature and society, *Nature*, 435(7043):814-818, 2005.

[17] Vendruscolo, M., Paci, E., Dobson, C. M., and Karplus, M., Three key residues form a critical contact network in a protein folding transition state, *Nature*, 409(6820):641-645, 2001.

[18] Dokholyan, N.V., Li, L., Ding, F., and Shakhnovich, E.I., Topological determinants of protein folding, *Proc. Natl. Acad. Sci. USA*, 99(13):8637-8641, 2002.

[19] Vendruscolo, M., Dokholyan, N.V., Paci, E., and Karplus, M., Small-world view of the amino acids that play a key role in protein folding, *Phys. Rev. E Stat. Nonlin. Soft. Matter Phys.*, 65(6 Pt 1):061910, 2002.

[20] Greene, L.H. and Higman, V.A., Uncovering network systems within protein structures, *J. Mol. Biol.*, 334(4):781-791, 2003.

[21] Wangikar, P.P., Tendulkar, A.V., Ramya, S., Mali, D.N., and Sarawagi, S., Functional sites in protein families uncovered via an objective and automated graph theoretic approach, *J. Mol. Biol.*, 326(3):955-978, 2003.

[22] Amitai, G., Shemesh, A., Sitbon, E., Shklar, M., Netanely, D., Venger, I., and Pietrokovski, S., Network analysis of protein structures identifies functional residues, *J. Mol. Biol.*, 344(4):1135-1146, 2004.

[23] Atilgan, A.R., Akan, P., and Baysal, C., Small-world communication of residues and significance for protein dynamics, *Biophys J.*, 86(1 Pt 1):85-91, 2004.

[24] Bagler, G. and Sinha, S., Network properties of protein structures, *Physica A*, 346(1-2):27-33, 2005.

[25] Brinda, K.V. and Vishveshwara, S., A network representation of protein structures: implications for protein stability, *Biophys J.*, 89(6):4159-4170, 2005.

[26] Kundu, S., Amino acid network within protein, *Physica A*, 346(1-2):104-109, 2005.

[27] Aftabuddin, M. and Kundu, S., Weighted and unweighted network of amino acids within protein, *Physica A*, 369(2):895-904, 2006.

[28] Aftabuddin, M. and Kundu, S., Hydrophobic, hydrophilic and charged amino acid networks within Protein, *Biophys J*, 93(1):225—231, 2007.

[29] Alves, N. A. and Martinez, A. S., Inferring topological features of proteins from amino acid residue networks, *Physica A*, 375(1):336—344, 2007.

[30] Batagelj, V. and Mrvar, A., Pajek - Analysis and Visualization of Large Networks, *Graph Drawing: 9th International Symposium*, 477, 2002.

[31] Noguchi, T. and Akiyama, Y., PDB-REPRDB: a database of representative protein chains from the Protein Data Bank (PDB) in 2003, *Nucleic Acids Res.*, 31(1):492-493, 2003.

[32] Huang, J., Gutteridge, A., Honda, W., and Kanehisa, M., MIMOX: a web tool for phage display based epitope mapping, *BMC Bioinformatics*, 7:451, 2006.

[33] Bulka, B., desJardins, M. and Freeland, S.J., An interactive visualization tool to explore the biophysical properties of amino acids and their contribution to substitution matrices, *BMC Bioinformatics*, 7:329, 2006.

[34] Hubbard, S. J. and Thornton, J. M., NACCESS, *Department of Biochemistry and Molecular Biology, University College London*, 1993.

[35] Sing, T., Sander, O., Beerenwinkel, N., and Lengauer, T., ROCR: visualizing classifier performance in R, *Bioinformatics*, 21(20):3940-3941, 2005.

[36] Swets, J. A., Measuring the accuracy of diagnostic systems, *Science*, 240(4857):1285-1293, 1988.

[37] Murzin, A. G., Brenner, S. E., Hubbard, T., and Chothia, C., SCOP: a structural classification of proteins database for the investigation of sequences and structures, *J. Mol. Biol.*, 247(4):536-540, 1995.

COMPUTATIONAL ANALYSIS OF PROTEIN-PROTEIN INTERACTIONS IN METABOLIC NETWORKS OF ESCHERICHIA COLI AND YEAST

CAROLA HUTHMACHER
carola.huthmacher@charite.de

CHRISTOPH GILLE
christoph.gille@charite.de

HERMANN-GEORG HOLZHÜTTER
hergo@charite.de

Medical Faculty of the Humboldt University, Charite, Institute of Biochemistry, Monbijoustr. 2, 10117 Berlin, Germany

Protein-protein interactions are operative at almost every level of cell function. In the recent years high-throughput methods have been increasingly used to uncover protein-protein interactions at genome scale resulting in interaction maps for entire organisms. However, biochemical implications of high-throughput interactions are not always obvious. The question arises whether all interactions detected by *in vitro* experiments also play a functional role in the living cell. In this work we systematically analyze high-throughput protein-protein interactions stored in public databases in the context of metabolic networks. Classifying reaction pairs according to their topological distance revealed a significantly higher frequency of enzyme-enzyme interactions for directly neighbored reactions (distance = 1). To determine possible functional implications for these interactions we examined randomized networks using original enzyme interactions as well as randomly generated interaction data. A functional relevance of enzyme-enzyme interactions could be demonstrated for those reactions that exhibit low connectivity. As this is a characteristic of enzyme pairs in metabolic channeling we systematically searched the literature and indeed recovered a certain fraction of enzyme pairs that has already been implicated in metabolic channeling. However, a substantial number of enzyme pairs uncovered by our large-scale analysis remains that up to now has neither been functionally nor structurally classified and therefore present novel candidates of the metabolic channeling concept.

Keywords: enzyme interactions; network randomization; randomly generated enzyme interactions; metabolic channeling.

1. Introduction

Protein-protein interactions play critical roles for almost all cellular processes. Permanent interactions are found in multimeric proteins such as the core and the regulators of proteasomes, the small and large subunits of ribosomes, or metabolic enzymes like fatty acid synthase. Transient interactions are important among others for signal transduction, splicing, cell motility, and cell cycle. For metabolic enzymes transient interactions have been described in the context of metabolic channeling, which is assumed to result in multiple catalytic advantages [16].

Recently, systematic searches for protein interactions have been conducted at large scale using high-throughput techniques such as yeast two-hybrid methods [9, 20] and affinity purification combined with mass spectroscopy [5, 7]. These experimental techniques yield enormous amounts of binary interaction data, which are valuable resources for protein function annotation. High-throughput data are thought to be free of bias from selection processes by the experimentator or the scientific community [15] since all data obtained from one experiment are measured under identical conditions and are recorded irrespectively of existing knowledge. This opens the possibility to relate binary interaction data to other biological properties by statistical analyses. Protein interaction data has been linked to a variety of other features like gene-fusion events [4, 19], gene expression [6, 21], and interactions of homologous proteins from different organisms [12]. The observed correlations provide evidence that a large percentage of interactions are biologically meaningful. On the other hand, there are increasing arguments for a considerable percentage of false positive and false negative rates inherent in high-throughput methods since the results of different techniques vary considerably [14]. Therefore the question arises whether high-throughput interactions are partially due to artefacts owing to limitations of the experiment or whether the biological meaning is simply not yet understood.
In this work we mapped high-throughput protein-protein interaction data onto metabolic networks of the model organisms *Escherichia coli* and *Saccharomyces cerevisiae*. We analyzed interaction probabilities for neighboring enzymes in original and randomized networks using real enzyme interactions as well as randomly generated data. Significantly higher interaction rates were observed for neighboring enzymes in original networks at least for sparsely connected reactions, suggesting functional relevance for such enzyme complexes. Additionally, we identified several putative novel examples of the metabolic channeling concept among the complexes involving neighboring enzymes.

2. Methods

2.1. *Database protein-protein interactions (PPI)*

For our analyses we used experimentally detected protein-protein interactions from the four public databases DIP (version of Jan 7th 2007), IntAct (version of Dec 4th 2006), MINT (version of Feb 6th 2007), and BioGRID (version 2.0.24). Only interacting proteins, which are both linked to SwissProt were included in our analyses in order to benefit from SwissProt annotations such as genetic locus information. The resulting set of more than 170,000 binary interactions includes high-throughput data as well as manually curated data. On the basis of SwissProt almost 18,000 interactions among proteins with assigned EC number were identified originating from more than 100 different organisms. Of these interactions 7,575 were observed in yeast and 3,288 in *E. coli*.

2.2. *Metabolic networks*

The obtained enzyme interactions were mapped onto the metabolic networks of the single-compartment organism *Escherichia coli* and the multi-compartment organism *Saccharomyces cerevisiae.* We used the metabolic network models iJR904 [18] (*E. coli*) and iND750 [3] (yeast) by Palsson and his colleagues. An important feature of these models are gene to protein to reaction associations, which allow the incorporation of further information such as protein-protein interaction data. Flux directions were assumed as proposed by Palsson. If reactions are defined as reversible we assumed a flux from left to right, meaning that metabolites on the left side of the reaction arrow are consumed. The *E. coli* set contains 931 reactions while the yeast set comprises 1,149 reactions. Genetic loci are assigned to 873 and 810 reactions, respectively. SwissProt annotation was mapped via locus information.

2.3. *Neighboring reactions*

Graphs representing metabolic networks were constructed where nodes correspond to reactions. Reactions having at least one metabolite in common were linked by edges regardless whether the metabolite is consumed or produced. Candidates were not defined as neighbors if the common metabolite is either abundant or unspecific like water and phosphate or acts as a cofactor like NAD+ and ATP. Otherwise this might result in linking reactions where a functional relevance is questionable in view of metabolic channeling. Reactions were coupled nevertheless if cofactor partners (NADH, ADP, etc.) do not occur in at least one of the reactions. By this means cofactor biosynthesis or degradation steps remain linked.

2.4. *Randomizing metabolic networks*

Networks were randomized in two different ways. In a first trial the skeleton of the original network was used and the reaction names assigned to nodes were permuted. In other words, node attributes were arbitrarily exchanged while edges remained unchanged. Since in this approach the original connectivity of reactions is ignored, we next generated networks preserving this characteristic. Reactions were grouped according to their overall connectivity and only reactions of similar connectivity were exchanged. To obtain few connectivity classes of similar size each class was defined to contain about one tenth of the reaction set. Only node labels corresponding to reactions of the same connectivity class were permuted, which roughly retains the number of neighbors for each reaction. For both types of randomization 1,000 networks were generated. The number of enzymes that form a complex and are located in direct neighborhood are counted in the original metabolic network and in the randomized ones. Homomeric interactions were not considered.

2.5. *Randomizing enzyme interactions*

We used two different approaches to generate random enzyme interactions. In the first approach interacting enzymes are drawn applying the same probability for all pairs. This probability was set to 0.5% in accordance to the observed interaction probability of *E. coli* enzymes. For the second approach we adjusted this probability to take into account the higher interaction probability of enzymes located at metabolic branchpoints [8]. If the connectivity of reactions was greater than 1 the base two logarithm of this value was multiplied to the interaction probability of reaction pairs resulting in higher probabilities for enzymes that are highly linked. 1,000 times random enzyme complexes were computed and mapped onto the original network as well as onto randomized networks as described above.

3. Results

For our analysis experimentally determined physical protein-protein interactions were obtained from the online databases DIP, IntAct, MINT, and BioGRID and mapped onto the metabolic networks of the model organisms *E. coli* and yeast. For the *E. coli* metabolic network model iJR904 enzymes corresponding to 1,759 reaction pairs were found in the interaction databases excluding those of overlapping genetic loci. In the case of the multi-compartment organism yeast enzymes corresponding to 2,508 reaction pairs were reported as interaction partners. Enzyme pairs were not included into the analysis if they belong to different cellular compartments. The focus of this work was set on enzymes that are direct neighbors within metabolic networks since such interactions have been found to play a functional role in metabolic channeling. We detected 104 pairs of reactions that are direct neighbors in the *E. coli* metabolic network and are catalyzed by enzymes recorded in protein interaction databases as partners. For yeast 103 such reaction pairs were found.

3.1. *Randomizing metabolic networks reduces the number of enzyme interactions involving neighboring reactions*

We compared the number of detected interactions in the metabolic networks of yeast and *E. coli* to that in randomized networks. In a first analysis step, reactions were randomly exchanged. On average 45 reaction pairs corresponding to interacting enzymes were located in direct neighborhood in the case of *E. coli* and 49 in the case of yeast, accounting for less than 50% of those in the original network (compare Table 1 run 1 with run 2). For both organisms all 1,000 randomized networks exhibited less interactions among neighboring enzymes as the original network (see Fig. 1 run 2). In a subsequent analysis we exchanged only reactions that are topologically equivalent with respect to connectivity. *E. coli* reactions were grouped into 10 classes of similar connectivity varying in size from 41 to 111 reactions and yeast reactions into 11 classes comprising minimal 66 reactions and maximal 130 reactions.

Enzymes catalyzing neighboring reactions in these randomized networks interact on average 92 times in *E. coli* and 88 times in yeast reaching almost the interaction probability of neighboring enzymes in the original metabolic networks (compare Table 1 run 1 with run 3). The number of interactions in the original networks rank among the 82.9^{th} (*E. coli*) and 87.9^{th} (yeast) percentile of the distribution obtained by randomizing the original networks (see Fig. 1 run 3). Considering only low connected reactions contained in classes 1 to 8 yields higher numbers of interactions in the original networks (see Table 2 run 4) than in both types of randomized networks (see Table 2 run 5 and 6). For *E. coli* 24 enzyme interactions were detected among direct neighbors in the original network while on average in arbitrarily randomized networks 12 interactions and in randomized networks preserving reaction connectivity 11 interactions were found. Only 4 and 2 randomized networks, respectively, exhibited as many interactions as the original networks (see Fig. 1 run 5 and 6). In the case of yeast 14 interactions among neighboring enzymes were detected in the original network and 8 and 6 interactions, respectively, in randomized networks (corresponding percentiles: 94.9 and 99.8).

Table 1. Overview on randomization runs 1-3 and results for *E. coli*. Run 3 preserves reaction connectivity.

	Run 1	Run 2	Run 3
Network	original network	permuted network	permuted network (preserving connectivity)
Enzyme interactions	original data	original data	original data
PPI involving neighboring enzymes (rel. amount)	1	0.43	0.89

3.2. *Randomly generated binary interaction data also yields higher frequency of protein interactions in original network than in permuted networks*

Relating network topology to the ability of enzymes to interact with each other reveals that enzymes are more likely to have interaction partners if they share metabolites with many other enzymes (data not shown). This tendency is much more pronounced for yeast than for *E. coli*. To assess the bias due to increased interaction probabilities of highly connected enzymes we repeated our analysis with randomly generated enzyme interactions. In a first approach enzyme pairs were randomly selected as interaction partners with equal probability. Mapping these

Table 2. Overview on randomization runs 4-6 and results for *E. coli*. Run 6 preserves reaction connectivity. Only those reactions were considered that are assigned to connectivity classes 1 to 8.

	Run 4	**Run 5**	**Run 6**
Network	original network	permuted network	permuted network (preserving connectivity)
Enzyme interactions	original data	original data	original data
PPI involving neighboring enzymes (rel. amount)	1	0.48	0.47

Table 3. Overview on randomization runs 7-9 for *E. coli* using randomly generated enzyme interactions and interaction probabilities correlated to connectivity of reactions. Run 9 preserves reaction connectivity.

	Run 7	**Run 8**	**Run 9**
Network	original network	permuted network	permuted network (preserving connectivity)
Enzyme interactions	artificial data (interact. probs. correl. to connectivity)	artificial data (interact. probs. correl. to connectivity)	artificial data (interact. probs. correl. to connectivity)
PPI involving neighboring enzymes (rel. amount)	1	0.45	0.96

Table 4. Overview on randomization runs 10-12 for *E. coli* using randomly generated enzyme interactions and interaction probabilities correlated to connectivity of reactions. Only those reactions were considered that are assigned to connectivity classes 1 to 8. Run 12 preserves reaction connectivity.

	Run 10	**Run 11**	**Run 12**
Network	original network	permuted network	permuted network (preserving connectivity)
Enzyme interactions	artificial data (interact. probs. correl. to connectivity)	artificial data (interact. probs. correl. to connectivity)	artificial data (interact. probs. correl. to connectivity)
PPI involving neighboring enzymes (rel. amount)	1	~1	~1

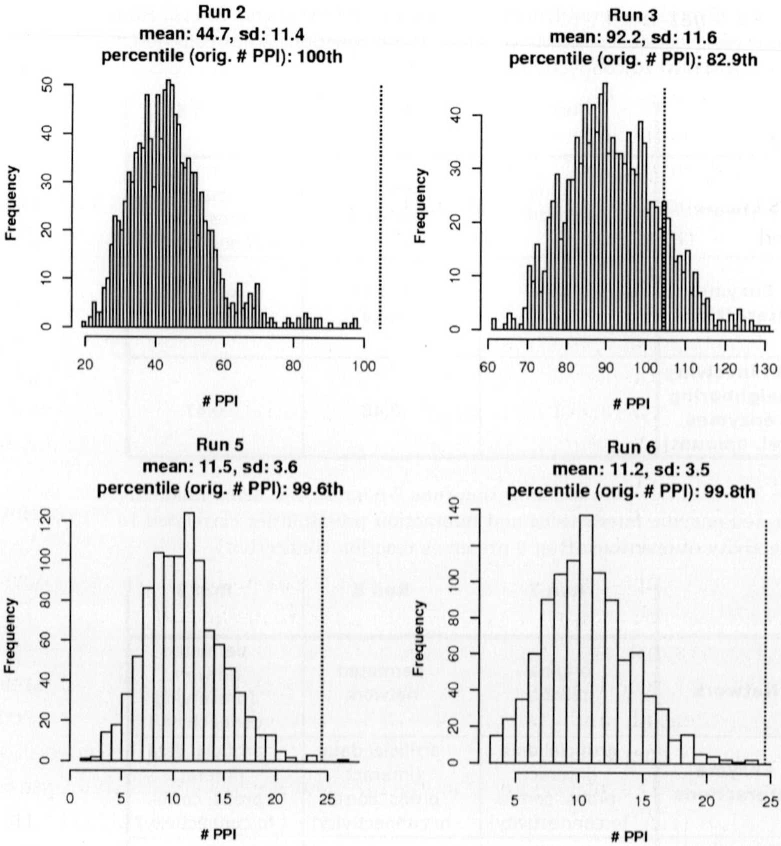

Fig. 1. PPI number distribution obtained by randomizing the *E. coli* network (see Table 1 and 2). Dotted lines represent number of interactions reported for neighboring enzymes in original network.

interaction data onto the original network and onto the two types of permuted networks resulted for all networks in similar numbers of interactions among neighboring enzymes (data not shown). In contrast, using in accordance to the real data higher interaction probabilities for those enzymes located at metabolic junctions yielded similar results as for experimentally observed interactions (compare Table 1 run 1-3 with Table 3 run 7-9). Significantly less interactions were found for neighboring enzymes in arbitrarily randomized networks than in the original network (compare Table 3 run 7 with run 8) but almost as high amounts in randomized networks where exchanges were restricted to reactions of similar connectivity (compare Table 3 run 7 with run 9). Considering only reaction pairs of low connectivity unveiled that interaction probabilities for neighboring enzymes in the original network and in the different types of randomized networks resemble (see Table 4 run 10-12).

3.3. *Known and yet uncharacterized enzyme complexes*

Our analyses of experimentally observed enzyme interactions as well as randomly generated interaction data mapped onto original metabolic networks and permuted networks suggest a functional role of interactions involving neighboring enzymes at least for those catalyzing sparsely connected reactions.

Reaction pairs corresponding to neighboring enzymes that are reported to interact can be grouped into three categories: both reactions consume the same metabolite (A), both reactions produce the same metabolite (B), or one reaction produces a metabolite that is consumed by the other reaction (C). Some reaction pairs have both common substrates and products and thus simultaneously belong to more than one category. Enzyme complexes of category C present candidates for metabolic channeling. Of the *E. coli* reaction set 37 reaction pairs were assigned to type A, 23 pairs to type B, and 48 reaction pairs to type C. In the case of yeast group A comprises 27 reaction pairs, group B 13 reaction pairs, and group C 73 pairs.

Scanning PubMed revealed that several previously known and characterized enzyme interactions were recovered by our large-scale analysis. This includes the tryptophan biosynthetic enzyme complex of anthranilate synthase and anthranilate phosphoribosyltransferase [1] as well as interactions involving the TCA enzymes malate dehydrogenase and citrate synthase [11]. In addition, enzyme complexes corresponding to 39 pairs of neighboring reactions in the *E. coli* network and 32 yeast reaction pairs were unveiled as potential candidates for metabolic channeling but have not been further experimentally characterized yet (for details see [8]). An example is given by the *E. coli* enzymes NAD synthase and the glycine cleavage system. Both enzymes produce metabolites (NAD and ammonia, respectively) that are consumed by the interaction partner. Furthermore, an interaction between the yeast enzymes fatty-acyl-CoA synthase and 1-acyl-glycerol-phosphate acyltransferase was observed. The first enzyme activates fatty acids of varying length with Co-A, which are substrates of the phosphatidate producing second enzyme thereby linking the fatty acid biosynthesis pathway to the phospholipid biosynthesis pathway.

Additionally, our analysis detected interactions that are not conform with the metabolic channeling concept (category A and B). This includes complexes comprising glutamine-fructose-6-phosphate transaminase (GFPT) from yeast, which is found to associate with several enzymes that use glutamine and produce glutamate like GFPT itself. The same holds for asparagine synthase, which uses the amido group of glutamine to generate asparagine from aspartate.

4. Discussion

Meanwhile, systematic and exhaustive analysis of protein-protein interactions in cells and tissues has become an important field of experimental cell biology and biochemistry. The protein-protein interaction networks resulting from these studies are on one hand impressive when depicted as complex network graphs but on the other hand they help only little to understand regulation and dynamics on the molecular level unless the physiological relevance of a protein-protein interaction established under *in vitro* conditions is substantiated in the context of the underlying reaction network. One useful validation may consist in mapping proteins with reported interactions onto the topology of the corresponding metabolic network. In this work we have carried out a systematic statistical analysis of enzyme-enzyme interactions in metabolic networks based on currently available data in protein-protein interaction databases. We focused on interactions involving direct neighbors of a network as such interactions are best understood in the context of metabolic channeling.

We previously found that adjacent enzymes are more likely to be recorded in protein interaction databases than distant enzymes [8]. This can be interpreted in the sense that a large proportion of such enzyme interactions plays indeed a functional role. However, one has to ascertain that this is definitively attributed to biochemical function and not due to a bias from network topology and the way pairs of reactions are iterated. Permutation of networks and randomization of interaction data are strategies to discriminate between both. We counted protein interactions in neighboring reactions and performed the same analysis for randomized networks. Arbitrary permutation resulted in significantly less interactions among neighboring enzymes than in the original network (compare Table 1 run 1 with run 2) while permutation restricted to exchanges of reactions that have similar connectivity, in order to preserve topological properties of the network, yielded only slightly less interactions (compare Table 1 run 1 with run 3). The observation could be misinterpreted in the sense that neighboring enzymes interact more frequently than expected due to functional relevance of these interactions, which is related to network topology. Repetition of this analysis using randomly generated enzyme interactions resulted in similar findings (compare Table 1 run 1-3 with Table 3 run 7-9) proving this prediction in its general form wrong. In this approach higher interaction probabilities were chosen for highly connected reactions in accordance to the real interaction data. However, when using randomly generated enzyme interactions and including only sparsely linked reaction pairs into the statistical analysis there we found practically no difference in the amount of next-neighbor interactions in the permuted networks and the original network (see Table 4 run 10-12) whereas for actually reported enzyme interactions the same permutation procedures resulted in a drop of more than 50% (compare Table 2 run 4 with run 5 and 6). This points to functional relevance of enzyme interactions for sparsely connected neighboring reactions. Although this statement is based on low numbers of observed interactions it is supported by the fact that the trend could be demonstrated for two distinct organisms. For enzyme interactions involving reactions with high connectivity functional importance cannot be concluded from our analysis. Anyhow, direct neighborhood of enzymes does not seem to play a role for these interactions. Since enzymes at metabolic branching points are often key enzymes controling whole pathways a regulatory function of interactions involving such enzymes is conceivable due to induced conformational changes, which alter enzyme activity.

Our result is conform with the concept of metabolic channeling, which often involves

consecutive reaction steps where intermediates are not or only little involved in other pathways. Examples of channeling where intermediates do not play important roles elsewhere are found in the synthesis pathways of pyrimidine [2], lumazine [10] and tryptophan [13, 17].
The number of well-studied cases is still small as biochemical evidences for a direct interaction of enzymes are difficult to obtain. Therefore we aimed to uncover putative novel examples of the metabolic channeling concept by analyzing high-throughput interaction data. Less than 3% of metabolic enzyme interactions reported for *E. coli* and yeast come into question for channeling of reaction intermediates, suggesting that metabolic channeling is not an ubiquitous phenomenon in cells. However, more than half of these candidate enzyme pairs have not been further experimentally characterized, in the case of *E. coli* even more than 80%, indicating that not all examples are known yet. Low connectivities of involved reactions might be a usefull discriminator for functional relevance of enzyme interactions as our results have shown. A putative metabolic channeling enzyme complex detected by our genome-wide analysis is that of fatty-acyl-CoA synthase and 1-acyl-glycerol-phosphate acyltransferase, which might have importance during cell devision as fatty acids are directly transfered to an enzyme of the phospholipid biosynthesis, which is needed for the synthesis of cellular membranes. By this means fatty acids are rather converted to phospholipids than being processed by competing enzymes. Intriguingly, our analysis revealed a substantial number of neighboring enzymes forming a complex for which the metabolic channeling concept does not apply as the two coupled reactions utilize a common substrate or generate the same product. It has to be noted that this finding has to be considered with care as it is based on *a priori* assumptions concerning the directionality of fluxes in the network. Nevertheless, it is worthwhile to ask for a possible functional relevance of such constellations deviating from the classical channeling concept. This includes several interactions of glutamine-fructose-6-phosphate transaminase (GFPT) with other enzymes utilizing amido nitrogen from glutamine as well. The functional relevance of these interactions might be the bundling of enzymes that either produce or consume the same metabolite. TAP pull-down assays revealed that GFPT furthermore interacts with glutamate synthase (precursor), which may form glutamine and 2-oxoglutarate from glutamate. Consequently, this interaction gives supply of glutamine for GFPT from which associated enzymes could participate.

References

[1] Browne, B. A., Itzel Ramos, A., and Downs, D. M., PurF-independent phosphoribosyl amine formation in yjgF mutants of Salmonella enterica utilizes the tryptophan biosynthetic enzyme complex anthranilate synthase-phosphoribosyltransferase, *J. Bacteriol*, 188(19):6786–6792, 2006.

[2] Christopherson, R. I., Traut, T. W., and Jones, M. E., Multienzymatic proteins in mammalian pyrimidine biosynthesis: channeling of intermediates to avoid futile cycles, *Curr. Top Cell. Regul.*, 1859–1877, 1981.

[3] Duarte, N. C., Herrgard, M J., and Palsson, B., Reconstruction and validation of

Saccharomyces cerevisiae iND750, a fully compartmentalized genome-scale metabolic model, *Genome Res.*, 14(7):1298–1309, 2004.

[4] Enright, A. J., Iliopoulos, I., Kyrpides, N. C., and Ouzounis, C A., Protein interaction maps for complete genomes based on gene fusion events, *Nature*, 402(6757):86–90, 1999.

[5] Gavin, A. C., Bosche, M. *et al.*, Functional organization of the yeast proteome by systematic analysis of protein complexes, *Nature*, 415(6868):141–147, 2002.

[6] Ge, H., Liu, Z., Church, G. M., and Vidal, M., Correlation between transcriptome and interactome mapping data from Saccharomyces cerevisiae, *Nat. Genet.*, 29(4):482–486, 2001.

[7] Ho, Y., Gruhler, A. *et al.*, Systematic identification of protein complexes in Saccharomyces cerevisiae by mass spectrometry, *Nature*, 415(6868):180–183, 2002.

[8] Huthmacher, C., Gille, C., and Holzhütter, H.-G., A comprehensive statistical analysis of protein interactions in metabolic networks reveals novel enzyme pairs potentially involved in metabolic channeling, *J. Theor. Biol.*, 2007 (submitted).

[9] Ito, T., Chiba, T., Ozawa, R., Yoshida, M., Hattori, M., and Sakaki, Y., A comprehensive two-hybrid analysis to explore the yeast protein interactome, *Proc. Natl. Acad. Sci. USA*, 98(8):4569–4574, 2001.

[10] Kis, K. and Bacher, A., Substrate channeling in the lumazine synthase/riboflavin synthase complex of Bacillus subtilis, *J. Biol. Chem.*, 270(28):16788–16795, 1995.

[11] Lindbladh, C., Rault, M., Hagglund, C., Small, W. C., Mosbach, K., Bulow, L., Evans, C., and Srere, P. A., Preparation and kinetic characterization of a fusion protein of yeast mitochondrial citrate synthase and malate dehydrogenase, *Biochemistry*, 33(39):11692–11698, 1994.

[12] Liu, Y., Liu, N., and Zhao, H., Inferring protein-protein interactions through high-throughput interaction data from diverse organisms, *Bioinformatics*, 21(15):3279–3285, 2005.

[13] Matchett, W. H., Indole channeling by tryptophan synthase of neurospora, *J. Biol. Chem.*, 249(13):4041–4049, 1974.

[14] von Mering, C., Krause, R. *et al.*, Comparative assessment of large-scale data sets of protein-protein interactions, *Nature*, 417(6887):399–403, 2002.

[15] Mrowka, R., Patzak, A., and Herzel, H., Is there a bias in proteome research?, *Genome Res.*, 11(12):1971–1973, 2001.

[16] Ovadi, J., Physiological significance of metabolic channelling, *J. Theor. Biol.*, 152(1):1–22, 1991.

[17] Pan, P., Woehl, E., and Dunn, M. F., Protein architecture, dynamics and allostery in tryptophan synthase channeling, *Trends. Biochem. Sci.*, 22(1):22–27, 1997.

[18] Reed, J. L., Vo, T. D., Schilling, C. H., and Palsson, B., An expanded genome-scale model of Escherichia coli K-12 (iJR904 GSM/GPR), *Genome Biol.*, 4(9):R54, 2003.

[19] Tsoka, S. and Ouzounis, C. A., Prediction of protein interactions: metabolic enzymes are frequently involved in gene fusion, *Nat. Genet.*, 26(2):141–142, 2000.

[20] Uetz, P., Giot, L. *et al.*, A comprehensive analysis of protein-protein interactions in Saccharomyces cerevisiae, *Nature*, 403(6770):623–627, 2000.

[21] Walhout, A. J. M., Reboul, J. *et al.*, Integrating interactome, phenome, and transcriptome mapping data for the C. elegans germline., *Curr. Biol.*, 12(22):1952–1958, 2002.

CONTEXT SPECIFIC PROTEIN FUNCTION PREDICTION

NAOKI NARIAI[1]
nariai@bu.edu

SIMON KASIF[1,2]
kasif@bu.edu

[1]*Bioinformatics Program, Boston University, 44 Cummington St., Boston, MA 02215, USA*
[2]*Department of Biomedical Engineering, Boston University, 44 Cummington St., Boston, MA 02215, USA*

Although whole-genome sequencing of many organisms has been completed, numerous newly discovered genes are still functionally unknown. Using high-throughput data such as protein-protein interaction (PPI) information to assign putative protein function to the unknown genes has been proposed, since in many cases it is not feasible to annotate the newly discovered genes by sequence-based approaches alone. In addition to PPI data, information such as protein localization within a cell may be employed to improve protein function prediction in two ways: 1) By using such localization information as a direct indicator of protein function (e.g. nucleolus localized proteins might be involved in ribosome biogenesis), and 2) by refining noisy PPI data by localization information. In the latter case, localization information may be used to distinguish different types of PPIs: Namely, interactions between co-localized proteins (more reliable), and interactions between differently localized proteins (potentially less reliable). In this paper, we propose a probabilistic method to predict protein function from PPI data and localization information. A Bayesian network is used to model dependencies between protein function, PPI data and localization information. We showed in our cross-validation experiment that in some cases, our method (conditioning PPI data by localization information) significantly improves prediction precision, as compared to a simple Naive Bayes method that assumes PPI data and localization information are conditionally independent given protein function. Finally, we predicted 57 unknown genes as "ribosome biogenesis" proteins.

Keywords: protein function prediction; protein-protein interaction; Bayesian networks.

1. Introduction

One of the challenges in computational biology is to annotate the thousands of unknown genes that are gleaned from the newly sequenced genome of many organisms. Sequence similarity based methods such as BLAST [1], and protein motif (domain) based approaches such as PFAM [2] have been widely used for protein function prediction. However, these sequence-based approaches often fail, when applied to the unknown proteins due to lack of orthologous proteins in other organisms or weak sequence similarity to other known proteins. Recently, high-throughput technologies have produced massive amount of genomic information, such as protein-protein interactions (PPIs), protein localization, and gene expression data. Several types of methods have been proposed to use these genome-wide data to predict protein function. One successful method uses PPI data to assign protein function, based on the assumption that interacting proteins tend to share the same function [3,4]. However, since PPI data produced from high-throughput analyses is known to be noisy, combining other types of genome-wide data may additionally improve protein function prediction methods.

Bayesian network methodologies have been proposed for integrating multiple types of genome-wide data, such as localization information, gene expression data, and co-essentiality to predict PPIs [5]. Moreover, combining such heterogeneous data to predict a functional linkage graph [6], in which an edge between two nodes (genes) represents functional similarity with a reliability score, has been extensively studied [3,7-10]. Instead of producing a functional linkage graph, assigning protein functions to each gene directly from genome-wide data has also been proposed, such as the methods based on Markov random fields (MRFs) [4,11], and other machine learning methods such as support vector machines [12]. A Bayesian method to combine different types of functional linkage graphs and other genomic features (e.g. protein localization, protein motif, etc.) has been shown to improve prediction coverage and accuracy significantly compared to using single source of data [13]. However, the majority of these Bayesian networks methodologies assume conditional independence between genomic features (i.e. PPI data, gene expression data, localization information, and protein motifs) given a class label (i.e. protein function).

In this paper, we use a more sophisticated Bayesian network structure to capture dependencies between genomic features (PPI data and localization information) and class label (protein function) for protein function prediction. Fig. 1 represents the difference between the proposed Bayesian network and Naive Bayes structure. In our context specific Bayesian network model, we condition PPI data by localization information. In other words, we differentiate PPIs into two types: 1) PPIs between co-localized proteins, and 2) PPIs between differently localized proteins. The assumption here is that PPIs between co-localized proteins should be more reliable than PPIs between differently localized proteins (which might be false positives). Hence, in our model, we can assign different weights probabilistically according to the PPI type when predicting protein function given localization information and interacting proteins (and their functions). Our method is applied to protein function prediction in the yeast *Saccharomyces cerevisiae.* In order to assign protein functions, we use the Gene Ontology [14] "biological process" terms as function labels. We show in our 5-fold cross-validation study that our method works significantly better than the Naive Bayes method, when predicting certain functional classes, such as the "ribosome biogenesis" GO term. However, in other cases such as the "mitotic cell cycle" GO term, we find that the simple Naive Bayes method works equally well or even much better than our proposed method. We analyze the results and hypothesize that the more sophisticated model works better when the assumptions made in our model are biologically appropriate for a specific function of interest: For example, when PPI patterns in a subset of co-localized proteins are characterized by a distinct topology or probability of interaction (e.g. "ribosome biogenesis" proteins tend to have PPIs within same localization). Finally, we annotated 57 unknown genes as "ribosome biogenesis" at the estimated precision of 50% (a complete gene list is available in Supplementary Information and available online at http://genomics10.bu.edu/nariai/context/).

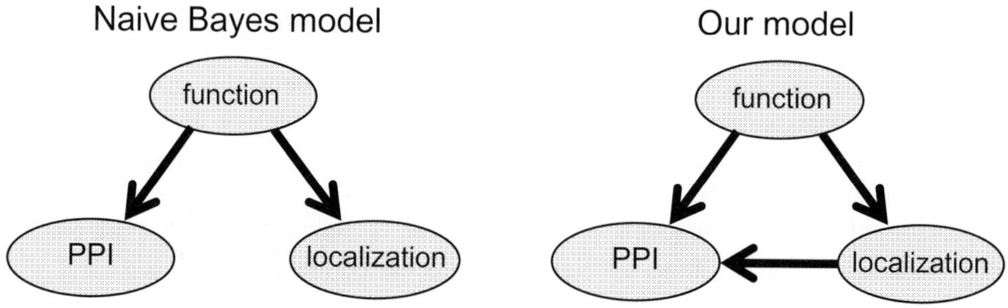

$$P(PPI, l \mid f) = P(PPI \mid f) \cdot P(l \mid f) \qquad P(PPI, l \mid f) = P(PPI \mid l, f) \cdot P(l \mid f)$$

Fig. 1. Context specific protein function prediction: Naive Bayes model (left) and our proposed model (right). In our model, there is a dependency between PPI data and localization information. In the equations, l, and f represent localization and function, respectively.

2. Methods

2.1. *Data preparation*

Physical protein-protein interactions (PPIs) of *Saccharomyces cerevisiae* are collected from the GRID database [15], as of 01/03/2007. After eliminating redundant interactions and self-self interactions, 31202 PPIs among 5151 genes are obtained. Protein localization information is obtained from the MIPS database [16], as of 11/14/2005. In total, 5191 protein-localization associations are obtained, in which 4076 proteins are associated with at least one of 41 cellular localization categories. For each protein, a feature vector $\mathbf{l} = (l_1, l_2, ..., l_L)^{\mathrm{T}}$ is defined, where l_i is a random variable to show localization ($l_i = 1$ if the protein localizes in l_i, and $l_i = 0$ otherwise), and L is the total number of localization features (41 in this case). The Gene Ontology (GO) "biological process" terms are obtained from the Yeast SGD database [17], as of 06/03/2006. For each gene-term association, we expanded and included all 'is-a' and 'part-of' ancestors of the GO label. In total, 107636 gene-term assignments are obtained, in which 6289 genes have at least one of 1965 GO terms. GO terms that appear more than 300 times or less than 5 times among the 6289 genes are subsequently discarded, since we believe that such overly broad or narrow functional terms are not very useful for further experimental validation. From the PPI data collected, we construct a functional linkage graph [6], in which nodes represent genes (proteins) and edges represent PPIs between nodes. From the protein localization information, edges (PPIs) can be divided into two types: 1) PPIs between proteins that share the same localization, and 2) PPIs between proteins that do not share any localization. More precisely, since some proteins do not have localization information at all, there is another type of PPIs: 3) neither 1) nor 2). We call each type of PPIs as 1) co-localized PPIs, 2) cross-localized PPIs, and 3) other PPIs. Generally speaking, it is expected that co-localized PPIs are more reliable than others, since it is usually the case that PPI occurs within the same localization, and other types of PPI might be rare cases or just false positives from high-throughput analyses. For each type of PPI and each GO term t, we calculate p_1, the probability that a protein has term t, given that the interacting partner has the label t, and p_0, the probability that a protein has term t, given that the interacting partner does not have the label t. Here, a χ^2 test is performed to ensure that p_1 and p_0 are statistically different using a Bonferroni-corrected p-value of 0.001. It is expected that p_1 is higher than p_0, which we are going to take advantage of to make function prediction, given a functional linkage graph. For convenience, we use notations $p_1^{(co)}$ and $p_0^{(co)}$ for the co-localized PPI, $p_1^{(\overline{co})}$ and $p_0^{(\overline{co})}$ for the cross-localized PPI, and $p_1^{(others)}$ and $p_0^{(others)}$ for the other PPI.

2.2. *Posterior probability of function given data*

For each protein and GO term, a Boolean random variable $f_{i,t}$ is associated, where $f_{i,t} = 1$ if the protein i is associated with the GO term t, and $f_{i,t} = 0$ otherwise. We calculate a posterior probability for all combinations of proteins and GO terms, given PPI data and localization information as $P(f_{i,t} = 1 \mid N_i, k_i, \mathbf{l}_i)$, where N_i is the total number of neighbors of protein i in the functional linkage graph (PPI network), k_i is the total number of neighbors of protein i which are annotated with t, and $\mathbf{l}_i$ is a feature vector for localization information of protein i. Applying Bayes' theorem (with omitting subscripts),

$$P(f \mid N,k,\mathbf{l}) = \frac{P(k,\mathbf{l} \mid f,N) \cdot P(f \mid N)}{P(k,\mathbf{l} \mid N)}$$

$$= \frac{P(k,\mathbf{l} \mid f,N) \cdot P(f \mid N)}{P(k,\mathbf{l} \mid f,N) \cdot P(f) + P(k,\mathbf{l} \mid \overline{f},N) \cdot P(\overline{f})}$$

$$= \frac{P(k \mid \mathbf{l},f,N) \cdot P(\mathbf{l} \mid f,N) \cdot P(f \mid N)}{P(k \mid \mathbf{l},f,N) \cdot P(\mathbf{l} \mid f,N) \cdot P(f \mid N) + P(k \mid \mathbf{l},\overline{f},N) P(\mathbf{l} \mid \overline{f},N) \cdot P(\overline{f} \mid N)}$$

$$= \frac{P(k \mid \mathbf{l},f,N) \cdot P(\mathbf{l} \mid f) \cdot P(f)}{P(k \mid \mathbf{l},f,N) \cdot P(\mathbf{l} \mid f) \cdot P(f) + P(k \mid \mathbf{l},\overline{f},N) P(\mathbf{l} \mid \overline{f}) \cdot P(\overline{f})},$$

where we assume that f and $\mathbf{l}$ are independent of the number of neighbors N, and hence, $P(f \mid N) = P(f)$, and $P(\mathbf{l} \mid f, N) = P(\mathbf{l} \mid f)$.

In the function above, $P(k \mid \mathbf{l}, f, N) = P(k_{co}, k_{\overline{co}}, k_{others} \mid f, N_{co}, N_{\overline{co}}, N_{others})$, where $N_{co}, N_{\overline{co}}, N_{others}$ are the number of co-localized neighbors, cross-localized neighbors, and others, respectively (please see Section 2.1. for the notations), and $k_{co}, k_{\overline{co}}, k_{others}$ are the those neighbors that are annotated with t. We assume a multinomial distribution and calculate this probability as

$$P(k_{co}, k_{\overline{co}}, k_{others} \mid f, N_{co}, N_{\overline{co}}, N_{others}) = \frac{N!}{\prod_{i=1}^{6} x_i!} \cdot \prod_{i=1}^{6} \theta_i^{x_i},$$

where

$$x_1 = k_{co},\ x_2 = N_{co} - k_{co}, x_3 = k_{\overline{co}},\ x_4 = N_{\overline{co}} - k_{\overline{co}}, x_5 = k_{others}, x_6 = N_{others} - k_{others},$$
$$\theta_1 = p_1^{(co)} \cdot P(PPI_{co} \mid f),\ \theta_2 = (1 - p_1^{(co)}) \cdot P(PPI_{co} \mid f),$$
$$\theta_3 = p_1^{(\overline{co})} \cdot P(PPI_{\overline{co}} \mid f),\ \theta_4 = (1 - p_1^{(\overline{co})}) \cdot P(PPI_{\overline{co}} \mid f),$$
$$\theta_5 = p_1^{(others)} \cdot P(PPI_{others} \mid f),\ \theta_6 = (1 - p_1^{(others)}) \cdot P(PPI_{others} \mid f),$$
$$\left(x_i \geq 0, \sum_{i=1}^{6} x_i = N,\ \sum_{i=1}^{6} \theta_i = 1. \right)$$

where $P(PPI_{co} \mid f), P(PPI_{\overline{co}} \mid f), P(PPI_{others} \mid f)$ are prior probabilities (fractions) of each type of PPI given a term t, which are pre-calculated from a training set.

Similarly,

$$P(k \mid \mathbf{l}, \overline{f}, N) = P(k_{co}, k_{\overline{co}}, k_{others} \mid \overline{f}, N_{co}, N_{\overline{co}}, N_{others}) = \frac{N!}{\prod_{i=1}^{6} x_i!} \cdot \prod_{i=1}^{6} \theta_i'^{x_i},$$

where

$$\theta_1' = p_0^{(co)} \cdot P(PPI_{co} \mid \overline{f}),\ \theta_2' = (1 - p_0^{(co)}) \cdot P(PPI_{co} \mid \overline{f}),$$
$$\theta_3' = p_0^{(\overline{co})} \cdot P(PPI_{\overline{co}} \mid \overline{f}),\ \theta_4' = (1 - p_0^{(\overline{co})}) \cdot P(PPI_{\overline{co}} \mid \overline{f}),$$
$$\theta_5' = p_0^{(others)} \cdot P(PPI_{others} \mid \overline{f}),\ \theta_6' = (1 - p_0^{(others)}) \cdot P(PPI_{others} \mid \overline{f}).$$
$$\left(\sum_{i=1}^{6} \theta_i' = 1. \right)$$

Finally, $P(\mathbf{l} \mid f)$ and $P(\mathbf{l} \mid \overline{f})$ are calculated as

$$P(\mathbf{l} \mid f) = \prod_{i=1}^{L} P(l_i \mid f), P(\mathbf{l} \mid \overline{f}) = \prod_{i=1}^{L} P(l_i \mid \overline{f}),$$

where $P(l_i \mid f)$ = (# of t–labeled genes that have a localization at l_i) / (# of t–labeled genes), $P(l_i \mid \overline{f})$ = (# of genes that are not labeled with t and have a localization at l_i) / (# of genes that are not labeled with t). $P(f)$ and $P(\overline{f})$ are prior probabilities, which are fractions of t-labeled genes, and genes that are not labeled with t in a training set, respectively.

3. Results

We applied our method to *Saccharomyces cerevisiae* protein function prediction (data preparation is described in Section 2.1.) and evaluate its performance through 5-fold cross validation. Since it is expected that the performance varies from one GO term to another, we choose "ribosome biogenesis" and "mitotic cell cycle" GO terms for our targets. In our 5-fold cross validation experiment, 6289 genes are first divided into five equally-sized sets. Following the standard conventions, four gene sets are selected and treated as a training set, and the remaining one is used as a test set. This second step is repeated until all gene sets have been chosen as a test set. Fig. 2 shows the prediction precision, #TP / (#TP+#FP), for varying posterior probability threshold by our method (conditioning), Naive Bayes method, and the method using PPI data alone. Error bars in graphs show the standard deviation of precision from 10 independent 5-fold cross validation experiments. In the case of predicting the "ribosome biogenesis" GO term, our method is significantly better than the Naive Bayes method and using PPI data alone (t-test, significance < 0.01). However, in the case of the "mitotic cell cycle" GO term, the proposed method is significantly worse than other methods (but Naive Bayes method is significantly better than using PPI data alone). Other than these cases, we found that for predicting the "generation of precursor metabolites and energy" GO term, our method works equally well compared to the Naive Bayes method (data not shown). These results show that whether our conditioning method works better than a Naive Bayes method or not depends on which GO term we are predicting. We explain why the prediction performance is so different depending on GO terms. Since our method weights PPI differently according to the localization (co-localized PPI, cross-localized PPI, and others), our method is most effective when positives (proteins that are annotated with the function of interest) have different values of $p_1^{(co)}, p_1^{(\overline{co})}, p_1^{(others)}$, and tend to have consistent frequencies for each type of PPI compared to negatives (proteins that are not annotated with the function). Note that $p_1^{(co)}, p_1^{(\overline{co})}, p_1^{(others)}$ are 0.62, 0.24, 0.41, respectively for "ribosome biogenesis", and 0.25, 0.20, 0.26, respectively for "mitotic cell cycle". We see that these values are quite different from each other for the "ribosome biogenesis" GO term, but not for "mitotic cell cycle". This means that co-localized PPI is more reliable for predicting the "ribosome biogenesis" GO term compared to others, hence is helpful to improve precision. Fig. 3 shows the number of neighbors annotated with the same function (*x*-axis) and the number of co-localized neighbors annotated with the same function (*y*-axis). A diagonal pattern is apparent for "ribosome biogenesis", but not for the "mitotic cell cycle" GO term. Since proteins annotated with "ribosome biogenesis" tend to have more co-localized PPI than other types of PPI compared to negatives, and $p_1^{(co)}$ is much higher than $p_1^{(\overline{co})}$, our method could successfully distinguish positives from negatives better than a Naive Bayes method.

From Fig. 2, we estimated the threshold probability 0.10 as the 50% precision point. We newly annotated 57 unknown genes as "ribosome biogenesis" (a complete gene list is available in Supplementary Information, http://genomics10.bu.edu/nariai/context/).

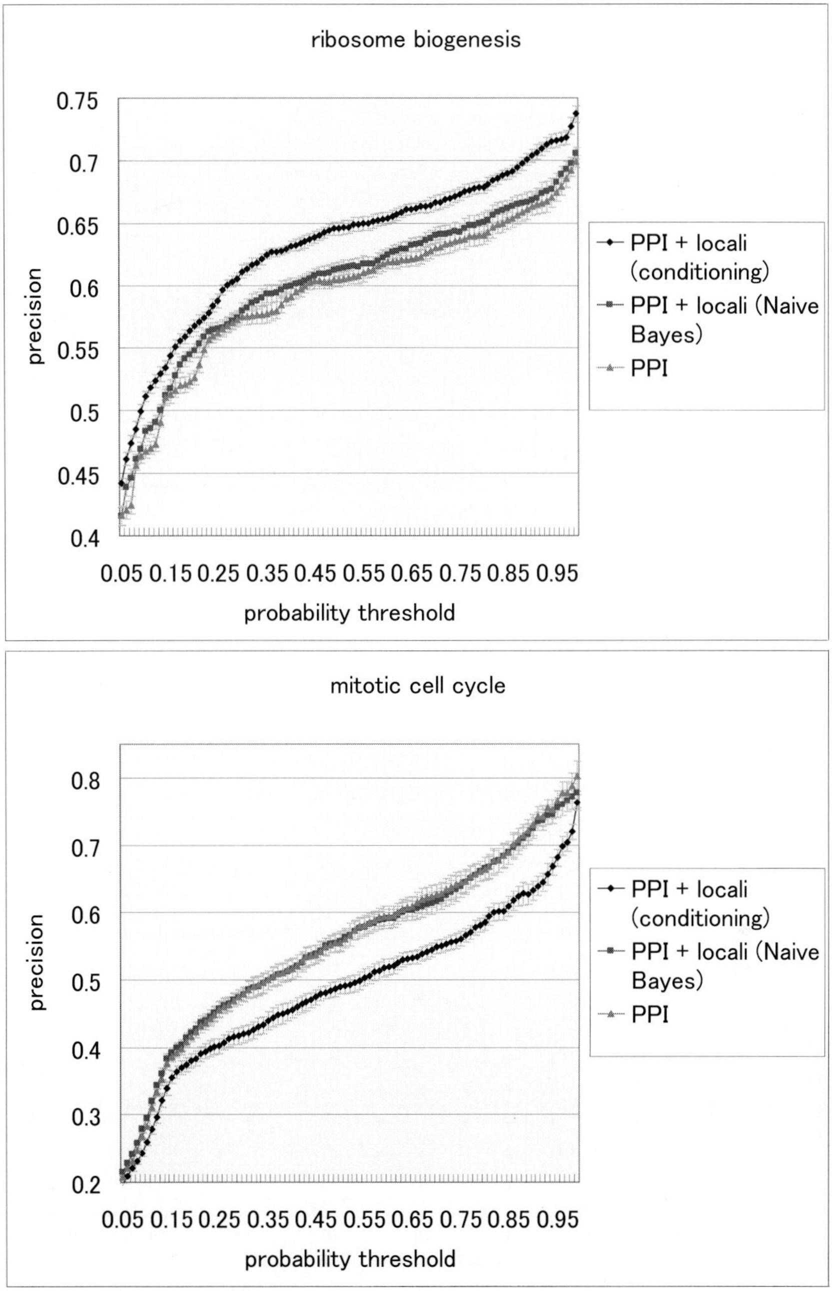

Fig. 2. Prediction performance for GO terms "ribosome biogenesis" (top) and "mitotic cell cycle" (bottom). Precision is defined as (#TP) / (#TP+#FP).

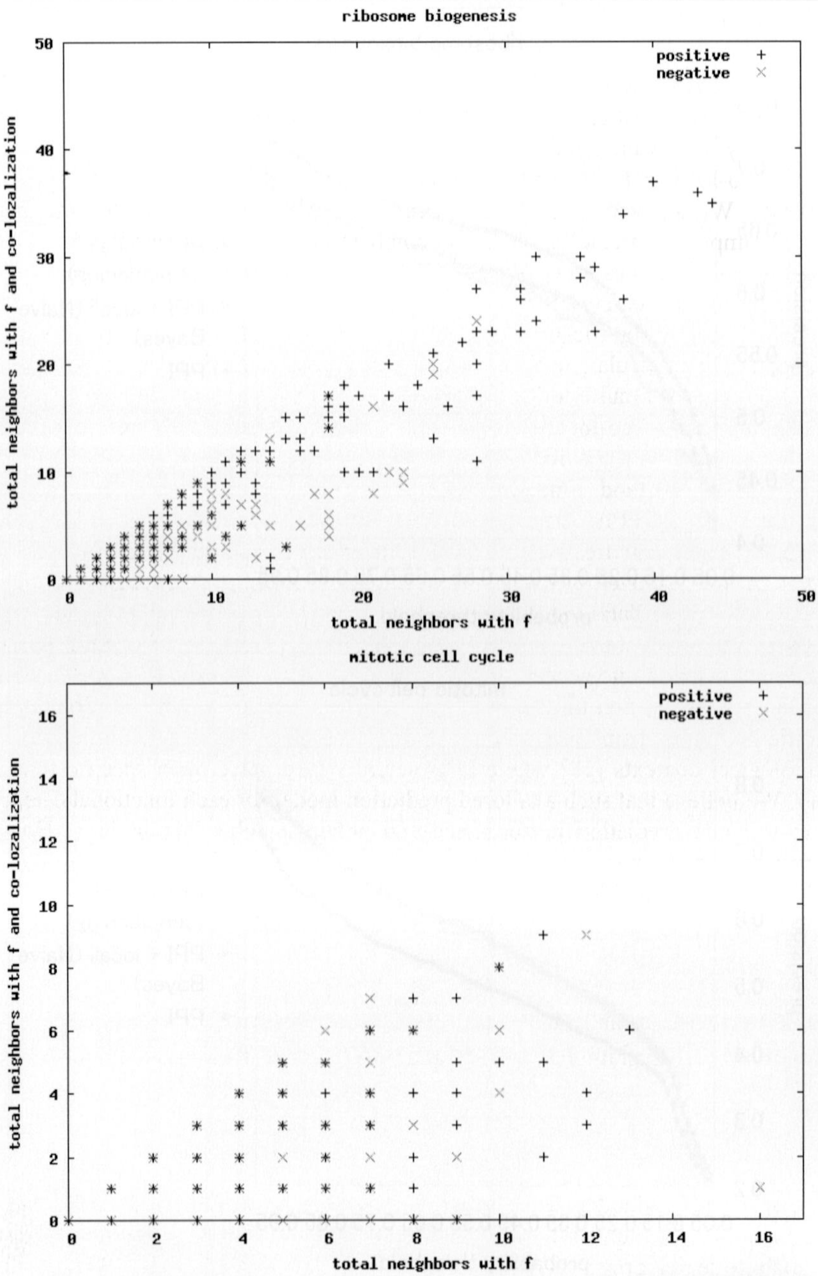

Fig. 3. The number of neighbors with the same function (x-axis) and the number of co-localized neighbors with the same function (y-axis). We can see a diagonal pattern for the GO term "ribosome biogenesis" (top), but not for "mitotic cell cycle" (bottom).

4. Discussion

In this paper, we propose a probabilistic method to predict protein function from PPI data and localization information under a Bayesian network structure, in which PPI data is conditioned by localization information. The assumption here is that treating PPI data differently (co-localized PPI and cross-localized PPI) will lead to better prediction performance. We showed in our 5-fold cross validation experiment that our method successfully improved prediction precision compared to a simple Naive Bayes method in some cases. In other cases, conditioning PPI data by localization did not improve prediction performance. However, even in these cases where the method does not provide a statistically significant improvement, it allows us to obtain deeper insight into gene function. In particular, it allows us to identify proteins that tend to interact in a similar fashion across multiple localizations. We analyzed the results and hypothesize that if the fraction of co-localized PPI and cross-localized PPI is not consistent for proteins annotated with a specific GO term, then a Naive Bayes method may work better than the proposed method. One limitation of our method is that we only distinguish between two types of PPIs: Co-localized PPIs and cross-localized PPIs. Ideally, every type of PPIs should be treated differently according to a specific localization: PPIs between a protein localized in A (such as nucleus) and a protein localized in B (such as ER). When more PPI data and localization information become available, our method can be extended to model additional types of PPIs. It might also be possible to take other contextual information into account, such as time (e.g. using time-series gene expression data during cell cycle) and biochemical context (e.g. interactions mediated or inhibited by specific protein domains or small molecules). We may then be able to determine the set of biological contexts [18] where PPIs actually take place for a specific functional category. We believe that such a tailored prediction model for each functional category is a key to improve prediction performance and obtain insights into biology. However, learning such a comprehensive context model would require a significant amount of data, while the currently available data remains sparse.

Acknowledgments

We thank Manway Liu and Dr. Martin Steffen for constructive comments. This work was supported by NSF grant ITR-048715 and NHGRI grant R01HG003367-01A1.

References

[1] Altschul, S.F., Madden, T.L., Schaffer, A.A., Zhang, J., Zhang, Z., Miller, W., and Lipman, D.J., Gapped BLAST and PSI-BLAST: a new generation of protein database search programs, *Nucleic Acids Res.*, 25(17):3389-3402, 1997.

[2] Finn, R.D., Mistry, J., Schuster-Bockler, B., Griffiths-Jones, S., Hollich, V., Lassmann, T., Moxon, S., Marshall, M., Khanna, A., Durbin, R., Eddy, S.R., Sonnhammer, E.L., and Bateman, A., Pfam: clans, web tools and services, *Nucleic Acids Res.*, 34(Database issue):D247-251, 2006.

[3] Karaoz, U., Murali, T.M., Letovsky, S., Zheng, Y., Ding, C., Cantor, C.R., and Kasif, S., Whole-genome annotation by using evidence integration in functional-linkage networks, *Proc. Natl. Acad. Sci. USA*, 101(9):2888-2893, 2004.

[4] Letovsky, S. and Kasif, S., Predicting protein function from protein/protein interaction data: a probabilistic approach, *Bioinformatics,* 19(Suppl 1):i197-204, 2003.

[5] Jansen, R., Yu, H., Greenbaum, D., Kluger, Y., Krogan, N.J., Chung, S., Emili, A., Snyder, M., Greenblatt, J.F., and Gerstein, M., A Bayesian networks approach for predicting protein-protein interactions from genomic data, *Science*, 302(5644):449-453, 2003.

[6] Yanai, I. and DeLisi, C., The society of genes: networks of functional links between genes from comparative genomics, *Genome Biol.*, 3: research0064, 2002.

[7] Marcotte, E.M., Pellegrini, M., Thompson, M.J., Yeates, T.O. and Eisenberg, D., A combined algorithm for genome-wide prediction of protein function, *Nature,* 402(6757): 83-86, 1999.

[8] Lee, I., Date, S.V., Adai, A.T., and Marcotte, E.M., A probabilistic functional network of yeast genes, *Science*, 306(5701):1555-1558, 2004.

[9] Lu, L.J., Xia, Y., Paccanaro, A., Yu, H. and Gerstein, M., Assessing the limits of genomic data integration for predicting protein networks, *Genome Res.*, 15(7):945-953, 2005.

[10] Troyanskaya, O.G., Dolinski, K., Owen, A.B., Altman, R.B., and Botstein, D., A Bayesian framework for combining heterogeneous data sources for gene function prediction (in Saccharomyces cerevisiae), *Proc. Natl. Acad. Sci. USA*, 100(14):8348-8353, 2003.

[11] Deng, M., Chen, T., and Sun, F., An integrated probabilistic model for functional prediction of proteins, *J. Comput. Biol.*, 11(2-3):463-475, 2004.

[12] Lanckriet, G.R., De Bie, T., Cristianini, N., Jordan, M.I., and Noble, W.S., A statistical framework for genomic data fusion, *Bioinformatics,* 20(16): 2626-2635, 2004.

[13] Nariai, N., Kolaczyk, E.D., and Kasif, S., Probabilistic protein function prediction from heterogeneous genome-wide data, *PLoS ONE*, 2(3):e337, 2007.

[14] Ashburner, M., Ball, C.A., Blake, J.A., Botstein, D., Butler, H., Cherry, J.M., Davis, A.P., Dolinski, K., Dwight, S.S., Eppig, J.T., Harris, M.A., Hill, D.P., Issel-Tarver, L., Kasarskis, A., Lewis, S., Matese, J.C., Richardson, J.E., Ringwald, M., Rubin, G.M., and Sherlock, G., Gene ontology: tool for the unification of biology. The Gene Ontology Consortium, *Nat. Genet.*, 25(1):25-29, 2000.

[15] Breitkreutz, B.J., Stark, C., and Tyers, M., The GRID: the General Repository for Interaction Datasets, *Genome Biol.*, 4(3):R23, 2003.

[16] Mewes, H.W., *et al.*, MIPS: analysis and annotation of proteins from whole genomes, *Nucleic Acids Res.*, 32(Database issue);D41-44, 2004.

[17] Dwight, S.S., *et al.*, Saccharomyces genome database: underlying principles and organisation, *Brief. Bioinform.*, 5(1):9-22, 2004.

[18] Rachlin, J., Cohen, D.D., Cantor, C., and Kasif, S., Biological context networks: a mosaic view of the interactome, *Mol. Syst. Biol.*, 2:66, 2006.

EVALUATION OF SEQUENCE ALIGNMENTS OF DISTANTLY RELATED SEQUENCE PAIRS WITH RESPECT TO STRUCTURAL SIMILARITY

AYSAM GÜRLER
guerler@chemie.fu-berlin.de

ERNST-WALTER KNAPP
knapp@chemie.fu-berlin.de

Freie Universität Berlin, Institut für Chemie und Biochemie, Takustr. 6, 14195, Berlin-Dahlem, Germany

We evaluate the performance of common substitution matrices with respect to structural similarities. For this purpose, we apply an all-versus-all pairwise sequence alignment on the ASTRAL40 [7] dataset, consisting of 7290 entries with a pairwise sequence identity of at most 40%. Afterwards, we compare the 100 highest scoring sequence alignments to their corresponding structural alignments, which we obtain from our structure alignment database. Our database consists of about 18.6 million pairwise entries. We calculated these alignments by applying the current version of GANGSTA [1], our non-sequential structural alignment tool, on about 26 million pairs. The results illustrate the difficulty of homology based protein structure prediction in cases of low sequence similarity. Further, the large fraction of structurally similar proteins in the ASTRAL40 dataset is quantitatively measured. Thereby, this investigation yields a new perspective on the topic of sequence and structure relation. Hence, our finding is a large-scale quality measure for any sequence based method, which aims to detect structural similarities.

Keywords: sequence alignment; protein structure prediction; substitution matrix; database comparison.

1. Introduction

Protein sequence alignment plays a key role in the investigation of protein functionality [4, 12]. The protein sequence determines the structure and through it the protein's function. Similar sequences often share similar structures. However, the opposite is not the case since similar structures can be encoded by dissimilar sequences [11]. Shakhnovich et al. analysed this issue in terms of a "free energy landscape" in sequence space. During evolution of a protein sequence, amino acid residues are deleted, inserted or replaced by others. This process of sequence altering can lead to cross "barriers" and to seed new local minima in sequence space. In some cases the new minima correspond to similar structures, which are conservative with respect to the protein's function. Here, the mutations in sequence do not cause an unsatisfactory structural change at functionally relevant protein sites. Hence, the structural conservation for specific sites is higher than the sequential conservation. These properties of sequence and structure coherence can lead to difficulties in the application of common sequence alignment methods. Current strategies are based on substitution matrices, which are applied for measuring sequence similarities [8, 9]. However, the most common substitution matrices like PAM (point accepted mutations) [2] and BLOSUM (blocks substitution matrix) [3] are based on preliminary sequence alignments of mainly similar protein sequence sections. Therefore, they are biased towards sequentially conserved regions. Despite these difficulties, many protein structure prediction methods apply a preliminary homology search in sequence databases [13]. In general, this process consists of four steps. First, the sequence homologue for a known sequence but unknown structure is searched. Then, both sequences are aligned. Afterwards, the backbone positions of the known structure are transferred to the other, based on the residue pairing on sequence level. Finally, the sidechains are added to the model. Certainly, this is a very effective and promising approach in case of high sequence similarity. Unfortunately, this search for structural properties based on sequence analysis becomes questionable when applied on distantly related sequences.

Sauder et al. performed an analysis with the structural alignment tool CE [9], the sequence alignment tool BLAST [8] and others. The quality of these methods on distantly related sequences is not known, yet [13]. In contrast to the current work, they measured the sequence alignment performance on sequence, instead of structure level. Further, the employed dataset was smaller. Sitbon et al. also applied an integrated analysis on sequence and structure information to determine the conservation of residues with respect to secondary structure elements. They found that helices and turns are underrepresented in conserved regions, in contrast to sheets, which are overrepresented. With respect to loops, they detected similar amounts in conserved and unconserved regions [4]. Further, Domingues et al. set up a benchmark protocol for sequence alignment algorithms with respect to threading. Thereby, they differ between local and global sequence alignment approaches. They claim that the alignments constructed with a combination of sequence alignment, atom pair interactions and protein solvent interactions are the most accurate. They evaluated the alignment quality by comparing the residue pairings between structure and sequence alignment results. Thereby, the local and global alignments performed quite similar. Additionally, they claim that the amount of incorrectly aligned residues with respect to the structural alignments is high for all algorithms [12].

In this paper, we evaluate the performance of common substitution matrices in detecting structural similarities. Therefore, we employ the ASTRAL40 dataset. The set consists of 7290 protein chains, which share less than 40% sequence identity. The sequences and the structures are available online [6]. In a first step, we align the sequences of each ASTRAL40 entry on the complete sequence set with FASTA [7]. Thereby, we retrieve the list of the 100 highest ranked protein pairs for each entry (as SCOP 1.69 codes [6]). Then, we select the corresponding structural scores (SC) of these pairs from our structure alignment database (SD). This procedure is applied in combination with BLOSUM50, BLOSUM62 and PAM120. The resulting structural scores (SC) are plotted in Fig. 6. Additionally, the 100 highest structural scores (SC) for each ASTRAL40 entry are selected from our structure alignment database and plotted as reference, respectively as upper performance limit. Since, our structure alignment method is able to detect non-sequential similarities between two protein structures we additionally plotted the sequential structure alignments separately.

2. Methods

2.1. *Sequence alignment*

Currently, the most popular sequence alignment tools are FASTA [7] and BLAST [8]. Both employ a set of substitution matrices to score the sequence alignment results. The most commonly used matrices are PAM and BLOSUM. Both matrix types are calculated on the basis of prior gapless sequence alignments. Initially the observed substitution frequencies q_{ij} are obtained by counting all of the aligned amino acid pairs *ij*. Further, the occurrence frequency p_i of each amino acid *i* is calculated. Finally, the log-odds ratio of the substitution frequencies against the background distribution of the amino acids is evaluated for each pair. The score s_{ij} is then written as

$$s_{ij} = (\ln \frac{q_{ij}}{p_i p_j}) / \lambda \tag{1}$$

with lambda [5] the scaling parameter. This procedure yields a symmetrical 20x20 substitution matrix. Sequence alignments are scored as summation of the s_{ij} values, corresponding to the aligned amino acid pairs *ij*. Since the scores employ a logarithmic scale, this is equivalent to the multiplication of amino acid occurrence probabilities against the background distribution under the independence assumption [5].

2.2. *Structure alignment scoring*

The basis of the structure alignment evaluation is the structure alignment score (SAS), which has been proposed by Kolodny et al. [10]. This score weights the RMSD of the C-alpha atoms by the number of structurally aligned residues $N_{aligned}$ (see equation 2).

$$SAS = \frac{RMSD * 100}{N_{aligned}} \tag{2}$$

Linear scaling yields the structural score (SC), which we define in this investigation to evaluate the structural similarity between two proteins. The structural score is defined as

$$\mathrm{SC} = 100 - 2 * \mathrm{SAS} \tag{3}$$

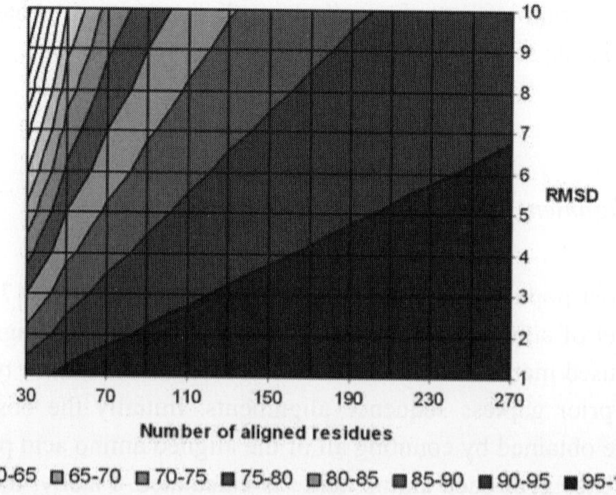

Fig. 1. The plot shows the range of structural scores (SC) as function of RMSD and the number of aligned residues (structural scoring scheme).

Fig. 1 shows the range of the structural score versus the RMSD and the amount of aligned residues.

2.3. *Structure alignment database*

Setting up the structure alignment database (SD) involved the evaluation of all ASTRAL40 (7290 entries) pairs, which leads to about 26 million structural alignments. These have been calculated with our non-sequential structure alignment method based on maximizing the GANGSTA score [1]. In contrast to sequence alignment methods the structural alignment does not incorporate amino acid identities, but crystallographic protein details. Our method is designed to ignore the sequential order of secondary structure elements in protein chains. Additionally, the method ensures that alignments are always topologically correct, such that only secondary structure elements of the same type are aligned on each other. Thereby, we attempt to capture the biologically relevant similarities between two proteins more accurately.

After evaluation, we kept the highest scoring alignment of each pair with a structural score (SC) above 30 and at least 50% of the secondary structure elements in the smaller of both proteins aligned. This amounts to about 18.6 million protein pairs. From them, about 450.000 pairs have a structural score above 90 (SC). Thus on average, each ASTRAL40 entry shares very high structural similarities with about 60 other proteins. About 7.15 million pairs score above 80 (SC), which indicates significant structural similarities between each ASTRAL40 entry and 980 other proteins in average (about 13% of the ASTRAL40 set). Fig. 2 shows the distribution of structural scores for the structure alignment database. About 7% (in total about 1.2 million) of all alignments are sequential, such that the secondary structure elements are aligned in sequence direction.

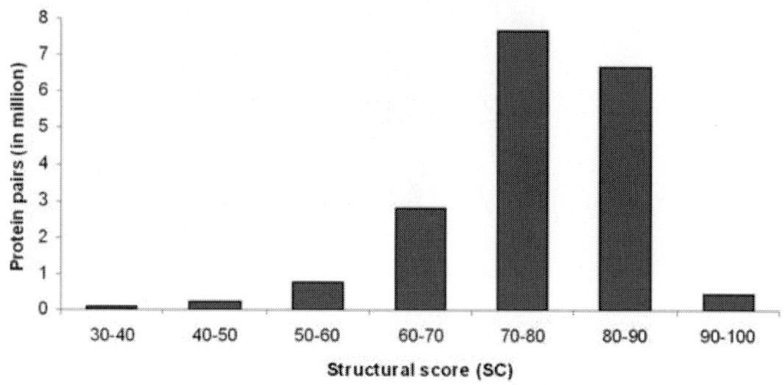

Fig. 2. Structural score (SC) histogram in our structure alignment database (SD).

The highest scoring pairs are **1mdah** (= SCOP code) with **2bbkh_** (SC = 0.99, $RMSD$ = 0.50 Å, $N_{aligned}$ = 337) in the sequential and **1fw8a_** with **1v6sa_** in the non-sequential entries (SC = 0.99, $RMSD$ = 1.27 Å, $N_{aligned}$= 323) (see Fig. 3). The highest amount of residues has been aligned in sequence direction between **1ogya2** and **2napa2** (SC = 0.99, $RMSD$ = 1.07 Å, $N_{aligned}$ = 512). Fig. 4 illustrates a case of non-sequential alignment by **1erja_** and **1m1xa4** (SC = 0.98, $RMSD$ = 1.59 Å, $N_{aligned}$= 180). About half of the secondary structure elements are aligned non-sequentially.

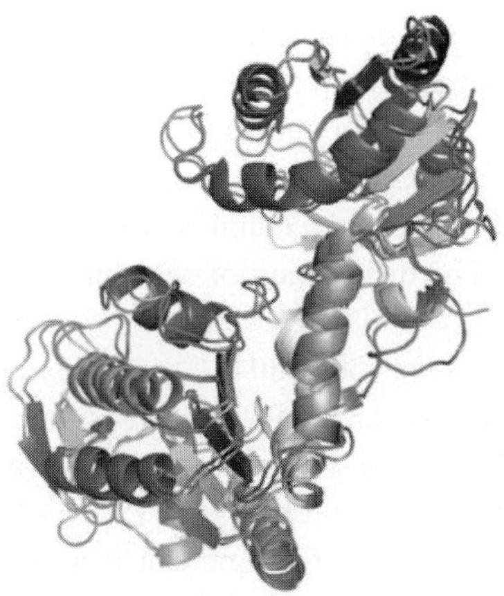

Fig. 3. Non-sequential alignment between 323 residues from **1fw8a_** and **1v6sa_** with a RMSD of 1.27Å. With respect to sequence direction, the initial three secondary structure elements (SSE) of 1v6sa_ are aligned on the last three elements of 1fw8a_. Secondary structure elements in dark, loops in light grey.

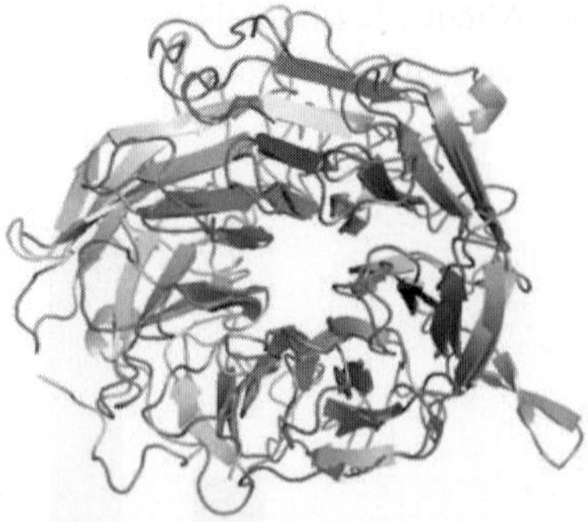

Fig. 4. Non-sequential structure alignment between **1erja_** and **1m1xa4** with 180 residues at 1.59Å. About half of the secondary structure elements are not aligned in sequence direction. Secondary structure elements in dark, loops in light grey.

3. Results

Initially, all-versus-all sequence alignments are performed on the ASTRAL40 dataset with FASTA. The highest ranking 100 sequences are kept for each entry. This yields 7290 sets of 100 sequentially high scoring entry pairs (=< 729000). Then, we select the structural scores (SC) for each of these pairs from our structure alignment database. Fig. 5 illustrates this data acquisition process and Fig. 6 shows the distribution of the corresponding structural scores plotted for FASTA with BLOSUM50. This evaluation has also been done with BLOSUM62 and PAM120. Since this gave almost identical results, only the BLOSUM50 plot is shown. Additionally, we plotted the 100 highest structural scores available for each entry from our database as reference. The reference plot is an upper performance limit for the sequence alignment. Since, the secondary structure elements can be disordered in terms of sequence direction (non-sequential alignments), we plotted the highest structural scores of the in-sequential structural alignment entries separately. The distribution of sequential entry scores has its mode at 85 (SC).

Most of the reference scores are above 80 (SC) and the mode (about 17%) is at about 92 (SC). As mentioned in section 2.3., this indicates significant structural similarities among the ASTRAL40 entries. The sequence alignment with FASTA was able to determine the structurally most similar protein pairs (SC >= 98). Furthermore, in most of these cases the corresponding structure alignment is arranged in sequence direction, more precisely these are sequential structure alignments (see dashed line in Fig. 6). However, only a small fraction of protein pairs scores in the range between 93 and 98 (SC). The mode (~ 4%) of accepted scores is at about 81 (SC). Unfortunately, for about

25% of the high ranking protein sequence pairs only very little structural similarity (SC < 30) could be detected by our structure alignment method.

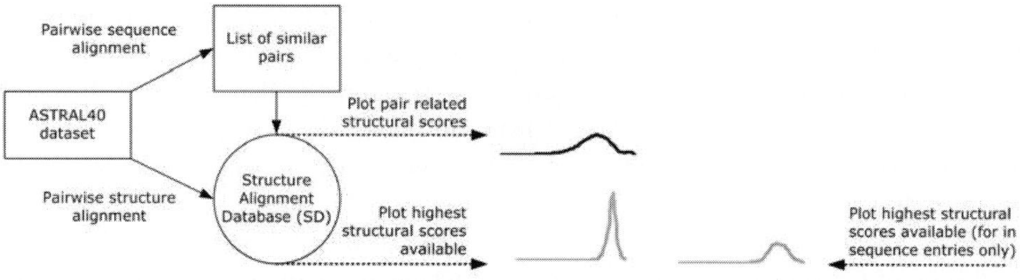

Fig. 5. This figure illustrates the data acquisition process by usage of sequence (dark) and structure (light) alignments. As result the structural score distributions, according to the structural alignment database (SD), are plotted. Additionally, the sequential structure alignment entries are plotted separately.

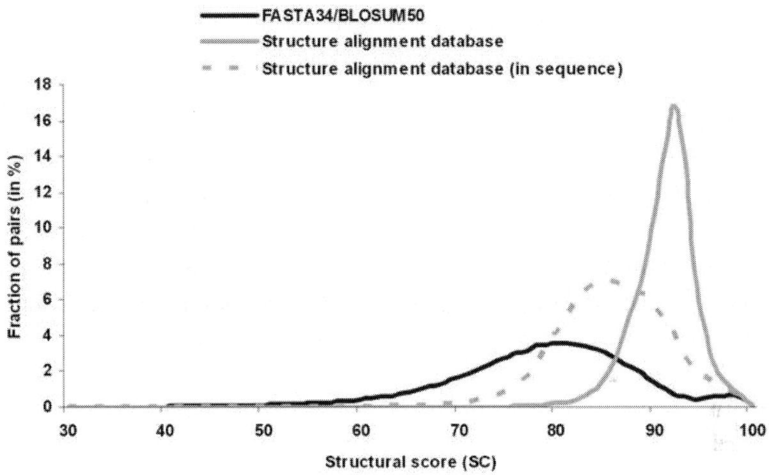

Fig. 6. Structural score distribution for similar protein pairs with respect to sequence (dark) and structure (light). The dashed line is related to the sequential structure alignments, in which the secondary structure elements of two proteins are aligned in sequence direction.

4. Discussion

The application of sequence alignment methods in protein science aims to reproduce structural similarities. Therefore, structure alignment methods, incorporating crystallographic details, are applied as a "gold standard" with respect to protein sequence alignment methods [14]. Since in many cases no crystal structure is known, sequence alignment is a promising and essential approach for the first step in protein structure prediction.

However, the results illustrate the difficulties of sequence alignment approaches in cases of low sequence similarity to already known protein structures. The sequence alignment method is able to reproduce the structurally most similar protein pairs, but in 25% of all high ranking FASTA results only very little structural similarity could be detected. This is related to the simplification of the model, since the sequence alignment method only incorporates the primary structure. Additionally, the sequence alignment method employs substitution matrices, which are biased towards conserved sequence segments. The structural alignment does not incorporate amino acid identities and the ASTRAL40 consists of distantly related sequences only. However, we applied the sequence alignment method only to produce pair lists of "similar" proteins. The evaluation of the similarities proceeded without taking any further information from the sequence alignment into account (e.g. score, residue assignment). Unfortunately, the recognition performance of structural similarities is low.

The fraction of sequential with respect to the non-sequential entries is at only about 7% (see details in 2.3.). Therefore, further investigations must be done to accurately measure the advantage of non-sequential versus sequential structure alignments. However, the results indicate a qualitative and quantitative gain through the non-sequential structure alignment approach. A reason for this can be the biochemical process of splicing. Furthermore, other genetic operations can reorder sequence segments [15]. Hence, our database incorporates relations between proteins and protein families, which are less constrained by these processes. Evaluating these relations can be useful to detect alternative structures and thereby support and improve protein structure prediction methods. Further, the database can be applied as reference for other sequence based approaches.

Acknowledgments

We like to thank Jorge Numata, Stephan Lorenzen and Jonas Maaskola for their comments and support. Furthermore, this study has been supported by the International Research Training Group (IRTG) on "Genomics and Systems Biology of Molecular Networks" (GRK1360, Deutsche Forschungsgemeinschaft (DFG)).

References

[1] Kolbeck, B., May, P., Schmidt-Goenner, T., Steinke, T., and Knapp, E.W., Connectivity independent protein-structure alignment: a hierarchical approach, *BMC Bioinformatics*, 7:510, 2006.

[2] Dayhoff, M.O., Schwartz, R.M., and Orcutt, B.C., A model of evolutionary change in proteins, *Atlas of Protein Sequence and Structure*, 5(3):345–352, 1978.

[3] Henikoff, S. and Henikoff, J.G., Amino acid substitution matrices from protein blocks, *Proc. Natl. Acad. Sci. USA*, 89(22):10915-10919, 1992.
[4] Sitbon, E. and Pietrokovski, S., Occurrence of protein structure elements in coserved sequence regions, *BMC Struct. Biol.*, 7:3, 2007.
[5] Altschul, S.F., Amino acid substitution matrices from an informaition theoretic perspective, *J. Mol. Biol.*, 219(3):555-565, 1991.
[6] Chandonia, J.M., Hon, G., Walker, N.S., Lo Conte, L., Koehl, P., Levitt, M., and Brenner, S.E., The ASTRAL compendium in 2004, *Nucleic Acids Res.*, 32(Database issue):D189-192, 2004.
[7] Pearson, W.R., Rapid and Sensitive Sequence Comparison with FASTP and FASTA, *Methods Enzymol.*, 183:63–98, 1990.
[8] Altschul, S.F., Madden, T.L., Schaffer, A.A., Zhang, J., Zhang, Z., Miller, W., and Lipman, D.J., Gapped BLAST and PSI-BLAST: a new generation of protein database search programs, *Nucleic Acids Res.*, 25(17):3389-3402, 1997.
[9] Shindyalov, I. N. and Bourne, P. E., A database and tools for 3-D protein structure comparison and alignment using the Combinatorial Extension (CE) algorithm, *Nucleic Acids Res.*, 29(1):228-229, 2001.
[10] Kolodny, R., Koehl, P., and Levitt, M., Comprehensive Evaluation of Protein Structure Alignment Methods: Scoring by Geometric Measures, *J. Mol. Biol.*, 346(4):1173-1188, 2005.
[11] Shakhnovich, E., Protein Folding Thermodynamics and Dynamics: Where Physics, Chemistry and Biology Meet, *Chem. Rev.*, 106(5):1559-1588, 2006.
[12] Domingues, F. S., Lackner, P., and Sippl, M. J., Structure Based Evaluation of Sequence Comparison and Fold Recognition Alignment Accuracy, *J. Mol. Biol.*, 297(4):1003-1013, 2000.
[13] Sauder, J.M., Arthur, J.W., and Dunbrack, R.L., Large-scale comparision of protein sequence alignment algorithms with structure alignments, *Proteins*, 40(1):6-22, 2000.
[14] Briffeuil, P., Baudoux, G., Lambert, C., De Bolle, X., Vinals, C., Feytmans, E., and Depiereux, E., Comparative analysis of seven multiple protein sequence alignment servers: clues to enhance reliability of predictions, *Bioinformatics*, 14(4):357-366, 1998.
[15] Cooper, D.N., Ball, E.V., and Krawczak, M., The human gene mutation database, *Nucleic Acids Res.*, 26(1):285-287, 1998.

CONFORMATIONAL ENTROPY OF BIOMOLECULES: BEYOND THE QUASI-HARMONIC APPROXIMATION

JORGE NUMATA[1]
numata@chemie.fu-berlin.de

MICHAEL WAN[1, 2]
mwan@fas.harvard.edu

ERNST-WALTER KNAPP[1]
knapp@chemie.fu-berlin.de

[1] *Macromolecular Modeling Group, Dept. of Chemistry and Biochemistry, Freie Universitaet Berlin, Takustr. 6, Berlin 14195 Germany*
[2] *Aspuru-Guzik Research Group, Harvard University, Dept. of Chemistry and Chemical Biology, 12 Oxford Street, Cambridge, MA 02138, USA*

A method is presented to calculate thermodynamic conformational entropy of a biomolecule from molecular dynamics simulation. Principal component analysis (the quasi-harmonic approximation) provides the first decomposition of the correlations in particle motion. Entropy is calculated analytically as a sum of independent quantum harmonic oscillators. The largest classical eigenvalues tend to be more anharmonic and show statistical dependence beyond correlation. Their entropy is corrected using a numerical method from information theory: the k-nearest neighbor algorithm. The method calculates a tighter upper limit to entropy than the quasi-harmonic approximation and is likewise applicable to large solutes, such as peptides and proteins. Together with an estimate of solute enthalpy and solvent free energy from methods such as MMPB/SA, it can be used to calculate the free energy of protein folding as well as receptor-ligand binding constants.

Keywords: conformational entropy; vibrational entropy; principal component analysis; quasi-harmonic approximation; anharmonicity; k-nearest neighbor entropy.

1. Introduction: thermodynamics of biological macromolecules

1.1. *Entropy of protein folding and ligand binding*

Protein folding and receptor-ligand binding occur in a spontaneous and specific way because the folded and bound states have a lower free energy than their unfolded and unbound counterparts, respectively. The Helmholtz free energy change ΔF or the Gibbs free energy change $\Delta G=\Delta F+P\Delta V$ predict the equilibrium constant (K_{eq}) for folding and binding. For incompressible fluids like water, the volume term $P\Delta V$ is negligible:

$$\Delta G \approx \Delta F = \Delta U - T\Delta S = -k_B T \ln K_{eq} \tag{1}$$

The net enthalpic (ΔH) and entropic ($T\Delta S$) contributions from all particles (solute and solvent) almost cancel out in natural or properly engineered proteins [1]. Stability against unfolding is typically around ΔG= 5 to 15 kcal/mol (K_{eq}= 10^{-4} to 10^{-11}). Upon folding, the solute becomes more rigid and loses conformational entropy. This unfavorable contribution is typically $T\Delta S_{\text{conformational}}$= 10 to 100 kcal/mol. Any estimation of free energy lacking this contribution will grossly overestimate the stability against unfolding.

A physically realistic combination of models to estimate the contributions to free energy can be found in the MM/PBSA method [2,3]:

$$\Delta G = \Delta U^{solute} + \Delta G^{solvent}_{electrostatic} + \Delta G^{solvent}_{hydrophobic} + \Delta U^{solu-solv}_{vdW} - T\Delta S^{solute}_{conformational} \tag{2}$$

The solute internal energy is evaluated by the molecular mechanics (MM) force field. Electrostatic solvation free energy can be obtained from the Poisson-Boltzmann (PB) equation, or its approximations. Hydrophobic free energy may be estimated as proportional to the solvent accessible surface (SA). An often neglected, but important term is the change in solute-solvent van der Waals interactions. Finally, the **solute conformational entropy** can be estimated with the method presented here.

Kollman, Case et al. say in a seminal article on MMPB/SA that "it would clearly be of interest to have better ways to estimate solute entropy, but there can be difficulties when the dynamics at room-temperature jump between basins." [3] The present method can deal with the multimodal probability distributions resulting from those jumps.

1.2. *Entropy quantifies conformational freedom*

Entropy is both a measure of disorder and of correlation. More formally, for a set of particles, entropy is a measure of the phase-space accessibility. Maximum entropy and accessibility in phase space for the non-interacting particles of an ideal gas are reached in a state of complete and random occupation of the container volume. For interacting particles, like the atoms in an organic molecule, entropy is still a measure of disorder (spread of the coordinates). But now the interactions between the atoms have to be included in the calculation by estimating the correlation between the atomic displacements. If we just add the entropy due to the conformational freedom of individual atoms, severe overestimation will occur. For macromolecules, correlation and dependence manifest themselves as concerted, delocalized motions spanning many atoms. [4] This phenomenon is incorporated in the form of Shannon information redundancy to provide a better estimate of entropy.

1.3. *Thermodynamics and information theory*

The information entropy of Claude Shannon [5] and the thermodynamic entropy of Rudolf Clausius [6] have the same functional form. This similarity alone makes it plausible that these quantities are the same. The formula for entropy (Eq. 3) contains a constant k and the natural or Naperian logarithm of the probability of the microstates of a molecular system. When dealing with statistics of coin tosses or genes, it is simpler to choose k=1 and other bases for the logarithm. As a fossil of the parallel historical development, the symbol for entropy is different in the Shannon context of communications theory (*H*) and thermodynamics (*S*). We will use *S*:

$$S = -k_B \sum_{i=1}^{t} p_i \ln p_i \tag{3}$$

The temptation is great to equate thermodynamic and information theoretical entropy. This temptation has been resisted in the present work, and each kind of entropy (quantum, classical and statistical) treated in its own mathematical framework.

1.4. *Novelty of the method*

The established protocol for estimating conformational entropy from molecular dynamics trajectories, the quasi-harmonic approximation [7, 8], provides a strict upper limit but is known to overestimate it considerably [9]. The method presented here takes the quasi-harmonic entropy and corrects it using a numerical method from information theory, the k-nearest neighbor entropy algorithm [10, 11]. The result is a method to estimate the entropy of solute molecules which accounts for anharmonicity (beyond Gaussians) and supralinear motion correlation (beyond covariance). This is achieved:

- Without omission of the mass-metric tensor [12, 13] to avoid deforming phase space. This is in contrast to other applications of numerical methods to calculate molecular entropy using dihedral angles, which will not yield the proper thermodynamic entropy without consideration of the Jacobian of the transformation from Cartesian to internal coordinates.
- Realizing that classical force-fields are fitted on quantum mechanical data. Quasi-harmonic frequencies higher than $k_B T / \hbar$ in the simulation correspond to quantum mechanical behavior, whose entropy should be calculated using all accessible states of a harmonic oscillator, and not just the ground state. Quantum entropy has the additional advantage of yielding an absolute value.
- By making use of an unbiased numerical method (the k-nearest neighbor entropy), which can give more precise results than simple binning of the trajectory.

2. Method for calculating conformational entropy of biomolecules

2.1. *Calculate principal components from covariance of a molecular dynamics trajectory*

Our starting point is a reasonably long and correctly set up molecular dynamics (MD) trajectory in the NVT ensemble. All particles except the solute atoms are deleted. Center-of-mass rotation and translation should be removed to calculate only the conformational entropy. If relevant, the rotational and translational components of entropy may be added later as ideal-gas S_{rot} and S_{trans} (See Chap 10 Ref. [14], Chap. 5 Ref. [15]).

The processed MD coordinate trajectory matrix **X** has dimensions ($3N_a$,n_f) where N_a is the number of atoms and n_f the number of simulation frames selected for analysis.

$$X = (x_1,\ x_2, x_3,\ ..\ ,\ x_{nf}) \tag{4}$$

We wish to apply principal component analysis (PCA) on this trajectory. It is desirable to calculate the principal components of variance directly in Cartesian space, without the matrix being singular [8]. Correct application of the mass-metric tensor is essential to diagonalization and uncoupling of the Hamiltonian in Cartesian coordinates. We thus apply mass weighting to each frame:

$$y_i = M^{1/2} x_i \qquad \text{for i=1,..,n}_f \tag{5}$$

Where **M** is a $(3N_a,3N_a)$ matrix with the atomic masses repeated three times in the diagonal elements m_{ii} and $m_{ij}=0$ for $i \neq j$. We thus obtain the $(3N_a,n_f)$ mass weighted trajectory matrix **Y**. To apply PCA on **Y**, we calculate its covariance matrix. The individual elements of the $(3N_a,3N_a)$ covariance matrix $\boldsymbol{\sigma}^2_{m.w.}$ are:

$$(\sigma^2_{m.w.})_{i,j} = \left\langle (y_i - \langle y_i \rangle)(y_j - \langle y_j \rangle) \right\rangle \tag{6}$$

with <...> denoting average across the considered simulation frames or trajectory. $\langle y_i \rangle$ is the precalculated average coordinate. Now $\boldsymbol{\sigma}^2_{m.w.}$ is diagonalized by an orthogonal transformation. After diagonalization, we obtain a new PCA coordinate system matrix with $3N_a$-6 eigenvalues or variances:

$$\lambda_{ii} = (m_{eff}\sigma^2_{PCA})_i \tag{7}$$

m_{eff} means that the PCA masses are combinations of the original ones. Of the total of $3N_a$ eigenvalues, 6 will have very high frequencies and may be discarded if translation and rotation were properly removed.

Also obtained are eigenvectors describing the PCA modes. Each PCA mode is a linear combination of the original coordinates. These combinations come from the weightings in the eigenvector matrix **W** of size (3Na,3Na-6).

PCA assumes that the particles have a Gaussian (normal) distribution with variance λ_{ii} in each mode i. PCA on the mass weighted MD-trajectory is also the called **quasi-harmonic approximation** because it implies fitting effective harmonic potentials on the observed coordinate covariance, smoothing over any anharmonicity. We may thus regard the method as producing the eigenvalues corresponding to a series of uncorrelated simple harmonic oscillators. We will later make use of this to calculate thermodynamic variables. Within the harmonic oscillator model we may connect the eigenvalues $\lambda_{ii}=m_{eff}\sigma^2_{PCA}$ to frequency ω through the **equipartition theorem.** Kinetic and potential energy are equal in the time average [15]:

$$\omega_i^2 \left(m_{eff}\sigma^2_{PCA}\right)_i = k_B T \tag{8}$$

PCA frequencies:

$$\omega_i = \sqrt{\frac{k_B T}{m_{eff}\sigma^2_{PCA}}} = \sqrt{\frac{k_B T}{\lambda_{ii}}} \tag{9}$$

The equipartition theorem is only valid in the classical limit $\omega << k_B T/\hbar$. Fortunately, high frequency quantum vibrations, where this approximation breaks down contribute less to molecular entropy. They also tend to be very close to Gaussian, thus not requiring corrections for anharmonicity and supralinear dependence beyond linear correlation.

We may now sort the $3N_a$-6 PCA frequencies. Large eigenvalues correspond to low frequency correlated motion. The lowest frequency modes are the most interesting collective motions, deemed important for enzymatic catalysis. [4] They are also the highest contributors to molecular conformational entropy. Still, it is not advisable to throw away higher frequency modes, as they may together make a sizeable contribution to entropy.

Using all the eigenvectors **W**, we may project the mass-weighted coordinates into the PCA collective coordinates **Z**:

All PCA modes: $$Z_{all} = W^T Y \tag{10}$$

Alternatively, we may select a subset of d eigenvectors, thus using a reduced eigenvector matrix of size (d,n_f):

Subset of d PCA modes: $$Z_d = W_d^T Y \tag{11}$$

2.2. *Absolute entropy of a harmonic oscillator as an upper limit*

Our objective is to provide an upper limit estimate of entropy. Assuming a Gaussian distribution is the safest bet because it has the highest entropy among all statistical distributions with the same given variance [16]. A statistical mechanical model that produces Gaussian-distributed coordinate displacements is the harmonic oscillator. For the quantum mechanical harmonic oscillator (including zero-point energy), the partition function is:

$$Q = \sum_{j=0}^{\infty} e^{-\alpha(j+1/2)} = e^{-\alpha/2} \sum_{j=0}^{\infty} \left(e^{-\alpha}\right)^j \quad \text{with} \quad \alpha = \beta\hbar\omega = \hbar\omega/(k_B T) \tag{12}$$

From it, we may directly calculate thermodynamic functions [14, 15] in the NVT ensemble, such as the Helmholtz free energy F and the internal energy U. Our main interest, however, is the quasi-harmonic entropy:

$$S_{Quantum,i} = \frac{F-U}{T} = k_B \frac{\alpha_i}{e^{\alpha_i}-1} - k_B \ln\left(1-e^{-\alpha_i}\right) \tag{13}$$

We now wish to use this result to calculate the entropy of all PCA modes. For many coordinates, we make a multivariate generalization of the formula. In other words, we create an approximation to molecular conformational entropy as a series of uncorrelated simple harmonic oscillators (SHO). Each SHO has frequency ω_i estimated from the classical mass weighted variance, and gives a total entropy [7]:

$$S_{Quantum} = k_B \sum_{i=1}^{3N-6} \left[\frac{\alpha_i}{e^{\alpha_i} - 1} - \ln(1 - e^{-\alpha_i}) \right] \quad (14)$$

with $\alpha_i = \frac{\hbar \omega_i}{k_B T} = \frac{\hbar}{\sqrt{k_B T}} \frac{1}{\sqrt{\lambda_{ii}}}$ in terms of the m.w. PCA eigenvalues λ_{ii}

This absolute entropy excludes the kinetic contribution (which is independent of conformation and may be added analytically from the equilibrated momenta [12]) and assumes distinguishable particles (Boltzmann statistics, as usual for systems of covalently bound atoms). Furthermore, zero-point energy does not add to the entropy because its contribution $\hbar\omega/2$ in F and U mathematically cancels out in Eq. 13.

The equipartition theorem (Eq. 8) is only strictly valid in the classical limit $\hbar\omega \ll k_B T$. The limiting frequency corresponds to $k_B T/\hbar$ = 4.06E13 Hz or f<< 216 cm^{-1} at T=310K.

Thermodynamic entropy of a classical harmonic oscillator

Using the classical partition function Q=1/α, we may calculate the classical thermodynamic entropy of a harmonic oscillator. This will become important to compute a correction to the entropy values within the classical region:

$$S_{Classical,thermo,i} = -k_B \left[\ln(\alpha_i) + 1 \right] \quad (15)$$

Classical entropy is a lower limit for quantum entropy at all points. At values $\alpha<0.6$, quantum and classical entropy agree better than 99%. At $\alpha=1$, the classical entropy still agrees to 96% with the quantum one. At $\alpha>1$ it incorrectly diverges towards $-\infty$ (See Fig. 1). Classical entropy calculations should be avoided in the quantum region, which is nonetheless relevant for molecular motion.

Statistical entropy of a Gaussian distribution

The classical statistical entropy for each eigenvalue can be calculated from the definition of differential entropy and the Gaussian probability density function. With the eigenvalues (variances) in terms of our dimensionless α:

$$S_{Class,Gaussian,i} = -k_B \left[\ln(\alpha_i) + \ln\left(\sqrt{2\pi e}\right) \right] \quad (16)$$

Comparing Eq. 16 to Eq. 15, one realizes that the additive constant is different for thermodynamic and statistical entropy. The important point is that Eq. 16 should be used to calculate the correction, because the k-nearest neighbors statistical method used to calculate non-Gaussian entropy in Sec. 2.4 produces comparable values.

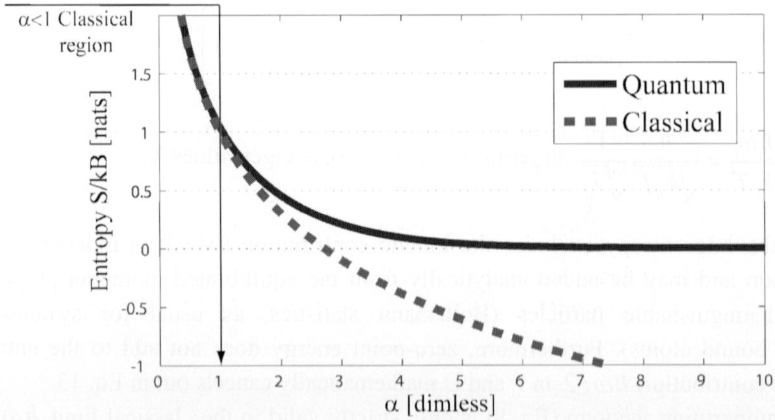

Fig. 1: Plot of thermodynamic entropy as a function of $\alpha = \hbar\omega/(k_BT)$, where ω is the harmonic oscillator frequency. Shown are the quantum (Eq. 14) and classical (Eq. 15) results. For large frequencies, quantum entropy displays the expected limiting behavior, such that $S_{Quantum} \rightarrow 0$. Classical entropy diverges $S_{Classical} \rightarrow -\infty$ as ω grows. The classical approx. breaks down in the quantum region because the spacings between energy levels become too small, and higher energy levels become significantly occupied.

2.3. *Pairwise non-Gaussian corrections for anharmonicity and supralinear dependence*

We now wish to provide a tighter upper limit to entropy than the quasi-harmonic approximation can afford. PCA decomposes the linear (Gaussian) correlation. If histograms are made of the principal components **Z**, most will be actual Gaussians. PCA modes with the largest classical eigenvalues are major contributors to entropy. They also tend to deviate from the Gaussian distribution.

An important property of Gaussians is that they have the largest entropy possible for a given variance [17]. This means that each principal component that displays **anharmonicity** will overestimate the entropy [9]. Furthermore, deviations from the Gaussian distribution also imply that higher order dependence in the trajectory data beyond linear correlation was missed. This supralinear dependence is quantified by **mutual information (M.I.)**. Missing higher order correlations causes a more severe overestimation of entropy than anharmonicity [12]. For instance, a kind of dependence that is systematically missed by PCA is that of 90° phase shifted atoms moving in parallel lines [18]. This is captured by the correction for the classical region.

We thus take the calculated absolute quantum entropy and correct it with numerical methods from information theory. This correction is strictly non-positive, because: a) a Gaussian has the maximum entropy for a given variance and b) two or more orthogonal Gaussians (after PCA) have zero Mutual Information. For a total of c modes in the classical regime, a **correction scheme for pairs of modes** (i,j) is proposed (see Fig. 2).

$$\Delta S_{Class,pairwise,correction} = -\left[\sum_{i=1}^{c} S_{anh,(i)} + \sum_{i=1}^{c} \sum_{j=i+1}^{c} M.I._{(i,j)} \right] \tag{17}$$

Anharmonic correction: Let $S_{non\text{-}Gauss,(i)}$ be the marginal (1-dimensional) entropy estimated by a non-Gaussian numerical method for PCA mode i:

$$S_{anh,(i)} = S_{Class,Gaussian,(i)} - S_{non-Gauss,(i)} \tag{18}$$

Supralinear dependence correction: Let $S_{non\text{-}Gauss(i,j)}$ be the joint (2-dimensional) entropy estimated by a non-Gaussian numerical method for PCA modes (i,j). The supralinear dependence correction is given by the mutual information (M.I.) [19]:

$$M.I._{(i,j)} = S_{non-Gauss,(i)} + S_{non-Gauss,(j)} - S_{non-Gauss,(i,j)} \tag{19}$$

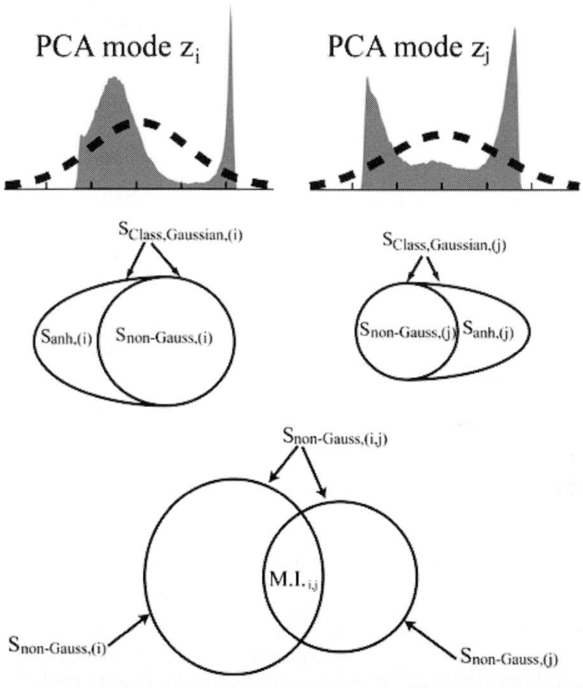

Fig. 2 The pairwise correction is the sum of anharmonicities and supralinear dependence (M.I.). Shown are two strongly anharmonic modes and their Gaussian fits. In the Venn diagrams, entropy is proportional to the area.

Because of numerical inaccuracies, S_{anh} and M.I. can be slightly negative. In the present implementation, a mode is taken as harmonic if $S_{anh}/S_{Quantum}<0.007$. Any M.I.<0 is taken as zero. If a large and negative M.I. is calculated for certain mode pairs, it is likely due to the frequency of data sampling. Both too low and too high resolution cause inaccuracies. For best results, use saved coordinates between every 10fs to 500fs.

In the next section, a practical method called the k-nearest neighbor algorithm ($S_{kNN,k}$) is used to calculate $S_{non\text{-}Gauss}$ for arbitrary dimensions. To guarantee numerical stability, the current implementation is restricted to pairwise corrections (d=1 and d=2).

2.4. *Estimation of non-Gaussian entropy*

The assumption of a Gaussian distribution is not necessary to estimate entropy. Indeed, one does not need to assume a particular distribution at all. Non-parametric methods avoid forcing a functional form for the distribution. One such method calculates the entropy from the k-nearest neighbor (kNN) points in the sample [10].

Mathematical foundation of k-nearest neighbor entropy

Let matrix $\mathbf{Z}_d$ of dimension (d,n_f) be a random sample of n_f observations of a d-dimensional random vector. In this case, the random vectors are a subset of d square-root-of-mass weighted principal components of the original MD trajectory of a solute molecule for n_f simulation frames:

$$Z_d = (z_1,\ z_2, z_3,\ ..\ ,\ z_{nf}) \tag{20}$$

Now we use a non-parametric estimate of the probability density around each observed PCA conformation. The conformational density for each frame of the simulation is assessed through its distance to the k most similar (nearest neighbor) frames. For the true underlying probability distribution function $f_{c\text{-true}}$ that produced the data, a non-parametric estimate for each d-dimensional sample vector z_i is given by f_c [20]:

$$f_c(z_i) = \frac{k}{n}\frac{1}{V_d\left(R_{i,k}\right)} \quad \text{with} \quad V_d\left(R_{i,k}\right) = \frac{\pi^{d/2} R_{i,k}^d}{\Gamma\left(\dfrac{d}{2}+1\right)} \tag{21}$$

k is defined below as the nearest-neighbor index and should not to be confused with k_B. V_d is the volume of a d-dimensional hyper-sphere with radius $R_{i,k}$:

$$R_{i,k} = \left\| z_i - z_{i,k} \right\|_2 \tag{22}$$

$R_{i,k}$ is the Euclidean distance between sample point $\mathbf{z}_i$ and its k-th nearest neighbor $\mathbf{z}_{i,k}$. The entropy of the probability distribution may now be estimated in an asymptotically unbiased form. The k-nearest neighbors entropy, following Hnizdo et al is [10, 11]:

$$\frac{S_{kNN,k}}{k_B} = \frac{d}{n}\sum_{i=1}^{n} \ln R_{i,k} + \ln \frac{n\pi^{d/2}}{\Gamma\left(\dfrac{d}{2}+1\right)} - L_{k-1} + \gamma \tag{23}$$

with $L_0 = 0$; $L_m = \sum_{i=1}^{m} \frac{1}{i}$ for $m \geq 1$; $\gamma = 0.5772157...$ (Euler-Mascheroni constant)

The first term of Eq. 23 may be understood intuitively as an estimate of entropy. The three last terms are an unbiasing correction. They arise to eliminate a known asymptotic bias in this estimator, as detailed by Hnizdo, Demchuk et al [10]. It is an extension to k-neighbors of a similar estimator proposed before by Kozachenko and Leonenko for k=1 [21]. Using higher neighbors, the method becomes less sensitive to the accuracy of the data. In practical tests by others [10] and our own with multidimensional random points shaped by statistical distributions of known entropy, the k=4th nearest neighbor has demonstrated to work well.

The k-nearest neighbors in a d-dimensional sample of n_f-points may be found using the ANN program, a C++ library by David Mount and Sunil Arya [22]. Specifically, the k-d tree nearest neighbors method implemented into ANN was used.

The kNN algorithm produces a relative, statistical and more importantly, a classical entropy. It is thus well suited to compute relative corrections in the classical region.

3. Practical application to small biomolecules

The main focus of the presented method is the calculation of conformational entropy of proteins and other macromolecules. As a first test, however, the combination of methods was applied to smaller molecules; glycine and alanine. Molecular dynamics simulations were done for 0.9 ns (after heating and equilibration for 0.170 ns). The C- and N-termini were charged, but the overall charge is zero. An explicit water box of 16Å^3 was used together with the Particle Mesh Ewald electrostatic method. The time steps were of 1fs, with freely moving hydrogens except for TIP3P waters (fixed with SETTLE). The Charmm22 force field was used with programs Charmm31a1 and NAMD 2.6b1. For analysis, solvent atoms were deleted. Translation and rotation were removed. Samples were taken every 100 fs for the entropy calculation, so that n_f=9000.

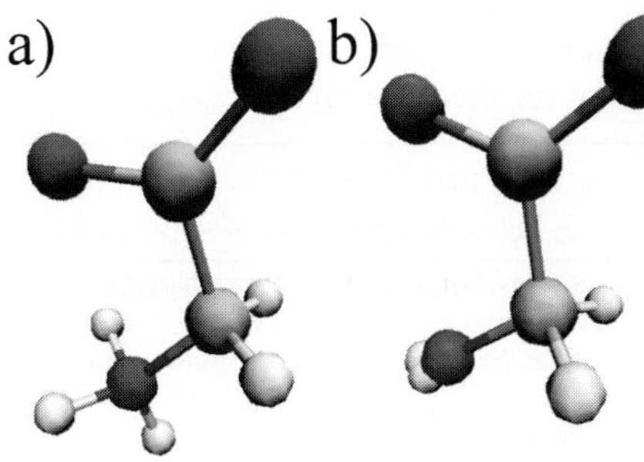

Fig. 3: Glycine molecule in a) its initial conformation and b) its average conformation during a 0.9 ns MD simulation. Notice that the amino group rotates freely, and this is manifested as an average close to the central nitrogen. It is not cause for worry that the average structure looks chemically unsound. In the words of Jaynes, "it is possible to make a sharp distinction in statistical mechanics: the physical and the statistical." [17] Our physical description is built into the MD simulation. Our statistical analysis happens in "shape and size" space, treating the conformational density with information theory.

The histograms shown in Fig. 4 represent each mass-weighted coordinate in the glycine trajectory. Fig. 4a presents the marginal distributions of the original trajectory. We may already calculate an entropy from the variance of these coordinates. Summing the marginal Gaussian entropies is tantamount to ignoring both anharmonicity and any correlations. This provides the highest upper limit to entropy (see Table 1). The severe overestimation speaks against proposals for additive entropy contributions, such as an "entropy per side chain" or "per chemical group".

After PCA is performed (Fig. 4b), the entropy may be calculated from the eigenvalues of the diagonalized mass-weighted covariance matrix. In total we obtain (3N-6)=24 eigenvalues for glycine. Only 3 collective coordinates are in the classical regime (bottom right in Fig. 4b), but together they contribute with 70% of the total entropy. The corresponding estimate can be seen in Table 1 as PCA, also known as the quasi-harmonic approximation [7]. It is lower because of the decomposition of linear correlation.

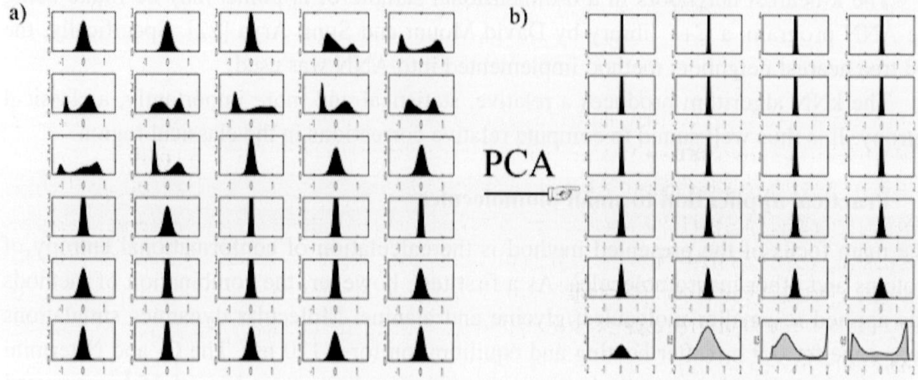

Fig. 4 Histogram from the molecular dynamics trajectory of glycine. a) 30 mass-weighted coordinates before principal component analysis (PCA). Notice how the coordinates 4 to 12 seriously deviate from Gaussians. These correspond to x,y,z fluctuations of the freely rotating hydrogens of the amino group. b) PCA groups motion types by decomposing the linear correlation. Shown are the 24 collective coordinates excluding translation and rotation. Note that although PCA implicitly assumes Gaussian collective coordinates, the actual projections may deviate. This is the case for the last 3 PCA coordinates (light gray), which are also in the classical region. For a) and b): The average of each coordinate has been subtracted for display purposes. Units: Horizontal axis in [a.m.u.$^{1/2}$Å] and vertical axis in histogram probability.

Table 1: Absolute molar entropy of two free-form amino acids. All estimates per column are based on the same 0.9 ns trajectory, except for the vacuum $S_{NormalModes}$.

Estimate type	Glycine $S_{conformational}$ cal/(mol K)	Alanine $S_{conformational}$ cal/(mol K)	Comments
Marginal	34.14	55.09	Sum of quantum entropy from variances of the marginal distributions of the original m.w. coordinates.
PCA	15.41	24.44	M.w. PCA, also called the Quasi-Harmonic Approximation (QHA). [7]
PCA-Anh	13.02	22.34	PCA with anharmonic correction.
PCA-Anh-M.I.	**9.44**	**14.88**	**Estimation from present method. PCA with anharmonic and pairwise supralinear (M.I.) corrections.**
Normal Modes	8.56	13.47	NMA After thorough minimization in vacuum. Lower limit benchmark.

In principle, a lower entropy estimate is better. The PCA value was corrected for anharmonicity (PCA-Anh), and then subsequently for anharmonicity and supralinear dependence (PCA-Anh-M.I.). Consistent with a recent study [12], the influence of supralinear dependence (M.I.) was much larger than that of anharmonicity.

For such small molecules, normal mode analysis is still a viable option and can be seen as a benchmark. It is thus encouraging that the corrected values are similar to it. It is not far-fetched to say that the corrected estimates are better gauges of entropy than normal modes, as they account for the entropy of the multimodal rotation of the amino and methyl groups as well as the influence of solvent particles. **For peptides and larger biomolecules, normal mode analysis of single conformations is definitely not appropriate**, as it cannot capture more complex potential energy surfaces. An estimate for the entropy of a protein as a whole can be obtained using the method presented here.

4. Conclusion

Estimation of absolute conformational entropy is important because it allows a detailed understanding of the thermodynamic driving forces at the molecular level [16]. While the quasi-harmonic approximation is known to systematically overestimate entropy [9], it was until recently the only option to calculate absolute entropy from MD simulations. A recent study [12] demonstrated that this overestimation is larger for the folded than for the unfolded state of peptides. Error cancellation is not exhibited because cooperativity and higher order motion correlations are more important in the compact, folded state than in an open, denatured state.

The current method provides a tighter upper limit to absolute entropy than the quasi-harmonic approximation. As a result, the method stays loyal to the E.T. Jaynes' spirit of maximum entropy as a "method of reasoning which ensures that no unconscious arbitrary assumptions have been introduced" [17].

Future directions include extending the method to more than pairwise supralinear dependencies. Also interesting is the combination with the permutation reduction method to calculate the entropy of diffusive systems (solvation shells and solvent entropy) [23].

A program to calculate the conformational entropy of peptides, proteins or nucleotides from Charmm and NAMD trajectories according to this algorithm will be made available on http://userpage.chemie.fu-berlin.de/~numata

Acknowledgments

This work was supported by the International Research Training Group "Genomics and Systems Biology of Molecular Networks" (GRK1360 of the DFG) and an internship grant for American students from the DAAD RISE program and German SFB498.

References

[1] Loladze, V.V. , Ermolenko, D.N., and Makhatadze, G.I., Thermodynamic consequences of burial of polar and non-polar amino acid residues in the protein interior, *J. Mol. Biol.*, 320(2):343-357, 2002.

[2] Zoete, V., Meuwly, M., and Karplus, M., Study of the insulin dimerization: Binding free energy calculations and per-residue free energy decomposition, *Proteins.*, 61(1):79-93, 2005.

[3] Srinivasan, J., Cheatham III, T.E., Cieplak, P., Kollman, P.A., and Case, D.A., Continuum solvent studies of the stability of DNA, RNA, and phosphoramidate-DNA helices, *J. Am. Chem. Soc.*, 120(37):9401-9409, 1998.

[4] Hammes-Schiffer, S. and Benkovic, S.J., Relating protein motion to catalysis, *Annu. Rev. Biochem.*, 75:519-541, 2006.

[5] Shannon, C.E. and Weaver, W., A mathematical theory of communication, *Bell Syst. Tech. J.*, 1948.

[6] Clausius, R., Über verschiedene für die Anwendung bequeme Formen der Hauptgleichungen der mechanischen Wärmetheorie, *Ann. der Physik*, 201:353-400, 1865.

[7] Andricioaei, I. and Karplus, M., On the calculation of entropy from covariance matrices of the atomic fluctuations, *J. Chem. Phys.*, 115:6289-6292, 2001.

[8] Schlitter. J., Estimation of absolute and relative entropies of macromolecules using the covariance matrix, *Chem. Phys. Lett.*, 215:617-621, 1993.

[9] Chang, C-E., Chen, W., and Gilson, M.K., Evaluating the accuracy of the quasiharmonic approximation, *J. Chem. Theory Compu.*, 1 (5):1017-1028, 2005.

[10] Hnizdo, V., Singh, H., Misra, N., Fedorowicz, A., and Demchuk, E., Nearest neighbor estimates of entropy, *Am. J. Math. Manage. Sci.*, 23:301-321, 2003.

[11] Hnizdo, V., Darian, E., Federowicz, A., Demchuk, E., Li, S., and Singh, H., Nearest-neighbor nonparametric method for estimating the configurational entropy of complex molecules, *J. Comput. Chem.*, 28(3):655-668, 2007.

[12] Baron, R., Biomolecular simulation: Calculation of entropy and free energy, polypeptide and carbopeptoid folding, simplification of the force field for lipid simulations (eth 16584). Zürich: ETH-Zürich; 2006.

[13] Knapp, E.W. and Hoffmann, D., Polypeptide folding with off-lattice monte carlo dynamics: The method, *Eur. Biophysics J.*, 24(6):387-403, 1996.

[14] Cramer, C.J., *Essentials of computational chemistry*, 2nd ed.; 2004.

[15] McQuarrie, D.A., *Statistical mechanics*, Harper & Row; 1973.

[16] Dill, K.A., Bromberg S: *Molecular driving forces* Garland Science; 2003.

[17] Jaynes, E.T., Information theory and statistical mechanics (part 1), *Phys. Rev.*, 106(4):620-630, 1957.

[18] Lange, O.F. and Grubmüller, H., Generalized correlation for biomolecular dynamics, *Proteins.*, 62(4):1053-1061, 2006.

[19] Cover, T.M. and Thomas, J.A, *Elements of information theory*; 1991.

[20] Loftsgaarden, D. and Quesenberry, C., A nonparametric estimate of a multivariate density function, *Ann. Math. Stat.,*1049-1051, 1965.

[21] Kozachenko, L. and Leonenko, N., Sample estimates of entropy of a random vector, *Problems of Information Transmission*, 23:95-101, 1987.

[22] Mount, D., Arya, S., Netanyahu, N., Silverman, R., and Wu, A., An optimal algorithm for approximate nearest neighbor searching fixed dimensions, *JACM*, 45(16):891-923, 1998.

[23] Reinhard, F. and Grubmüller, H., Estimation of absolute solvent and solvation shell entropies via permutation reduction, *J. Chem. Phys.*, 126(1):014102, 2007

DETECTING NEAR-NATIVE DOCKING DECOYS BY MONTE CARLO STABILITY ANALYSIS

STEPHAN LORENZEN
lorenzen@chemie.fu-berlin.de

Macromolecular Modelling Group, Free University Berlin, Takustr. 6, 14195 Berlin, Germany

Since protein complex crystallization is expensive and time-consuming, computational docking tools provide a valuable method to investigate protein interactions. While the sampling of possible docked conformers of two proteins can be performed efficiently by Fast Fourier Transform (FFT) methods, the selection of near-native decoys from the pool of thousands of possible decoys is still far from being solved. Here, a new approach for docking decoy selection by Monte Carlo stability analysis is presented. In the course of replica exchange Monte Carlo simulations (REMC), replica from near-native decoys show a significantly lower structural diversity than replica from non-native decoys. The effect is successfully applied to rank docking decoys in a benchmark set of 59 protein complexes.

Keywords: protein–protein docking; Monte Carlo; decoy selection; structural stability; ROTAFIT.

1. Introduction

Despite the rising number of protein structures deposited in the Protein Data Bank (PDB [1]), structural information about protein complexes is still sparse, since especially transient protein complexes are hard to crystallize. However, the function of most proteins depends on their interaction with other proteins. Since in many cases the individually crystallized structures of the monomers are available, computational protein–protein docking tools are of rising interest. The search for the structure of a protein complex basically consists of two steps: First, all possible orientations of the two monomers towards each other have to be scanned (*sampling*), resulting in hundreds to thousands of possible docked conformers (*decoys*). In a second step, the near-native structures in the decoy pool have to be identified, and the structure of the proposed complex has to be refined.
The application of the Fast Fourier Transformation technique for docking by Katchalski-Katzir [16] made the efficient scanning of the six-dimensional translational and rotational space possible for the first time, and today a variety of FFT-based docking programs are used [5, 11, 13, 29]. Basically, both monomers are represented in a grid which distinguishes between the interior and the surface of the proteins. The geometrical complementarity of a complex can then be efficiently cal-

culated as a correlation function rewarding surface contacts and penalizing interior overlap.

The original FFT method described by Katchalski-Katzir was extended tremendously in the following years. A pair-wise shape complementarity function favoring continuous patches of high curvature over several small contact patches [8] as well as terms for desolvation [7] and electrostatics [11] have been introduced.

To rerank the decoys produced by FFT methods, a number of further scoring functions have been developed. Besides electrostatic interactions and surface complementarity, common scoring functions include the Atomic Contact Energy (ACE [33]), an atomic level extension of the quasichemical potential of Miyazawa and Jernigan [23]. The term describes desolvation energy and loss of side chain entropy upon binding ($\Delta G - T\Delta S$). As a residue based potential, the RP score [24] compiled from frequencies of amino acid contacts in protein interfaces is used. A variety of different combinations of scoring functions has been described [25, 26].

Another common procedure in decoy selection is the clustering of decoys generated by FFT methods. The basic assumption behind the clustering approach is that the free energy landscape exhibits a broader well near the native structure than near nonnative structures, which often lie in local minima [4]. Vakser *et al.* introduced clustering of docking decoys in low-resolution docking studies [30, 31], and the docking server ClusPro [9] depends on clustering of FFT docking decoys. When clustering a diverse set of docking decoys, 5 Å has been shown to be a reasonable RMSD cutoff [20].

A common problem in scoring docking decoys from unbound monomers are incorrect conformations of key side chains or even the backbone: unbound protein structures superposed on the bound complex often show several clashes and are thus hard to rank by common energy functions. Due to these structural inaccuracies, the useage of common energy functions, especially the 'hard' van der Waals potential, in scoring docked unbound monomers is not feasible. A refinement of docked structures can make the scoring by common energy functions possible. Methods to allow side chain flexibility include multicopy approaches [14, 19, 32] or quick energy minimizations prior to scoring [2, 3, 18]. In most cases, these methods do not substantially change the conformation of the complex. Extensive Molecular Dynamics (MD) refinement of docked conformers has been shown to increase the fraction of native contacts [17]. However, due to the high computational cost, this approach is not feasible for scoring and refining large decoy sets.

A third way of structure refinement besides minimization and Molecular Dynamics are Monte Carlo simulations. The principle of the method is to apply random moves and decide on their acceptance according to the change in energy caused by the move. It could be shown hat a move acceptance probability of $p = \min(1, e^{-\frac{\Delta E}{kT}})$ leads to a Boltzmann-like energy distribution of structures rather than a single local minimum like in a minimization [22]. Since the acceptance of moves depends on the temperature chosen, an extension of the method [28] uses several *replicas* simulated at different temperatures. In regular time intervals, the replicas simulated in neighboring temperature slots are exchanged with a probability of $p = \min(1, e^{(\frac{1}{kT_1} - \frac{1}{kT_2})\Delta E})$. In this way, low-temperature replicas are used to refine structures of already favorable energies, whereas high-temperature replicas are used to cross energy barriers by also accepting some energetically unfavorable moves.

A Monte Carlo approach to protein–protein docking is implemented in the Rosetta program [12, 27]. Starting from random orientations, the monomers are first assembled to generate glancing contact and then subjected to 500 rigid body MC steps rated by a low-resolution residue–residue potential using a side chain centroid representation. After adding explicit side chains, 50 cycles of rigid-body perturbation, side chain optimization and rigid-body minimization are carried out. The acceptance of the moves is decided by the Metropolis criterion [22]. After the minimization, the 200 best-scoring conformations from 10^5 independent runs are clustered. The algorithm has been succesful in recent rounds of CAPRI [15], an international protein–protein docking contest.

The algorithm presented here, ROTAFIT, uses initial rigid-body docking decoys generated by an FFT method and subjects them to 1000 steps of replica-exchange Monte Caro (REMC) simulation with 20 replica. The protocol combines the power of an efficient sampling of the six-dimensional translational and rotational space by the FFT method with subsequent side chain and rigid body refinement. By the generation of a large number of structures resulting from each decoy, the structural stability of the initial decoy in the course of the simulation could be analyzed. Based on this criterion alone, near-native decoys could be identified in the first 10 predicted decoys in 20 of the 35 cases where the initial decoy set contained near-native decoys.

2. Methods and Results

2.1. *Benchmark set and initial decoy generation*

The protein–protein docking benchmark of Chen *et al.* [6] was used throughout this study. The set consists of 59 protein complex structures of several types (22 enzyme-inhibitor, 19 antibody-antigen, 11 other, 7 difficult). For 31 of the 59 complexes, unbound structures of both monomers are known; in the other cases, one of the proteins is provided in its bound form. Initial decoys were generated by ZDOCK2.3 [5] with sparse rotational sampling.

For each of the 59 structures, *interface* residues are defined as residues of both partners which have at least one heavy atom below 10 Å of a heavy atom of the partner protein in the *co-crystallized* structure. The interface C_α *RMSD* is calculated by superposing the respective residues of the docked *unbound* proteins on the native interface. Docking results with an interface RMSD of ≤ 2.5 Å are considered sufficiently close to the native structure and defined as *hits*. In 35 of the 59 cases, the first 100 structures generated by ZDOCK contained a hit (Tab. 1). For enzyme–inhibitor cases (20/22), ZDOCK performed remarkably better in this range than for antibody–antigen (11/19) and 'other' (4/7) cases. In none of the seven difficult cases, a hit was placed within the first 100 decoys.

2.2. *Replica Exchange Monte Carlo simulation*

The first 100 decoys of each of the complexes were used as input decoys for ROTAFIT, a novel Monte Carlo algorithm. Each decoy is copied to 20 replica which are simulated by different temperatures (250 to 2 000 K) for 1 000 steps. Each step consists of a Gaussian distributed random translation of one protein with a standard deviation of 0.5 Å and a rotation around a random axis through the center of the protein. The rotation angle is also randomly distributed, so that its standard deviation leads to a mean displacement of 0.5 Å of interface residues. A move step also contains a change of side chain residues in the interface using the rotamer library by Lovell *et al.* [21]. The library consists of 7328 rotamers of the 20 amino acids and their probabilities to occur in proteins. In each step, five interface amino acids (heavy atoms within 10 Å of a heavy atom of the binding partner) of each of the monomers are selected. For each of these 10 amino acids, one random rotamer for each is chosen from the library, according to its probability. The energies of the resulting 2^{10} possible rotamer combinations (current or new rotamer at 10 positions) are calculated, and the combination of side chains with the lowest energy is chosen for the actual step. The decision about the acceptance of a move consisting of a translation, a rotation and side chain changes is made by using the Metropolis criterion [22]. The acceptance rate of moves varied from 5 to 50%, dependent on the temperature of the replica (data not shown). In each step, replica exchange [28] is performed between the different replica.

2.3. *Energy function*

The energy function used is comprised of a sum of a Lennard-Jones potential describing van der Waals interactions, a Coulomb term using a distant dependent dielectric (ϵ=4r) and and an Atomic Contact Energy (ACE [33]) term. Parameters for the Lennard-Jones potential and partial charges of atoms are taken from the extended atom CHARMM19 potential [10]. To direct the simulation towards interface minimization rather than overestimating intra-chain energy changes by side chain movements, intra-chain energies are weighted with a factor of 2.5% to prevent unfavorable side chain conformations rather than to guide the simulation.

2.4. *Stability analysis*

To analyze the trajectories after the ROTAFIT run, the pair-wise C_α RMSDs of ligands after superposition of the receptors were calculated. The last 500 steps of the 'coldest' 15 replica of the ROTAFIT run are analyzed. For each decoy, 'structural neighbors' are defined as replica within the same trajectory (descending from the same ZDOCK decoy) with a ligand C_α RMSD below the specified threshold.

Fig. 1 shows the average number of structural neighbors for each decoy in dependence of the RMSD threshold for all cases where a near-native structure was included in the decoy set generated by ZDOCK. In many cases, near-native decoys

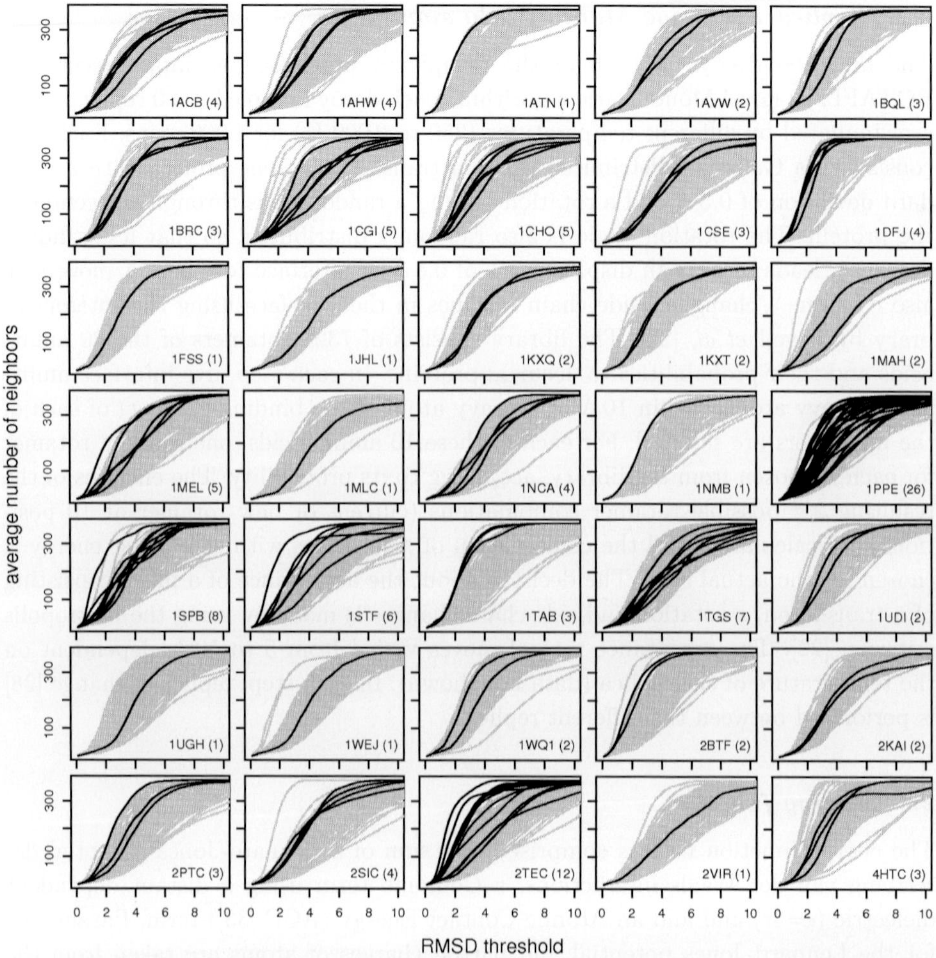

Fig. 1. Structural stability of docking decoys in the second half of the simulation. The graphs show the number of structural neighbors in dependence of the RMSD threshold. Black lines: decoys with interface C_α RMSD of ≤ 2.5Å; Inset: PDB code of the complex, number of decoys with interface C_α RMSD of ≤ 2.5Å

(black lines) show a significantly higher degree of structural stability than average decoys. The results are summarized in Tab. 1. As a measure of structural stability, the integral under the curves plotted in Fig. 1 is calculated. Ranking by the value of this integral identifies hits within the first 10 best-scoring decoys in 20 cases, compared to 16 cases for ZDOCK (Tab. 1). In most cases, the integral value for hits is significantly higher than for non-hits (see z-scores in Tab. 1). Interestingly, the ranking by integral values is able to clearly define hits in some cases where the ZDOCK score fails (e. g., 1ATN, 1TAB, 2PTC, 2SIC) and vice versa (e. g. 1KXT, 4HTC). However, the complexes where the different scoring algorithms fail or succeed fall in different classes, and there is no correlation with interface sizes or other characteristics of the binding sites (data not shown).

Table 1. Number and rank of near-native decoys based on ZDOCK score and structural stability.

PDB		hit	Rank of first hit		z-score of hits		Best RMSD in top 10	
code	class[a]	count	ZDOCK	integral	first	average	ZDOCK	integral
1ACB	E	4	12	9	1.21	0.63	4.68	1.03
1AHW	A	4	1	5	1.68	1.21	1.22	1.43
1ATN	O	1	49	2	2.32	2.32	15.89	1.19
1AVW	E	2	12	18	1.00	0.79	6.87	9.94
1BQL	A	3	4	3	1.82	1.46	1.60	1.04
1BRC	E	3	29	15	1.05	0.93	2.67	2.67
1CGI	E	5	14	6	1.33	0.14	3.03	1.96
1CHO	E	5	1	7	1.33	0.67	0.90	0.90
1CSE	E	3	24	12	1.18	0.83	5.78	5.78
1DFJ	E	4	2	2	1.80	1.51	2.17	1.49
1FSS	E	1	24	19	0.79	0.79	7.84	6.99
1JHL	A	1	41	54	-0.10	-0.10	4.91	5.29
1KXQ	A	2	13	53	-0.04	-0.16	6.98	8.72
1KXT	A	2	2	36	0.43	0.38	1.23	12.48
1MAH	E	2	40	35	0.47	-0.45	5.85	6.13
1MEL	A	5	5	1	2.02	1.08	1.13	1.13
1MLC	A	1	61	65	-0.37	-0.37	6.28	4.55
1NCA	A	4	1	1	1.82	0.82	1.52	1.17
1NMB	A	1	2	2	1.94	1.94	1.02	1.02
1PPE	E	26	1	2	1.90	0.77	0.75	0.75
1SPB	O	8	2	2	2.52	1.15	0.78	0.78
1STF	E	6	1	1	3.25	1.46	0.93	0.93
1TAB	E	3	14	2	2.09	1.96	6.32	1.56
1TGS	E	7	3	8	1.22	-0.22	2.16	2.16
1UDI	E	2	6	10	1.12	0.81	1.37	1.11
1UGH	E	1	12	51	0.13	0.13	5.35	10.45
1WEJ	A	1	69	36	0.40	0.40	10.79	4.60
1WQ1	O	2	11	7	1.21	0.48	3.26	1.54
2BTF	O	2	1	2	2.32	2.10	0.85	0.85
2KAI	E	2	16	42	0.22	-0.45	4.91	4.93
2PTC	E	3	47	1	2.06	1.18	4.79	0.88
2SIC	E	4	13	3	1.51	1.01	2.53	1.40
2TEC	E	12	1	1	2.67	1.30	0.85	0.78
2VIR	A	1	75	53	0.04	0.04	8.73	12.33
4HTC	E	3	3	20	0.89	0.42	1.32	2.96

Note: [a] A: antibody–antigen; E: enzyme–inhibitor; O: other; D: difficult

2.5. *Selection and evaluation of final predictions*

To select the final predictions for each of the 59 docking cases, the last 500 time steps of the 'coldest' 15 replica were clustered. In contrast to clustering a diverse set of docking decoys, where 5 Å prooved to be a reasonable threshold [20], for replica originated from the same decoy, a 1 Å RMSD threshold is appropriate. Using this threshold, an average of 5% of the replica descending from one decoy build the biggest cluster. The cluster center is defined here as the structure with the most structural neighbors within the trajectory of the 15 replica. Fig. 2(a) shows the interface C_α RMSD of the selected decoy to the natice structure before and after refinement. In the cases where our integral score finds a hit within the first 10 predicted decoys (solid circles), the ZDOCK decoys are in most cases brought closer to the native structure, if their RMSD was sufficiently low before (less than about 3 Å). Above this threshold, the changes in C_α RMSD are small but undirected. As an example, Fig. 2(b) shows the 60th decoy (ZDOCK ranking) of 1TGS (ranked 8th by the integral score), which improves from 2.64 to 1.85 Å interface RMSD. In the case of 1FBI, the refinement generated a hit where no hit was present in the original decoy set (RMSD decreases from 2.8 to 2.35).

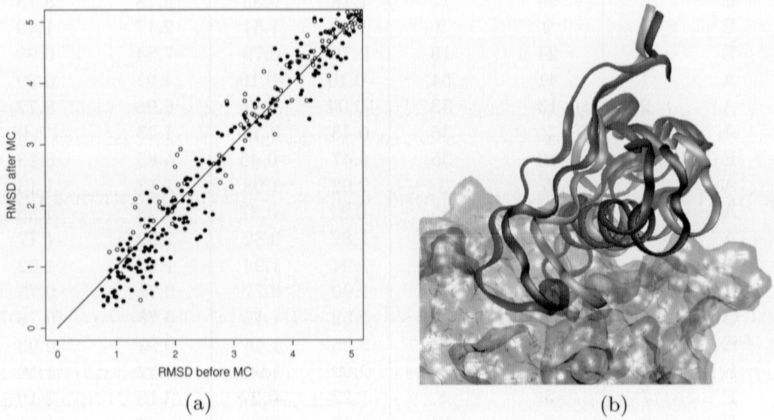

Fig. 2. Refinemnent of initial ZDOCK decoys by ROTAFIT. (a) Interface C_α RMSD before and after refinement for *all* decoys. Filled circles: Integral score ranks hit in first 10 decoys; Empty circles: No hit in first 10 ecoys ranked by integral score. (b) Decoy 60 (ZDOCK rank) of 1TGS (ranked 8 with integral score): Surface representation: bound receptor; green: ZDOCK decoy; blue: co-crystallized ligand; red: ROTAFIT refinement result.

3. Discussion

A new method of scoring and refining protein–protein docking decoys is presented. The refinement is achieved by a replica exchange Monte Carlo approach, rather than by minimizing independent decoys separately. While clustering of independently generated docking decoys is widely used, the stability analysis of trajectories generated from a single decoy is a novel approach and performs well in the examined benchmark set. This also corresponds to the more physical distribution of decoys generated by Metropilos Monte Carlo runs compared to energy minimizations [22]. A possible explanation of the higher structural stability of near-native decoys is that they lie in broader energy minima, so that random perturbations still result in a conformation lying in this same energy well and the perturbed decoy thus easily retains back its native state, rather than to slip in a new, local energy minimum. With a computation time of 2–3 minutes per decoy, ROTAFIT provides an interesting alternative to quick scoring methods on one side and extensive simulations on the other side. The integral score provided as output can be an important contribution to novel scoring functions.

Acknowledgments

I wish to thank Ernst-Walter Knapp, Aysam Gürler and Stefan Günther for comments and fruitful discussions. The work was supported by the International Research Training Group (IRTG) "Genomics and Systems Biology of Molecular Networks" (GRK1360, Deutsche Forschungsgemeinschaft (DFG)).

References

[1] Berman, H. M., Westbrook, J., Feng, Z., Gilliland, G., Bhat, T. N., Weissig, H., Shindyalov, I. N., and Bourne, P. E., The protein data bank, *Nucleic Acids Res.*, 28(1):235–242, 2000.

[2] Camacho, C. J., Gatchell, D. W. , Kimura, S. R., and Vajda, S., Scoring docked conformations generated by rigid-body protein-protein docking, *Proteins*, 40(3):525–537, 2000.

[3] Camacho, C. J., Ma, H., and Champ, P. C., Scoring a diverse set of high-quality docked conformations: a metascore based on electrostatic and desolvation interactions, *Proteins*, 63(4):868–877, 2006.

[4] Camacho, C. J., Weng, Z., Vajda, S., and DeLisi, C., Free energy landscapes of encounter complexes in protein-protein association, *Biophys J.*, 76(3):1166–1178, 1999.

[5] Chen, R., Li, L., and Weng, Z., ZDOCK: An initial-stage protein-docking algorithm, *Proteins*, 52(1):80–87, 2003.

[6] Chen, R., Mintseris, J., Janin, J., and Weng, Z., A protein-protein docking benchmark, *Proteins*, 52(1):88–91, 2003.

[7] Chen, R., and Weng, Z., Docking unbound proteins using shape complementarity, desolvation, and electrostatics, *Proteins*, 47(3):281–294, 2002.

[8] Chen, R. and Weng, Z., A novel shape complementarity scoring function for protein-protein docking, *Proteins*, 51(3):397–408, 2003.

[9] Comeau, S. R., Gatchell, D. W., Vajda, S., and Camacho, C. J., Cluspro: A fully automated algorithm for protein-protein docking, *Nucleic Acids Res.*, 32:W96–W99, 2004.

[10] Neria, E., Fischer, S., and Karplus, M., Simulation of activation free energies in molecular systems, *J. Chem. Phys.*, 105:1902–1921, 1996.

[11] Gabb, H. A., Jackson, R. M., and Sternberg, M. J., Modelling protein docking using shape complementarity, electrostatics and biochemical information, *J. Mol. Biol.*, 272(1):106–120, 1997.

[12] Gray, J. J., Moughon, S., Wang, C., Schueler-Furman, O., Kuhlman, B., Rohl, C. A., and Baker, D., Protein-protein docking with simultaneous optimization of rigid-body displacement and side-chain conformations, *J. Mol. Biol.*, 331(1):281–299, 2003.

[13] Heifetz, A., Katchalski-Katzir, E., and Eisenstein, M., Electrostatics in protein-protein docking, *Protein Sci.*, 11(3):571–587, 2002.

[14] Jackson, R. M., Gabb, H. A., and Sternberg, M. J., Rapid refinement of protein interfaces incorporating solvation: application to the docking problem, *J. Mol. Biol.*, 276(1):265–285, 1998.

[15] Janin, J., Assessing predictions of protein-protein interaction: The CAPRI experiment, *Protein Sci.* 14(2):278–283, 2005.

[16] Katchalski-Katzir, E., Shariv, I., Eisenstein, M., Friesem, A. A., Aflalo, C., and Vakser, I. A., Molecular surface recognition: determination of geometric fit between proteins and their ligands by correlation techniques, *Proc. Natl. Acad. Sci. USA*, 89(6):2195–2199, 1992.

[17] Król, M., Tournier, A. L., and Bates, P. A., Flexible relaxation of rigid-body docking solutions, *Proteins*, 68(1):159–169, 2007.

[18] Li, L., Chen, R., and Weng, Z., RDOCK: Refinement of rigid-body protein docking predictions, *Proteins*, 53(3):693–707, 2003.

[19] Lorber, D. M., Udo, M. K., and Shoichet, B. K., Protein-protein docking with multiple residue conformations and residue substitutions, *Protein Sci.*, 11(6):1393–1408, 2002.

[20] Lorenzen, S. and Zhang, Y., Identification of near-native structures by clustering protein docking conformations, *Proteins*, 68(1):187–194, 2007.

[21] Lovell, S. C., Word, J. M., Richardson, J. S., and Richardson, D. C. The penultimate rotamer library, *Proteins*, 40(3):389–408, 2000.

[22] Metropolis, N., Rosenbluth, A. W., Rosenbluth, M. N., Teller, A. H., and Teller, E., Equations of state calculations by fast computing machines, *J. Chem. Phys.*, 21:1087–1092, 1953.

[23] Miyazawa, S. and Jernigan, R., Estimation of effective interresidue contact energies from protein crystal structures: quasi-chemical approximation, *Macromolecules*, 18:534–552, 1985.

[24] Moont, G., Gabb, H- A., and Sternberg, M. J., Use of pair potentials across protein interfaces in screening predicted docked complexes, *Proteins*, 35(3):364–373, 1999.

[25] Murphy, J., Gatchell, D. W., Prasad, J. C., and Vajda, S., Combination of scoring functions improves discrimination in protein-protein docking, *Proteins*, 53(4):840–854, 2003.

[26] Pierce, B. and Weng, Z., ZRANK: Reranking protein docking predictions with an optimized energy function, *Proteins*, 67:1078–1086, 2007.

[27] Schueler-Furman, O., Wang, C., and Baker, D., Progress in protein-protein docking: Atomic resolution predictions in the CAPRI experiment using rosettadock with an improved treatment of side-chain flexibility, *Proteins*, 60(2):187–194, 2005.

[28] Swendsen, S. and Wang, C., Replica monte carlo simulation of spin glasses, *Phys. Rev. Lett.*, 57:2607–2609, 1986.

[29] Tovchigrechko, A. and Vakser, I. A. Development and testing of an automated approach to protein docking, *Proteins*, 6082):296–301, 2005.

[30] Vakser, I. A., Low-resolution docking: Prediction of complexes for underdetermined structures, *Biopolymers*, 39(3):455–464, 1996.

[31] Vakser, I. A., Matar, O. G., and Lam, C. F., A systematic study of low-resolution recognition in protein–protein complexes, *Proc. Natl. Acad. Sci. USA.*, 96(15):8477–8482, 1999.

[32] Zacharias, M., Protein-protein docking with a reduced protein model accounting for side-chain flexibility, *Protein Sci.*, 12(6):1271–1282, 2003.

[33] Zhang, C., Vasmatzis, G., Cornette, J. L., and DeLisi, C., Determination of atomic desolvation energies from the structures of crystallized proteins, *J. Mol. Biol.*, 267(3):707–726, 1997.

AUTOMATICALLY GENERATED MODEL OF A METABOLIC NETWORK

SIMON BORGER
borger@molgen.mpg.de

WOLFRAM LIEBERMEISTER
lieberme@molgen.mpg.de

JANNIS UHLENDORF
uhlndorf@molgen.mpg.de

EDDA KLIPP
klipp@molgen.mpg.de

Max Planck Institute for Molecular Genetics, Berlin, Germany

We demonstrate an approach to automatically generating kinetic models of metabolic networks. In a first step, the metabolic network is characterised by its stoichiometric structure. Then to each reaction a kinetic equation is associated describing the metabolic flux. For the kinetics we use a formula that is universally applicable to reactions with arbitrary numbers of substrates and products. Last, the kinetics of the reactions are assigned parameters. The resulting model in SBML format can be fed into standard simulation tools. The approach is applied to the sulphur-glutathione-pathway in Saccharomyces cerevisiae.

Keywords: Metabolic networks; systems biology; parameter estimation; data integration; sulphur-glutathione pathway; Bayesian data analysis.

1. Introduction

Studying the regulation of metabolic reaction networks is an important task in systems biology and functional genomics. A complete understanding of metabolic regulation requires quantitative information about kinetic laws and the concentrations of metabolites and enzymes. This quantitative knowledge in combination with the known network of metabolic pathways allows the construction of mathematical models that describe the dynamic changes in metabolite concentrations over time. The models are high-dimensional systems of ordinary, non-linear differential equations. The main problems of the approach are the setup of the equations that describe the metabolic pathways in form of kinetic rate equations and the identification of the system parameters. To solve these problems, a variety of pathway modeling tools such as Copasi [6], CellDesigner [15], and others have been developed which simplify model construction and analysis. Most of these tools are able to store and exchange models in the Systems Biology Markup Language (SBML, [21]) and to fit parameters for a given set of experimental data. A long-term goal is the construction of genome-scale metabolic models. For various model organisms stoichiometric genome-scale models have been constructed. They have been shown to be useful for the investigation of steady state fluxes in wild type cells and in mutants. Large-scale dynamic models would be very useful to predict the effect of transient perturbations, for instance by gene regulation, or to apply powerful analysis tools such as metabolic control analysis (MCA), but is still hampered by a lack of systematically retrieved data. Facing the need of a more systematic construction and parameterisation of metabolic models, we present an approach to (i) automatically construct an SBML model from a list of reactions, (ii) automatically associate kinetic expressions to all reactions and (iii) automatically assign the parameter values based on available information and on statistically based estimates for missing information.

2. Methods

We have set up a workflow to automatically generate models of metabolic networks from a given set of reactions; it consists of the following steps:

(i) set up a structural model
(ii) assign a kinetic law to each reaction
(iii) collect kinetic and thermodynamic data
(iv) determine a feasible set of parameters
(v) construct the kinetic model in SBML format

The parameter estimation is based on Bayes statistics; the result is a posterior distribution of parameter sets that reflects the most probable values and their remaining uncertainties. This result can serve as a prior in further modelling steps. We shall now describe the single steps of the workflow; a more detailed description of it is given in [4].

Structural model

We base our models on knowledge contained in the Kyoto Encyclopedia of Genes and Genomes (KEGG) [16]. KEGG is a set of databases that constitute a computer representation of biological knowledge at different levels, i.e. pathways, reactions, enzymes, compounds and genes. These levels are interconnected.

Our metabolic networks are built from chemical reactions stored in KEGG. We start with a set of reactions; in practice we map the reactions to their identifiers in KEGG. With these reaction identifiers we retrieve information on the relevant compounds (in the form of KEGG compound indentifiers) and enzymes (EC numbers). Each reaction in the model has a reaction identifier, an enzyme identifier and a metabolite identifier associated with it. In the resulting SBML file, each element is described by a MIRIAM-compliant annotation [8, 18], which points to the respective KEGG identifer.

Kinetic laws

In a next step each reaction is assigned a kinetic expression. We use the convenience kinetics, a rate law that assumes a random-order enzyme mechanism and is applicable to reactions with any number of substrates and products [9]:

$$v = E \cdot \left(\prod_a \frac{\bar{c}_a}{\bar{c}_a + k_a^A}\right) \cdot \left(\prod_i \frac{k_i^I}{\bar{c}_i + k_i^I}\right) \cdot \frac{k_+^{cat} \prod_s \bar{c}_s^{\,n_s} - k_-^{cat} \prod_p \bar{c}_p^{\,n_p}}{\prod_s \left(\sum_{m=0}^{n_s} \bar{c}_s^{\,m}\right) + \prod_p \left(\sum_{m=0}^{n_p} \bar{c}_p^{\,m}\right) - 1} \,. \quad (1)$$

The index variables a, i, s and p run over the sets of activators, inhibitors, substrates and products of the reaction, respectively. The concentration of the enzyme catalysing the reaction is denoted by E. The variables k_+^{cat} and k_-^{cat} stand for the turnover rates of the enzyme in the forward (+) and the backward (−) direction, $\bar{c}_s = c_s/k_s^M$ and $\bar{c}_p = c_p/k_p^M$ are the ratios of the substrate and product concentrations with their k^M values, K_i^I denotes the inhibition constant of inhibitor i and K_a^A the activation constant of activator a. Finally, n_s and n_p are the stoichiometric coefficients of substrate s and product p, respectively.

This formula is directly applicable once the stoichiometric structure of the model - i.e., the substrates and products of all reactions - and the regulatory structure - i.e. activators and inhibitors of an enzyme - are known.

The parameters that enter this kinetic expression are a k^M value for each reactant, a k^A value for each activator of the enzyme, a k^I value for each inhibitor and two turnover rates $k_\pm^{cat}$ for the enzyme. The total number of kinetic parameters entering the formula is $N_s + N_p + N_i + N_a + 2$ as indicated in Table 1. In order to ensure thermodynamic consistency of the parameter set, we actually regard the turnover rates $k_\pm^{cat}$ as dependent quantities and express them by two different kinds of parameters, one k^V value for each reaction and one k^G value for each metabolite [9].

Table 1. Types of kinetic parameters and their numbers entering the kinetic formula Eq.(1).

Parameter type	number required
k^M	$N_s + N_p$
k^I	N_i
k^A	N_a
k^{cat}	2

Note: N_s: number of substrates, N_p: number of products, N_i: number of inhibitors, N_a: number of activators of a reaction.

Data collection

We use two types of data available. First, we search literature and databases for thermodynamic, kinetic, metabolomic and proteomic data. The thermodynamic data include Gibbs free energies of formation [2, 5, 11], and equilibrium constants [19]. The kinetic data comprise k^M values [14, 17], k^I values [14, 17], $ic50$ values [17] and turnover rates k^{cat} [14, 17]. Metabolomic data sources are metabolite concentrations [1]. Protein concentrations come from [23].

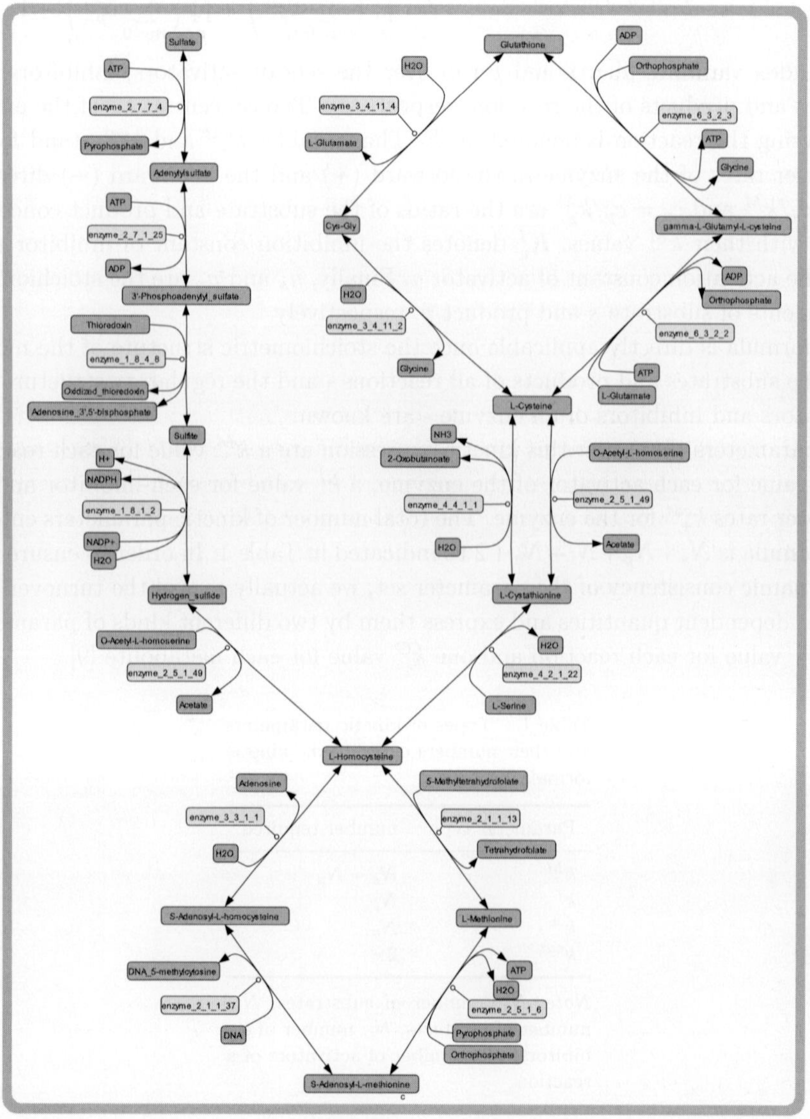

Fig. 1. An automatically generated metabolic network set up starting with the KEGG reaction identifiers R00529, R00509, R02021, R00858, R01287, R00192, R00380, R00177, R00946, R01290, R01001, R03217, R00894, R00497, R00494, R00899.

Furthermore, we use predicted k^M values based on a statistical linear model [3]. The idea behind is that there are different statistical factors that explain the logarithm of a k^M value, $\ln(k^M)$. The first of the three factors is the substrate contribution μ determined by the substrate's chemical properties. Secondly, there is the substrate-enzyme contribution α reflecting evolutionary conservation across organisms. Finally, there is the substrate-organism contribution β stemming from the adjustment of k^M values to typical concentrations of the respective metabolite in a certain organism. The sum will be the value of the logarithm of the k^M value: $\ln(k^M) = \mu + \alpha + \beta$.

We map all the collected experimental data to one of the entities `reaction`, `enzyme` and `compound` from KEGG, or to a combination of them, according to the type of data. The entities are represented by their respective KEGG identifiers. With the KEGG identifiers the data are written into a database. By searching the database for the KEGG identifiers present in the model, data can be retrieved for the kinetic expression Eq.(1) of each reaction.

The data for the model are searched in the following way: first we look for data that are identified by a reaction identifier like equilibrium constants k^{eq} or Michaelis Menten constants k^M. If for a certain reaction no such data are found by the reaction identifier, we search by the enzyme identifier that is associated with the reaction. An equilibrium constant is already completely determined by a reaction identifier, for a k^M value the metabolite identifier also has to be extracted. In the next step, data that require either only a metabolite identfier, like concentrations and Gibbs free energies of formation, are searched by the the respective identifier.

Table 2. Prior distributions.

quantity	no. of data	5% quantile	median	95% quantile
thermodynamic				
equilibrium constant	2088	0.000001	0.119	162.0
Gibbs energy of formation (kJ/mol)	9804	-1522.6	-331.0	324.3
kinetic				
k^M (mM)	90240	0.00098	0.14	20
k^I (mM)	21092	0.000003	0.016	14.0
$ic50$ (mM)	8324	0.000002	0.002	0.67
k^{cat} (mM)	22587	0.008	6.0	1100
metabolomic				
metabolite concentration (mM)	225	0.0018	0.122	4.9
proteomic				
protein abundance (molecules/cell)	10141	279	2939	33502

Note: The parameter types used for data collection for the model. Shown are the number of available data and properties of their distributions.

Distribution of feasible parameter sets

The collected experimental data cannot be directly written into the kinetics of the reactions of the metabolic network. The reasons are: (i) experimental data are noisy because of measurement errors. Sources of noise are biological variability, measurement errors, measurement in vitro or in other species; (ii) we may find different contradicting values for the same parameter; (iii) due to thermodynamical constraints [9] (iv) many parameters will not be available and therefore remain undetermined.

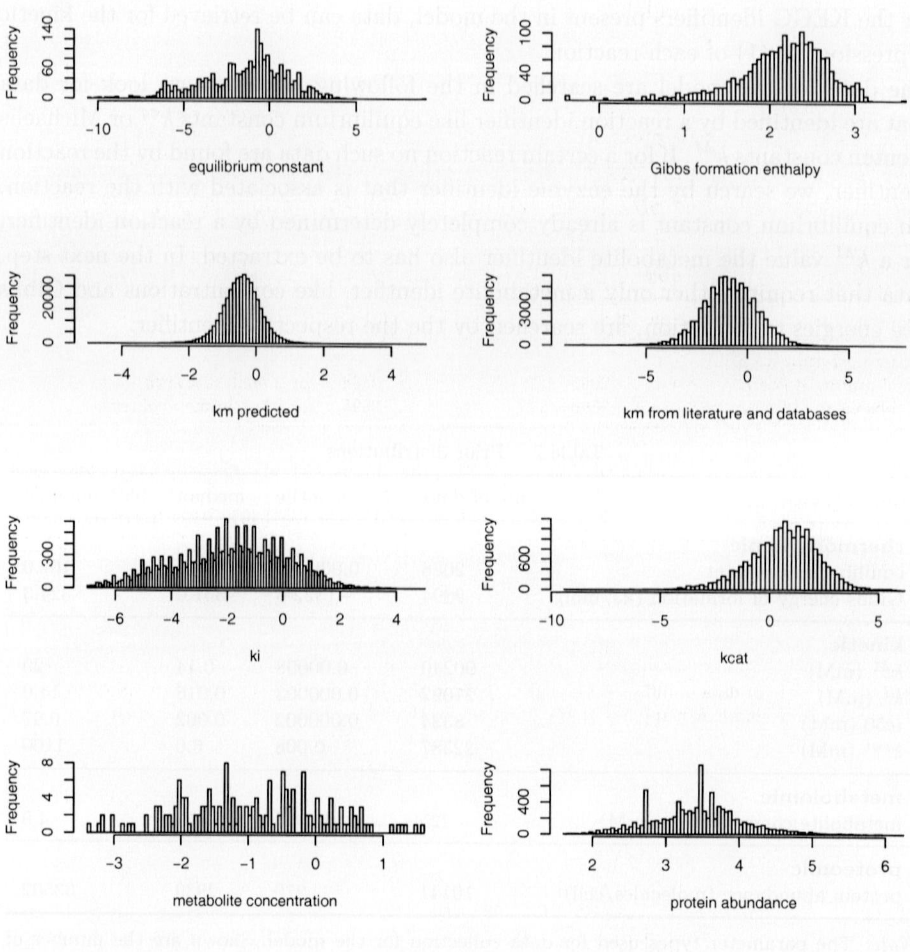

Fig. 2. Prior distributions of logarithms of different data types used in the Bayesian parameter estimation approach.

We thus regard the information gained for the parameters as uncertain data and take

a Bayesian approach to find a complete set of parameters that is thermodynamically consistent [10]. From a statistics over all collected data of certain parameter types, we derive prior distributions for these parameter types as indicated in Table 2. The distributions of the logarithmic parameters are also shown in Fig. 2. For instance, a log-normal distribution fitted to k^M values in the database Brenda [14] is used as a prior for each k^M value in a model. The parameter values we retrieve for our specific model are used as data that have to be explained by the parameter set of the model; this determines a likelihood function. The prior and the likelihood function combined yield a posterior distribution of the model parameters given the found data points for the model [10].
By random sampling from the posterior distribution we can find distributions of the behaviour of the model.

Kinetic model

Once the entities of the metabolic network, i.e. reactions, metabolites and enzymes, are assigned their parameters, the result is written to an output file in SBML format. The annotations in the SMBL file accord with the MIRIAM standard [8, 18].

Table 3. Statistics of data retrieved for model.

quantity	no. of data	thereof for Sacc. Cerev.	no. of data retrieved for model
thermodynamic			
equilibrium constant	2088	2088	for 2 of 16 reactions
Gibbs energy of formation	9696	9696	for 22 of 36 metabolites
kinetic			
k^M	90240	2475	for 34 of 36 substrate-reaction-pairs
k^I	21092	158	2
$k^{cat}_{\pm}$	22587	144	3 of 74 metabolite-enzyme-pairs
metabolomic			
metabolite concentration	225	30	for 7 of 36 metabolites
proteomic			
protein abundance	10141	10141	for 13 of 15 enzymes

Note: The number of data of different parameter types in the database, how many of them apply to Saccharomyces cerevisiae and how many have been extracted for the model (not only kinetic parameters).

3. Results

As a test case, we applied the described automatic model generation to the sulphur assimilation and the glutathione biosynthesis pathways in the yeast Saccharomyces cerevisiae. These pathways play an important role in the buffering of arsenic in order to avoid toxic effects: the cell increases the uptake of sulphur, leading to a raised glutathione level. Glutathione, having a high reduction potential, forms a complex

with arsenic and the complex then is disposed in the vacuoles. The expression of the enzymes involved in these pathways is enhanced upon exposure to arsenic [12]. From a manually sketched metabolic network of the sulphur assimilation and the glutathione biosynthesis pathways, an enhanced version of the model in [12], we looked up the KEGG reaction identifiers. With these identifiers, information about the reactants and enzymes is fetched from the KEGG database. The result is the metabolic network shown in Fig. 1.
In the data retrieval step we could find 131 entries; some of them refer to the same parameter of the model and are averaged. After averaging and balancing we are left with 49 parameters.
The prior distributions of the logarithms of the parameters are shown in Fig. 2. In some cases they can be well described by a normal distributions (i.e. the parameter itself is log-normal distributed). Especially where the number of collected data (see. Table 2) are large the normal distribution is a good description of the respective actual distribution. This is especially the case for the k^M, k^{cat}, k^i values and the Gibbs formation enthalpies.

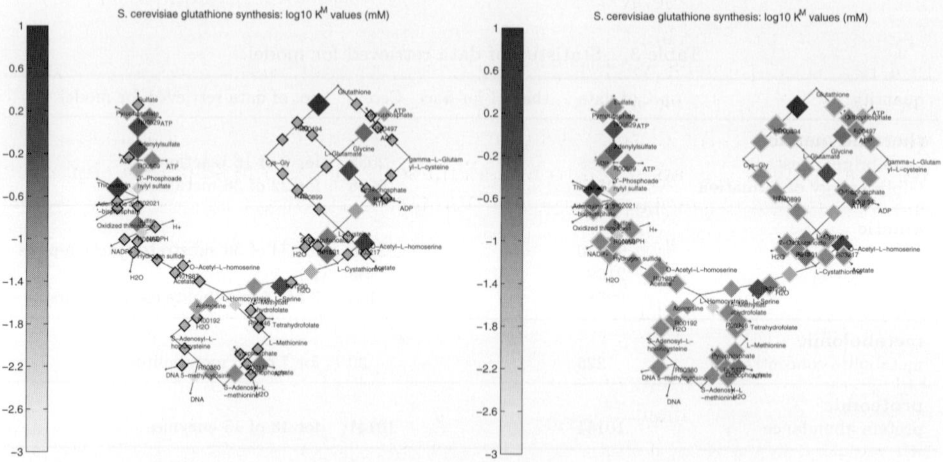

Fig. 3. Michaelis-Menten constants in the sulphur-glutathione model. Left: k^M values retrieved from the database Brenda [14]. Some of the values are missing (grey diamonds with black border). Right: balanced, complete set of k^M values for the model.

After averaging we are left with 49 experimental values useful for assigning values to the kinetic parameters of the model (s. Table 3). The kinetic parameters of the model are determined by 127 independent kinetic and thermodynamic values. Those values that cannot be extracted from the database are mainly determined by the mean values of their prior distributions (seeTable 2) and then undergo the thermodynamical adjustment and Bayesian procedure. In Fig. 3 we show as an

example the k^M values of the model. To the left we display the number of extracted k^M values from the database (missing data are indicated by black borders of the diamonds) and their numerical values. To the right we show the model parameters after the adjustment to thermodynamical constraints and the replacement of missing values in the course of the Bayesian procedure. High numerical values tend to stay high, missing ones are replaced by "average" numerical values.
When simulated with initial concentration values of the metabolites in the range of 0.1 to 10 mM, and holding the concentrations of the cofactors constant, the model yields concentrations in the range of 1μM to 1mM. The fluxes obtain values in the range of 1nM/s to 1μM/s.

4. Discussion

We have presented a workflow to automatically generate metabolic models. We begin with a set of reactions, set up the structure of the model, search data that determine the kinetic parameters of the model as well as concentrations of enzymes and metabolites. These data cannot simply be written into the kinetic expression of Eq.(1), but have to be averaged and adjusted to thermodynamic constraints [9]. The data are taken as hints that determine distributions of the model parameters: In a Bayesian approach we draw values from a posterior distribution of the model parameters given the values extracted from the database [10]. In further analyses we can thus assess the distribution of models and their dynamic behaviours.
We have applied this approach to a medium-scale model, the sulphur assimilation and glutathione biosynthesis pathways. We fed an enhanced version of the model in [12] into the workflow. The outcome is a parametrised model in the SBML format with annotations according to the MIRIAM standard. The model can be fed into simulation tools for further analysis of its dynamical properties, for instance for comparison with metabolite time courses.
The parameter balancing, i.e. the adjusting to thermodynamical constraints and combining prior and likelihood, results in a joint posterior distribution of all parameters; in particular, it ensures that we obtain a mean value for each parameter, even if no data about this parameter are available. The posterior distribution of an "unknown" parameter will reflect two sources of knowledge: (i) its prior (e.g., the posterior of an unknown k^I value is just its prior, that is, the distribution of all known k^I values); (ii) its relationships with other, possibly known, parameters (e.g., as $v^{max} = Ek^{cat}$, experimental data of E and v^{max} will affect the posterior of k^{cat}).
The posterior parameter distribution can be used as a prior for subsequent modelling in which new data, e.g., metabolic timecourses, are incorporated (see [10]). Hence, our approach takes the form of an iterative learning tool.
The fact that we could not find data for all the parameters of the model is partly due to the simple lack of available data. Another reason is incomplete mapping of reaction names and metabolite names to the appropriate KEGG entities. Further-

more, also in KEGG there are cases that two entities appear to the knowledgeable user as the same, but not to computer tools. Hence, a standard for names of biological entities is desirable. It would greatly improve approaches like the presented automated modelling approach and help in paving the way towards genome scale models.

References

[1] Albe, K.R., Butler, M.H., and Wrigth, B.E., Cellular concentrations of enzymes and their substrates, *J. Theor. Biol.* 143(2):163–195, 1990.

[2] Alberty, R.A., Equilibrium compositions of solutions of biochemical species and heats of biochemical reactions, *Proc. Natl. Acad. Sci. USA*, 88(8):3268–3271, 1991.

[3] Borger, S., Liebermeister, W., and Klipp, E., Prediction of Enzyme Kinetic Parameters Based on Statistical Learning, *Genome Inform*, 17(1): 80–87, 2006.

[4] Borger, S., Uhlendorf, J., Helbig, A., and Liebermeister, W., Integration of enzyme kinetic data from various sources, *In Silico Biol.*, 7(S1):09, 2007.

[5] Hartmann, K. and Schomburg, D., GibbsPredictor: Predicting Gibbs energies from molecular structures, *Bioinformatics*, submitted.

[6] Hoops, S., Sahle, S., Gauges, R., Lee, C., Pahle, J., Simus, N., Singhal, M., Xu, L., Mendes, P., and Kummer, U., COPASI - a COmplex PAthway SImulator, *Bioinformatics*, 22(24):3067–3074, 2006.

[7] Klipp, E., Herwig, R., Kowald, A., Wierling, C., and Lehrach, H., *Systems Biology in Practice. Concepts, Implementation and Application*, Wiley-VCH Verlag GmbH and Co. KGaA, Weinheim, 2005.

[8] Le Novere, N., Finney, A., Hucka, M., Bhalla, US., Campagne, F., Collado-Vides, J., Crampin, E.J., Halstead., M., Klipp, E., Mendes, P., Nielsen, P., Sauro, H., Shapiro, B., Snoep, J.L., Spence, H.D., and Wanner, B.L., Minimum information requested in the annotation of biochemical models (MIRIAM), *Nat. Biotechnol.*, 23(12):1509–1515, 2005.

[9] Liebermeister, W. and Klipp, E., Bringing metabolic networks to life: convenience rate law and thermodynamic constraints, *Theor. Biol. Med. Model.*, 3:41, 2006.

[10] Liebermeister, W. and Klipp, E., Bringing metabolic networks to life: integration of kinetic, metabolic, and proteomic data, *Theor. Biol. Med. Model.*, 3:42, 2006.

[11] Mavrovouniotis, M.L., Estimation of standard Gibbs energy changes of biotransformations, *J. Biol. Chem.*, 266:14440–14445, 1991.

[12] Thorsen M., Lagniel, G., Kristiansson, E., Junot, C., Nerman, O., Labarre, J., and Tamas, M.J., Quantitative transcriptome, proteome and sulfur metabolite profiling of the Saccharomyces cerevisiae response to arsenite, *Physiol Genomics*, 30(1):35–43, 2007.

[13] Schomburg, I., Chang, A., and Schomburg, D., BRENDA, enzyme data and metabolic information, *Nucleic Acids Res.* 30(1):47–49, 2002.

[14] `http://www.brenda.uni-koeln.de/`

[15] `http://www.celldesigner.org/`

[16] `http://www.genome.ad.jp/kegg/`

[17] `http://sysbio.molgen.mpg.de/KMedDB`

[18] `http://www.ebi.ac.uk/compneur-srv/miriam/`

[19] `http://xpdb.nist.gov/enzyme_thermodynamics/`

[20] `http://www.r-project.org/`

[21] `http://sbml.org/`

[22] `http://sysbio.molgen.mpg.de/KMedDB/`

[23] `http://yeastgfp.ucsf.edu/`

CONVERSION FROM BIOPAX TO CSO FOR SYSTEM DYNAMICS AND VISUALIZATION OF BIOLOGICAL PATHWAY

EUNA JEONG
eajeong@ims.u-tokyo.ac.jp

MASAO NAGASAKI
masao@ims.u-tokyo.ac.jp

SATORU MIYANO
miyano@ims.u-tokyo.ac.jp

Human Genome Center, Institute of Medical Science, University of Tokyo, Tokyo 108-8639, JAPAN

The vast accumulation of biological pathway data scattered in various sources presents challenges in the exchange and integration of these data. Major new standards for representation of pathway data and the ability to check inconsistency in pathways are inevitable for the development of a reliable pathway data repository. Within the context of biological pathways, the cell system ontology (CSO) had been developed as a general framework to model system dynamics and visualization of diverse biological pathways. CSO provides an excellent environment for modeling, visualizing, and simulating complex molecular mechanisms at different levels of details. This paper examines whether CSO addresses the integration capability of pathway data with system dynamics. We present a conversion tool for converting BioPAX to CSO. Transforming the data from BioPAX to CSO not only allows an analysis of the dynamic behaviors in molecular interactions but also allows the results to be stored for further biological investigations, which is not possible in BioPAX. The conversion is done using simple inference algorithms with the addition of view- and simulation-related properties. We demonstrate how CSO can be used to build a complete and consistent pathway repository and enhance the interoperability among applications.

Keywords: Cell System Ontology (CSO); biological pathway integration; ontology mapping; system dynamics; BioPAX.

1. Introduction

In the current post-genomic era, the interactions among biological entities and networks are being uncovered by molecular biologists at an accelerating pace. Understanding individual biological entities and networks is not sufficient to explain how a cell works. There is a growing need for developing environments that enable us to describe complex and dynamic biological pathways at the system level. In order to address this requirement, we had developed a new system-dynamics-centered ontology called the cell system ontology (CSO) [3]. The three main features of CSO are as follows:

First, CSO allows the manipulation of different levels of granularities and abstraction of pathways, e.g., metabolic pathways, regulatory pathways, signal transduction pathways, and cell-cell interaction. CSO has a hierarchical structure to explicitly define the classes and relationships among those classes. It ensures that the relations between the classes are treated in a correct and consistent way.

Secondly, CSO can capture both quantitative and qualitative models by using the hybrid functional Petri net with extension (HFPNe) [6]. CSO can explain not only the qualitative aspects of a model such as the biological functions and behavior of the networks but also the quantitative features such as reaction priority and kinetics.

Thirdly, CSO can encode information related to visualization and simulation of biological pathways. A well-designed representation will reduce the development time of special applications and enhance communication between software tools. In addition, CSO provides mature core vocabularies for annotating biological properties and standard icons for easy modeling and accelerating the exchangeability among applications.

Recently, a large amount of biological pathway data has been generated. This data is available in several formats such as BioPAX [1] and SBML [2]. Unfortunately, these formats neglect the system dynamics behavior or lack the formal definitions of each term. In order to facilitate data integration, we have made an effort to convert the existing pathway representations, particularly BioPAX, to CSO. BioPAX is based on a formal ontology, and many pathway databases export their data to the BioPAX format. BioPAX level 2 represents only metabolic pathway and molecular interactions. Additional types of pathway data such as signal transduction pathways and genetic regulatory networks are yet to be captured. Furthermore, the BioPAX format does not support dynamic models for simulation.

In this paper, we investigate the capability of CSO with regard to integration of pathway data with system dynamics. Transforming the data from BioPAX to CSO not only allows an analysis of the dynamic behaviors along with a visualization of the molecular interactions but it also allows the results to be stored for further biological investigations, which is not possible in BioPAX. The conversion allows other pathway data represented in the BioPAX format to benefit from CSO tools such as Cell Illustrator for visualization and simulation [5] and BioGraphLayout for automatic layout [4].

2. Comparison of Two Biological Pathway Representations – CSO and BioPAX

This section compares two representations–CSO and BioPAX–for biological pathways in the Web Ontology Language (OWL) [7]. In order to avoid confusion among terms used in the different formats, we use a namespace prefix `cso:` for CSO and `bp:` for BioPAX. CSO is a comprehensive representation for dynamic cell systems based on HFPNe [6] and consists of 195 classes. For the complete description and specifications of CSO, refer [3] and [10], respectively. BioPAX is a data exchange format for metabolic and molecular interactions, consisting of 41 classes. For the specifications of BioPAX, refer [1]. In this section, we compare only the two formats of the main data model.

Biological pathway model CSO is a general framework to understand the behavior of cell systems in an integrated manner. Therefore, the ontology has to represent not only a biological model itself but also a complete environment of the model, such as the results of model simulation, graphical representation of a model, and literature citations. All data in CSO is structured around `cso:Project` that represents the comprehensive environment of a pathway model. A project has one `cso:Model`, which describes the pathways via a set of simulatable biological processes based on HFPNe and biological facts. The class `cso:Fact` is designed to represent information that is not related to dynamic simulation but important for understanding the pathway functionalities such as the effect of a drug's efficacy in terms of the degree to which it binds to plasma protein.

BioPAX defines `bp:pathway` that consists of a set of interactions. Some interactions can be grouped to convey any meaning as pathway steps. A pathway consists of subpathways and interactions; alternatively, a pathway can be defined without specifying the interactions within it. In a pathway, the pathway steps can be listed in `bp:pathwayStep` and order relationships between pathway steps may be established to describe the overall flow of a pathway by using NEXT-STEP and STEP-INTERACTIONS slots. However, the temporal order may not be significant for specific steps.

Since the BioPAX classes `bp:pathway` and `bp:pathwayStep` provide insights into the underlying pathways, this information is mapped to `cso:Fact`. If the slot PARTICIPANTS of `bp:interaction` includes `bp:pathway`, i.e. a pathway catalyzed by an enzyme, this `bp:interaction` is converted to `cso:Fact`.

Biological interaction CSO has a process-centered structure to represent biological pathways. The class `cso:Process` represents the simulatable interactions among physical entities via `cso:Connector`. The same entity may play different roles along with the involved process as an activator, an inhibitor, or a reactant. Depending on its role, a different connector is needed.

BioPAX describes a catalyzed interaction by using `bp:control` and `bp:conversion` as subclasses of `bp:interaction`. The `bp:control` interaction must have one controller as a physical entity and one controlled interaction as a conversion. If a biochemical reaction is catalyzed by multiple enzymes, a separate catalysis interaction is required for each enzyme. If uncatalyzed, the `bp:conversion` interaction can be defined without `bp:control`.

Figure 1 shows how an enzyme-mediated trimerization of a protein complex is represented in (a) BioPAX and (b) CSO. In the figure, the boxes depict instances of classes and the arrows indicate the relationships between instances, i.e., slot names. BioPAX uses two classes `bp:catalysis` and `bp:biochemicalReaction` for the catalyzed interaction, while CSO requires only one–`cso:ProcessBiological`. The participants of `bp:catalysis` are an enzyme and an interaction described via CONTROLLER and CONTROLLED slots, respectively. In turn, `bp:biochemicalReaction` has two participants, as shown in the LEFT and RIGHT slots. On the other hand, in CSO, three entities are involved in trimerization, and each of them is linked to trimerization via different connectors. The stoichiometric coefficient included in `bp:physicalEntityParticipant` is represented as the simulation property of `cso:Connector` rather than that of the physical entity.

Biological entity BioPAX defines several slots such as CONTROLLER, COFACTOR, RIGHT, and LEFT for describing the participants in interactions. The range of these slots are all `bp:physicalEntityParticipant`, which holds a physical entity as `bp:physicalEntity` and other information such as cellular location, stoichiometric coefficient, and sequence features.
In CSO, `cso:Entity` contains properties for location and sequence features related to an entity as well as the entity itself; it is not separated into two classes. Therefore, the combination of `bp:physicalEntityParticipant` and `bp:physicalEntity` corresponds to a single `cso:Entity`.
Figure 1 also shows how the participants involved in the trimerization of a protein complex are represented in (a) BioPAX and (b) CSO. In BioPAX, each LEFT and RIGHT is `bp:physicalEntityParticipant` that wraps `bp:complex`. In turn, the COMPONENTS slot of complex_R3 has another physical entity participant whose physical entity is complex_R1. In (b), each connector links a trimerization process to each entity. The three ENTITY slots of complex_R3 indicate that complex_R3 consists of three complex_R1s. The stoichiometric coefficient is represented as the connector's concentration to transfer from the entity to the process in CSO. The same `cso:Entity` can participate in two different processes via different connectors, which may have different kinetics.

Core vocabulary BioPAX refers to external controlled vocabulary to annotate biological pathway data such as cell types, cellular locations, evidence, and experimental forms. CSO also defines a class called `cso:ControlledVocabulary` (`CV`) for annotating biological properties.
However, there are several differences between BioPAX and CSO. First, in CSO, `cso:CV` is further divided into several subclasses for distinctive usage and rapid parsing, for example, `cso:BiologicalEvent` for a biological process and `cso:BiologicalRole` for the entity's primitive role, e.g. cofactor and enzyme. Second, `cso:CV` provides a predefined common vocabulary for annotation as instances of each subclass, rather than refers external sources. The terms are selected from freely available sources and reorganized to meet the objective of CSO. Lastly, in order to avoid losing the relationship already defined in the external sources, we place this information in CSO. It will reduce the time to parse and query external sources.

How to represent visualization and simulation The functionality of modeling system dynamics and visualization is unique to CSO. For the mathematical simulation of biological pathways, `cso:Entity`, `cso:Process`, and `cso:Connector`

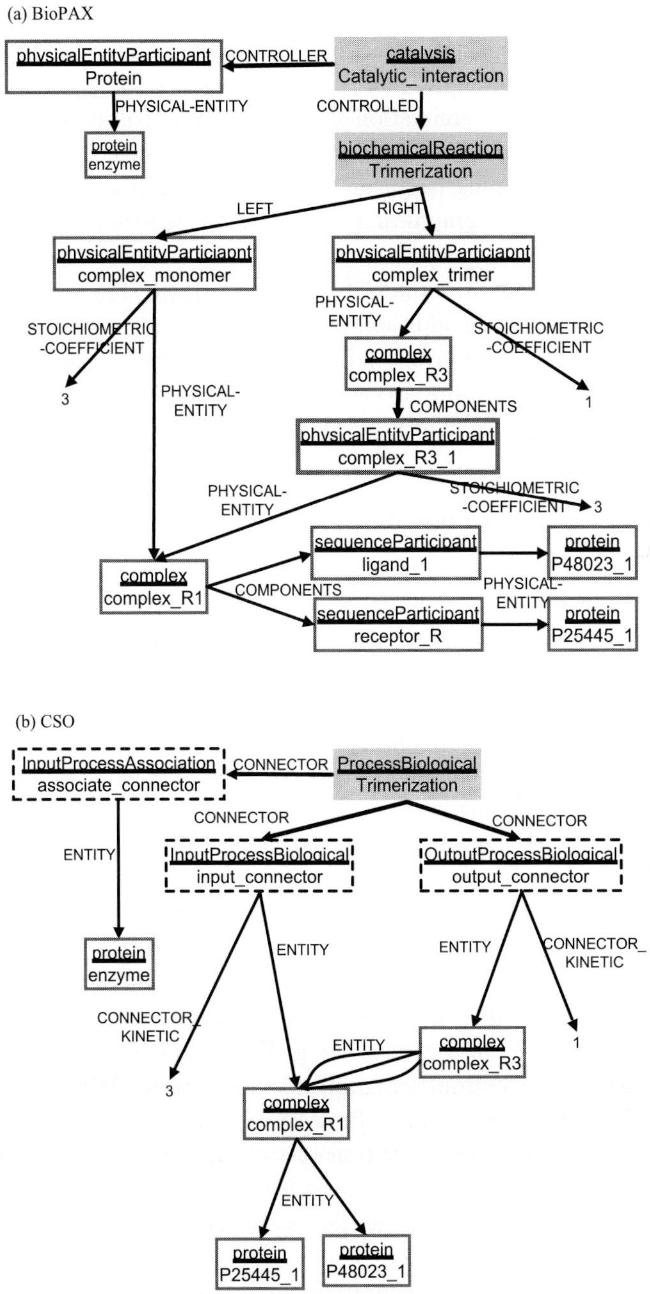

Fig. 1. Graphical view of (a) BioPAX and (b) CSO classes representing trimerization of a protein complex. The boxes show classes: grey for interaction, dotted for `cso:Connector`, and others for participants. The arrows connect two instances (classes) via slots.

all have different slots to describe simulation properties such as initial value, evaluatable script, kinetics, and delay (for details, see [3]). The conversion procedure assigns default values to the simulation-related properties. The simulation tools will use these values to enable the simulation of a time-interval-based discrete event, continuous events performed by differential equations, and more complicated events by using object-like programming language. CSO defines several graphical properties such as geometric position, graphical shape, and image-file-related properties for visualization. The layout of the biological components in networks and the corresponding image files can be stored in CSO, following which they will be used by the pathway visualization tools. In addition, the corresponding icons for all core terms of biological events and cell components are embedded in CSO. As an image file, both non-scalable and scalable coding are acceptable. For example, Portable Network Graphics (PNG) [12] and Scalable Vector Graphics (SVG) [14] are recommended.

3. Inference for Transforming BioPAX to CSO

For ontology mapping between BioPAX and CSO, the main focus is on the inference of `cso:Connector`, which defines the role of the participating entities and the simulation conditions. The comprehensive rules for mapping between BioPAX and CSO are given in Supplementary material.

In the first step, we identify `cso:Entity` from BioPAX data. From the definition of BioPAX, the physical entity participating in a process is stored in PHYSICAL-ENTITY of `bp:physicalEntityParticipant`. The same physical entity can be used in multiple `bp:physicalEntityParticipants`. Hence, a pair of `bp:physicalEntityParticipant` and its `bp:physicalEntity` is mapped to one `cso:Entity`.

In the next step, we identify `cso:Process` from BioPAX data. The pair of `bp:control` and `bp:conversion` is mapped to `cso:Process`. If a control interaction controls a pathway, it will be mapped to `cso:Fact` and not to `cso:Process`. If `bp:conversion` is uncatalyzed, it is mapped to `cso:Process`. The algorithms for mapping `cso:Entity` and `cso:Process` are given in Supplementary material.

Here, we need to generate appropriate connectors for participant properties, namely, CONTROLLER, LEFT, and RIGHT. BioPAX contains five slots to influence the direction of a reaction and the regulation role of an enzyme: CONTROL-TYPE, DIRECTION, SPONTANEOUS, LEFT, and RIGHT. The LEFT and RIGHT slots denote the reaction direction as well as the participants. Algorithm 3.1 shows the inference procedure for the connectors.

The inference of CSO connector is simply divided into two parts based on whether or not a conversion interaction is controlled by a control interaction. The priority is given in the following order: CONTROL-TYPE, DIRECTION, and SPONTANEOUS. The CONTROL-TYPE property of `bp:control` defines the control relationship, which is described in lines 2 to 6 in the algorithm. Depending on the value, i.e., activation or inhibition, the controller entity in BioPAX is connected to

Algorithm 3.1 inferCSOConnector

```
if bp:control has bp:conversion then
  if (CONTROL-TYPE eq "activation") then
    create cso:InputAssociation for CONTROLLER
  else if (CONTROL-TYPE eq "inhibition") then
    create cso:InputInhibitor for CONTROLLER
  end if
  if (c_i is bp:catalysis and defines DIRECTION) then
    if (DIRECTION eq "reversible") then
      create cso:InputProcess for LEFT and cso:OutputProcess for
      RIGHT {for one process P1}
      create cso:OutputProcess for LEFT and cso:InputProcess for
      RIGHT {for another process P2}
    else if (DIRECTION eq "physiol-left-to-right" or "irreversible-left-to-
    right") then
      create cso:InputProcess for LEFT
      create cso:OutputProcess for RIGHT
    else if (DIRECTION eq "physiol-right-to-left" or "irreversible-right-to-
    left") then
      create cso:OutputProcess for LEFT
      create cso:InputProcess for RIGHT
    end if
  end if
end if
if bp:conversion is uncatalyzed then
  if (SPONTANEOUS eq "L-R" or empty) then
    create cso:InputProcess for LEFT
    create cso:OutputProcess for RIGHT
  else if (SPONTANEOUS eq "R-L") then
    create cso:OutputProcess for LEFT
    create cso:InputProcess for RIGHT
  end if
end if
```

a process via an association connector or an inhibitor connector, respectively. The "create" function in the algorithm is accompanied by a referral to an entity already defined in the first step, thereby identifying `cso:Entity`.

BioPAX uses the DIRECTION property to indicate the directionality and reversibility of the reaction, using values such as reversible, irreversible-left-to-right, irreversible-right-to-left, physiol-left-to-right, and physiol-right-to-left. If the DIRECTION slot of `bp:catalysis` is defined, then the entities included in the LEFT and the RIGHT slots will be mapped to different connectors in CSO, as shown in lines 7 to 18. The reversibility of the reaction is specified, for example, "reversible" for the interaction occurring in both directions, and "irreversible-left-to-right" and "irreversible-right-to-left" for the interactions occurring only in the specified direction. In the "reversible" case, the participants in the interaction may be either reactants or products. In CSO, two processes are required for a reversible interaction. For each direction, every participant is linked to one process as an input and to another process as an output, as shown in lines 8 to 10. We consider that "physiol-left-to-right" and "irreversible-left-to-right" imply the same direction from left to right regardless of the reversibility shown by lines 11 to 13. The same assumption is used for the opposite direction, "physiol-right-to-left" and "irreversible-right-to-left," in lines 14 to 16. The class `bp:catalysis` has a slot for the cofactor (not shown in the algorithm). If the cofactor is defined, it will be mapped to an association connector whose biological role is annotated as a cofactor in CSO.

The SPONTANEOUS slot is used to represent a process if it occurs without any external intervention in BioPAX. The lines from 20 to 28 show the case where the conversion interaction is uncatalyzed and the SPONTANEOUS slot is defined. The possible values indicate the direction of the interaction: "L-R" for left-to-right, "R-L" for right-to-left, and "not-spontaneous" for not at all. If the slot value is left empty, the spontaneity is not known. In this case, we assume that the direction is left-to-right. If a conversion is catalyzed, "not-spontaneous" can be used to confirm whether the interaction is controlled by a control interaction.

In BioPAX, LEFT and RIGHT are not used to indicate the direction of a conversion. As shown above, depending on the directionality, LEFT may constitute either substrates or products of a conversion. If the directionality is not specified, we consider that LEFT stores substrates and RIGHT stores products, following the conventions for LEFT and RIGHT of BioPAX [1].

4. Experimental Results

This section describes the preliminary experiments based on the inference algorithms described in Section 3. We have selected the G1/S DNA damage checkpoints pathway in BioPAX from Reactome (ID=69615). For the purpose of reasoning, the Pellet OWL reasoner [8] is used. For obtaining a detailed mapping table between classes in two ontologies, see Supplementary material.

Since the OWL format is intended to be machine-readable, an intuitive, graphical visualization of the interaction is desirable to analyze the given model. In the following figures, we have used simplified notations for brevity. For example, the components of 26S_proteasome, which consists of 41 proteins, are omitted. If the Reactome model uses a complete text-based description for the name, the name is abbreviated; for example, "ubiquitanation" instead of "Ubiquitination_of_phosphorylated_Cdc25A." The layout is modified manually to improve an understanding of the figures.

Figure 2 shows the Reactome BioPAX model loaded in Cytoscape [11]. Cytoscape shows the binary relation between two nodes along with the direction.

The nodes represent instances of `bp:catalysis`, `bp:biochemicalInteraction`, and `bp:physicalEntity`. The edges indicate the slots connecting two instances such as CONTROLLED and LEFT. Since Cytoscape shows the participants at the `bp:physicalEntity` level, the information about cellular location and post-transformation stored in `bp:physicalEntityParticipant` is lost in this view. For example, Unitprot:P30304-1, a boxed protein shown in the top left of Figure 2, is related to three participants: P30304 located in the cytoplasm, P30304 located in the nucleoplasm, and P30304 phosphorylated located in the nucleoplasm. Each participant plays a different role: LEFT of ubiquitination, LEFT of phosphorylation, and RIGHT of phosphorylation, respectively. Similar situations are shown for UniProt:P04637-1 and UniPort:Q00987-1 in the bottom left of the figure.

The converted model in CSO is visualized via the Cell Illustrator [9] in Figure 3. Since CSO encodes physical entities and their related information simultaneously, Unitprot:P30304-1 is correctly represented as three entities depending on their cellular location and modified state in the Cell Illustrator view. Cell Illustrator uses the cellular locations to correctly locate the entities and processes in subcellular compartments. The used icons for the processes are the default images provided by CSO.

In addition, the initial values for simulation are assigned for each process and entity because they are not available in the BioPAX format. The changes in the concentration of the interesting entities along with time are charted. The results of the simulation with charts can be stored in CSO. Any changes such as replacement of images with user-defined images and adjustments to parameters are recorded in CSO. The information encoded in CSO can be reused by any associated tool for pathway analysis, visualization, and simulation with less effort.

During conversion, we detected several problems such as ambiguous and missing information in the Reactome BioPAX. BioPAX has a restriction in that some classes are defined for organizational purposes and should not have instances. Unfortunately, when pathway databases are exported to BioPAX, some concepts could not be converted into BioPAX. Furthermore, there are no proper guidelines for incorporating external databases to BioPAX. In the Reactome example, ubiquitin ligase and cyclin E are defined as `bp:physicalEntity` and not `bp:protein`. Although CSO provides a more detailed hierarchy of classes, it is not easy to identify the entity type, e.g., protein or DNA, without some kind of human intervention.

We detected another ambiguity in a catalyzed interaction. In the lower right region of Figure 2, p21/p27 plays two different roles: CONTROLLER of catalysis and LEFT of inactivation. This case cannot be simulatable because an enzyme does not consume its concentration during an interaction, but a substrate does it. In CSO, the catalyzed inactivation is mapped to one process, and p21/p27 is involved in inactivation via two connectors: an association connector and an input connector. In this case, the interaction can be mapped to a biological fact in CSO because it is not clear which one is inactivated. On the other hand, we interpreted this interaction as two processes by adding one more binding process. In the first process, p21/p27

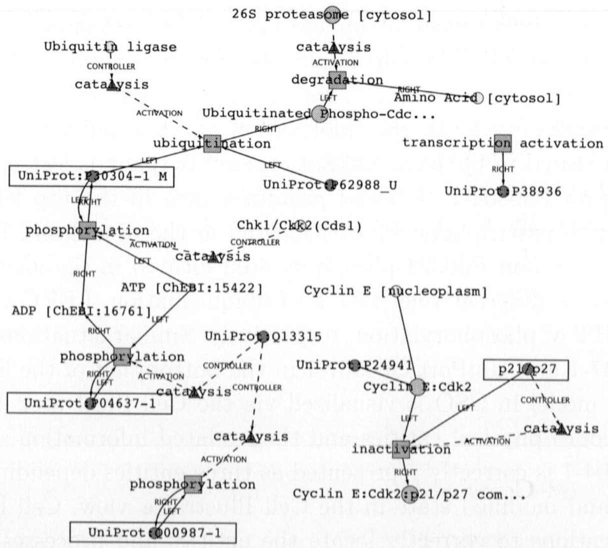

Fig. 2. Cytoscape view. The names of nodes are abbreviated after modification. The visual styles are set manually. Catalysis is denoted by the triangles; conversion, by the rectangles; and physicalEntity, by the circles.

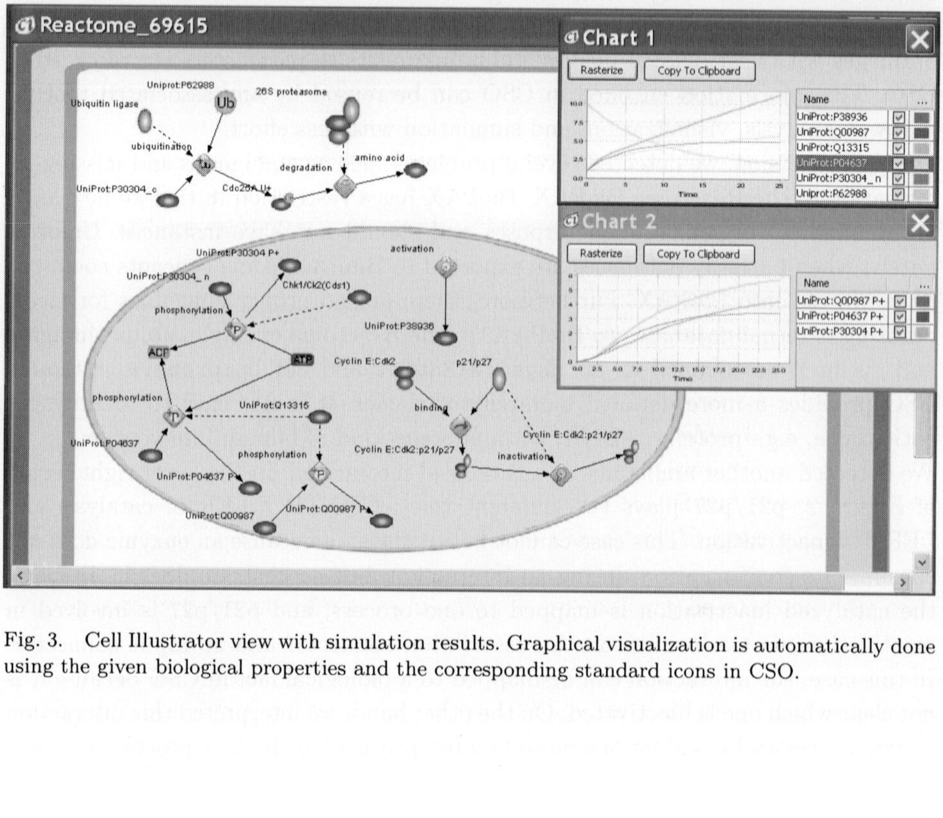

Fig. 3. Cell Illustrator view with simulation results. Graphical visualization is automatically done using the given biological properties and the corresponding standard icons in CSO.

and Cyclin E:Cdk2 bind together and a complex is generated, while in the second process, the resulting complex Cyclin E:Cdk2:p21/p27 is inactivated by p21/p27, as shown in Figure 3.
Another problem is that INTERACTION-TYPE is not defined in the Reactome BioPAX. This property is important to decide the type of process that occurs. In particular, CSO provides the corresponding image for each process and supports its visualization. Generally, if the type of entity or process is not given, an unknown type is assigned. For this experiment, we assign the interaction type from the given instance ID that describes the detailed interaction.

5. Discussion and Conclusion

In this study, we have explored the potential of the cell system ontology (CSO) to integrate biological pathway data with system dynamics and visualization. CSO captures the quantitative aspects of biological functions and equips mature core vocabularies and the corresponding standard icons. These functionalities are important to enhance semantic validation for a given model. The experimental results show how CSO can be used to generate a complete and consistent pathway repository. The conversion generates dynamic models with improved visualization from static ones. The simulations allows to explore the possible dynamic behavior of pathway components and these results might be useful for further investigation of biological systems.
During conversion to CSO, we faced several problems caused by incomplete BioPAX data. First, some important data for visualization, such as the interaction type, is often missing. Therefore, human intervention becomes necessary in order to obtain the correct value, which is time consuming, as the comments have to be read. Secondly, the used terms that have no external references may cause problems. BioPAX uses PSI-MI [13] as controlled vocabulary for the interaction type. For example, "trimerization" is used to describe the interaction that binds three identical molecules in CSO, which is not defined in PSI-MI. However, it is used to describe a binding of three nonidentical molecules in BioPAX. As visualized with the CSO standard icon, this makes a model invalid. Thirdly, as described in Section 3, some concepts could not be converted into BioPAX because BioPAX only supports metabolic pathways and molecular interactions. In the conversion from BioPAX to CSO, it might be necessary to recover the primitive, original meaning used in the external sources.
These abovementioned problems are the challenges we face in the next step of conversion. The ability to check inconsistency and incompleteness in pathways is inevitable for the development of a reliable pathway data repository. In a future work, we will develop more comprehensive inference methods for missing or ambiguous data to reduce human intervention as much as possible.

References

[1] Bader, G. and Cary, M., BioPAX – biological pathways exchange language level 2, version 1.0 documentation, 2005.

[2] Finney, A. and Hucka, M., Systems biology markup language (SBML) Level 2: structures and facilities for model definitions, 2003.

[3] Jeong, E., Nagasaki, M., Saito, A., and Miyano, S., Cell system ontology: representation for modeling, visualizing, and simulating biological pathways, accepted in *In Silico Biology*, 2007.

[4] Kojima, K., Nagasaki, M., Jeong, E., Kato, M., and Miyano, S., An efficient grid layout algorithm for biological networks utilizing various biological attributes, *BMC Bioinformatics*, 8, 76, 2007.

[5] Nagasaki, M., Doi, A., Matsuno, H., and Miyano, S.,Genomic Object Net: I. A platform for modeling and simulating biopathways, *Applied Bioinformatics*, 2:181-184, 2003.

[6] Nagasaki, M., Doi, A., Matsuno, H., and Miyano, S., A versatile Petri net based architecture for modeling and simulation of complex biological processes, *Genome Inform.*, 15(1):180–197, 2004.

[7] Smith, M., Welty, C., and McGuinness, D., OWL Web Ontology Language Guide, 2004

[8] `http://pellet.owldl.com/`, Pellet: The open source OWL DL reasoner.

[9] `http://www.cellillustrator.com/`, Cell Illustrator 3.0.

[10] `http://www.csml.org/`, Cell System Markup Language (CSML).

[11] `http://www.cytoscape.org/`, Cytoscape: analyzing and visualizing network data.

[12] `http://www.libpng.org/pub/png/`, PNG (Portable Network Graphics).

[13] `http://www.psidev.info/index.php?q=node/60`, Molecular interaction XML format documentation.

[14] `http://www.w3.org/TR/SVG11/`, Scalable vector graphics (SVG) 1.1 specification.

AN IMPROVED SCORING SCHEME FOR PREDICTING GLYCAN STRUCTURES FROM GENE EXPRESSION DATA

AKITSUGU SUGA
suga@kuicr.kyoto-u.ac.jp

YOSHIHIRO YAMANISHI
yoshi@kuicr.kyoto-u.ac.jp

KOSUKE HASHIMOTO
khashimo@kuicr.kyoto-u.ac.jp

SUSUMU GOTO
goto@kuicr.kyoto-u.ac.jp

MINORU KANEHISA
kanehisa@kuicr.kyoto-u.ac.jp

Bioinformatics Center, Institute for Chemical Research, Kyoto University, Gokasho, Uji, Kyoto 611-0011, Japan

The prediction of glycan structures from gene expression of glycosyltransferases (GTs) is a challenging new area in computational biology because the biosynthesis of glycan chains is under the control of GT expression. In this paper we developed a new method for predicting glycan structures from gene expression data. There are two main original aspects of the proposed method. First, we proposed to increase the number of predictable glycan structure candidates by estimating missing glycans from a global glycan structure map, which enables us to predict new glycan structures that are not stored in the database. Second, we proposed a more general scoring scheme based on real-valued gene expression intensity rather than converting it into binary information. In the result we applied the proposed method to predicting cancer-specific glycan structures from gene expression profiles for patients of acute lymphocytic leukemia (ALL) and acute myelocytic leukemia (AML). We confirmed that several of the predicted glycan structures successfully correspond to known cancer-specific glycan structures according to the literature, and our method outperforms the previous methods at a statistically significant level.

Keywords: glycosyltransferase; glycan structure; DNA microarray; gene expression.

1. Introduction

Glycans are carbohydrate chains attached to lipids or proteins and are notable as the third type of biological chain next to DNA and proteins, since they have a huge variety of structures and play key roles in a wide variety of biological processes, such as immunity and disease pathogenesis. Pathogens have evolved to exploit host lineage-specific glycans and are constantly shaping the glycomes of their hosts [3]. It is well known that some N-linked glycans are necessary in proper protein folding in eukaryote and specific glycan structures are expressed in carcinoma samples [8]. In addition, some glycans are involved in cell adhesion [1]. Understanding glycan functions requires determining glycan structures, as well as genome and amino-acid sequences.

Some powerful experimental instruments for glycan purification and analysis have been developed and successively improved, such as high-performance liquid chromatography, capillary electrophoresis, mass spectrometry and nuclear magnetic resonance technology [9]. In addition, a variety of computational tools have recently been developed, such as automatic annotation tools for mass spectrometry [4], glycan structure matching methods [2], glycan composite structure maps [6] and glycan structure prediction methods [7]. However, even with these advances, the experimental determination and computational analysis of glycan structures is still difficult. This is because glycans have more complicated structures than DNA and proteins. While nucleotide and amino acid chains are linear and consist of 4 and 20 elementary components, respectively, glycan chains are branched structures and consist of a number of monosaccharides. In addition, they are multivalent, and linkages have anomeric configurations (alpha and beta).

Recently, Kawano *et al.* developed a method for predicting glycan structures based on microarray gene expression data [7]. The basic idea of their method stems from the fact that glycan biosynthesis is under the control of the expression of glycosyltransferases (GTs). If the expression level of GTs is known in the transcriptome or in the proteome of a given organism, it should be possible to predict the repertoire of glycan structures related with the experimental conditions of expression data such as tissues, organs, and diseases. In their method, the gene expression information of GTs is used in the prediction process. However, there are some limitations in Kawano's method. First, the number of predictable glycans depends on the number of glycans stored in the database, because their prediction is based on a database search. Secondly, the prediction accuracy is far from ideal at practical levels, because their method can treat only binary value information of microarray gene expression data.

In this study, we propose a new method to predict glycan structures from gene expression profiles by improving on the framework of Kawano's method. First, we introduced a strategy of predicting missing glycans, which are not stored in the glycan database, in order to add new glycan structures into our candidate set using the glycan composite structure map. Next, we proposed a new scoring scheme to use the original real-valued expression values, the so-called 'signal', from the microarray data, rather than using binary values, the so-called 'detection', because gene expression levels are observed with real-valued signals in most microarray data in nature. Finally, we applied the proposed method to an experimental gene expression dataset from acute lymphocytic leukemia (ALL) and acute myelocytic leukemia (AML) in order to predict cancer specific glycan structures. As a result, we found that the proposed method outperform Kawano's method in terms of the number of correctly predicted cancer-specific glycan structures.

2. Materials and Methods

2.1. Data

2.1.1. Glycosyltransferase reaction

To construct a GT reaction pattern library, GT genes were obtained from the human genome in the KEGG GENES database based on their annotations [1]. The reaction specificity of each GT was determined according to the published literature and was characterized by the following three features: (1) the acceptor monosaccharide residue in the glycan chain, (2) the donor monosaccharide residue, and (3) the linkage between them. Fig. 1 shows an illustration of GT-related reactions. 186 GT genes are currently annotated in the human genome. The reaction pattern library consists of nine kinds of monosaccharides: glucose (Glc), galactose (Gal), mannose (Man), N-acetyl-galactosamine (GalNAc), fucose (Fuc), xylose (Xyl) glucuronic acid (GlcA) and N-acetyl-neuraminic acid (sialic acid, Neu5Ac).

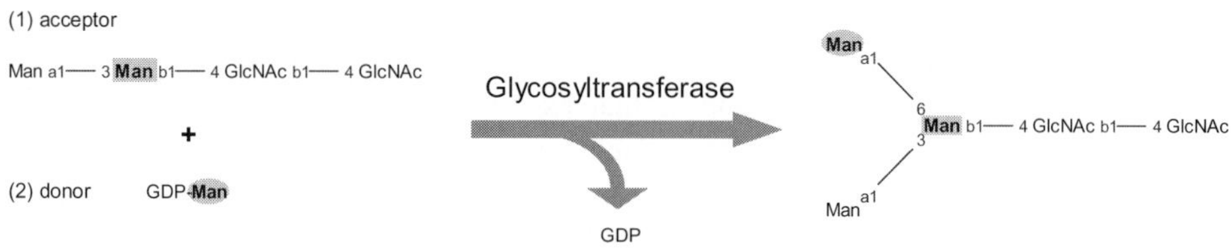

Fig. 1. An example of GT related reactions. This reaction is catalyzed by a GT. The substrates are shown in the left and the product is shown in the right. The acceptor monosaccharide is represented by a gray square and the donor monosaccharide is represented by a gray oval. In this case, the reaction component is 'Man a1-4 Man'. The numbers in the reaction component represent the positions of covalent bonding in the monosaccharides, and 'a' and 'b' represent anomeric configurations 'alpha' and 'beta', respectively.

2.1.2. Glycan structures

All glycan structures were collected from the KEGG GLYCAN database which contains 13022 entries as of this writing. Non-carbohydrate residues in the entries, such as Cer (ceramide), Asn, Ser/Thr, S (sulfate) and P (phosphate) were deleted to obtain glycan entries consisting of only carbohydrates, and duplicate structures were merged. Furthermore, glycan entries, including monosaccharides that are not present in the reaction library, were removed. In this study we focused on the analysis of N-glycans stored in the database, where the number of N-glycans is 1723.

2.1.3. Microarray expression data

Human DNA microarray expression data was obtained from the previous study [5]. The expression dataset consists of 5357 genes for 48 ALL and 21 AML patients. Note that gene expression is measured using real-valued intensity data. We prepared two types of gene expression datasets: ALL and AML datasets. In this study, the gene expression values in ALL and AML datasets are averaged over 48 and 21 patients, respectively. We used these datasets in the prediction process.

The previous method cannot handle real-valued gene expression data [7]. To carry out the previous method, we also prepared a binary type of gene expression dataset in the following manner. The gene expression values were transformed into binary values using a threshold of 1. If the gene expression signal is greater than the threshold, we assign 1 to the corresponding gene. Otherwise, if the gene expression signal is not greater than the threshold, we assign 0 to the corresponding gene. The binary-valued version is used as predictor data in the previous method.

2.2. Methods for predicting glycan structures

2.2.1. Previous method

The previous method is based on a combination of the co-occurrence score of the GT-related reactions and candidate score for the database search [7]. Here we make a brief review of the previous method.

The co-occurrence score is designed to represent how frequently two GT-related reactions occur simultaneously. Suppose that we have a candidate set consisting of n glycans in the database as $G = \{g_i\}_{i=1}^{n}$. Each glycan g_i is broken down into reaction components, where a reaction component consists of two adjacent monosaccharides and their linkage. For example, the reaction component in Fig. 1 is 'Man a1-4 Man'. Here a set of all possible reaction components is represented by $R = \{r_j\}_{j=1}^{m}$, where m is the number of reaction components. Let us define a reaction pattern vector $\mathbf{x}_j$ for the j-th reaction component r_j as $\mathbf{x}_j = (x_{1j}, x_{2j}, \ldots, x_{nj})^T$, where x_{ij} is the frequency of r_j in the i-th glycan g_i. Then, we represent all the reaction components by the reaction pattern vector as $\{\mathbf{x}_j\}_{j=1}^{m}$. The co-occurrence score between two reaction components r_j and r_k is obtained by computing the correlation coefficient between the corresponding reaction pattern vectors $\mathbf{x}_j$ and $\mathbf{x}_k$, for example, using Pearson's correlation coefficient defined as follows:

$$C(r_j, r_k) = \frac{\mathrm{cov}(\mathbf{x}_j, \mathbf{x}_k)}{\sqrt{\mathrm{var}(\mathbf{x}_j)\,\mathrm{var}(\mathbf{x}_k)}}.$$

This score represents how likely it is that two GT genes involving r_j and r_k are co-expressed. The prediction of glycan structures is performed by looking for glycans associated with the GT gene expression. Suppose that we are given a gene expression profile $\mathbf{z} = (z_1, z_2, \ldots, z_p)^T$ from a microarray experiment, where p is the number of GT genes in the microarray experiment. Note that the gene expression profile consists of binary-valued gene expression data in the previous method, where the presence or absence of GT gene transcripts is coded as 1 or 0, respectively. Then, we represent the corresponding reaction components of GT genes in the microarray as $R^* = \{r_l^*\}_{l=1}^{p}$.

Glycans associated with gene expression data are selected from our candidate set G. To select the most appropriate glycan out of the candidate set, Kawano *et al.* proposed to compute the following candidate score for all the candidate glycans g_i $(i = 1,2,\ldots,n)$:

$$S(g_i) = \frac{1}{\sum_{j=1}^{m} I(r_j \in g_i)} \sum_{l=1}^{p}\sum_{j=1}^{m} I(z_l = 1) I(r_j \in g_i) C(r_l^*, r_j),$$

where $C(\cdot,\cdot)$ is the co-occurrence score function, $r_j \in g_i$ means that glycan g_i contains the reaction component r_j, and $I(\cdot)$ is an indicator function which returns 1 if the event is true and returns 0 if the event is false. After the candidate scores are calculated for all glycans stored in the database, high scoring glycan structures are predicted.

2.2.2. *Proposed method*

There are several limitations and drawbacks in the previous method. First, we cannot predict glycan structures which are not stored in the database, because the prediction procedure is only able to select high scoring glycans from the set of glycans stored in the database. Therefore, the number of predictable glycan candidates is very limited in practical applications. Secondly, we have to transform the original expression values into binary values by taking an appropriate threshold. However, the gene expression is observed with real-valued intensity in most microarray experiments, so the discretization process might lead to a loss of information. Therefore, it is more natural to use the original real-valued intensity rather than converting it into binary information. In this study, we develop a new prediction method in order to overcome the two problems.

First, we propose to increase the number of glycan structure candidates to be predicted. A global glycan composite structure map enables us to suggest missing glycan structures and to estimate possible intermediate structures for the corresponding missing glycans [6]. Fig. 2 shows an illustration of missing glycans and possible intermediate structures. The structures of such intermediate glycans have not been determined experimentally so far, but they are likely to exist in nature. We propose to add such intermediate glycan structures as candidates for predictable glycans. Therefore, our candidate glycan set is composed of not only known glycan structures stored in the

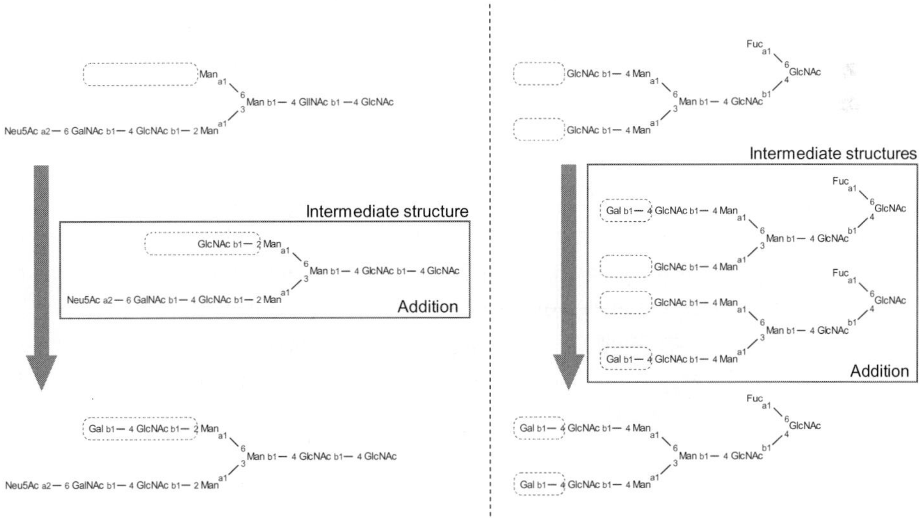

Fig. 2. An illustration of missing glycans and possible intermediate structures. When the intermediate glycans are not stored in the database, they are added into our candidate set for the prediction.

database but also newly identified glycan structures. We represent the number of known glycans and newly identified glycans by n and N, respectively.

Second, we propose a new scoring scheme based on real-valued intensities of microarray gene expression data. Suppose that we are given a GT gene expression profile $\mathbf{z}^* = (z_1^*, z_2^*, \dots, z_p^*)^T$ from a microarray experiment. Unlike the previous method, the gene expression profile consists of real-valued expression intensities in this context. For the database search, we propose the following candidate score based on the original real-valued expression intensities for all the candidate glycans g_i $(i = 1, 2, \dots, n+N)$:

$$S(g_i) = \frac{1}{\sum_{j=1}^{m} I(r_j \in g_i)} \sum_{l=1}^{p} \sum_{j=1}^{m} z_l^* I(r_j \in g_i) C(r_l^*, r_j),$$

where $C(\cdot,\cdot)$ is the co-occurrence score function, p is the number of GT genes in the microarray experiment, and m is the number of all GT reaction components in a similar manner to the previous method. After the candidate scores are calculated for all glycans in the database and for newly identified glycans, then high scoring glycan structures are predicted.

3. Results

3.1. Amplification of predictable glycan structure candidates

We constructed a global composite structure map for all the glycans in the database, which enabled us to identify many missing glycans [6]. We estimated the intermediate glycan structures for the corresponding missing glycans from the global composite structure map in order to increase the number of predictable glycan structure candidates. As a result, we identified 1291 new intermediate glycans which contain N-glycan core structures. Table 1 shows the numbers of N-glycan structures before and after applying

Table 1. The number of predictable glycan structures before and after applying the global composite structure map.

	Before	After
All N-Glycan	1723	3014
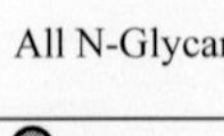	1164	2094
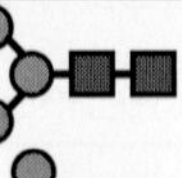	218	372
Others	341	548

GlcNAc, Man

this process. There are two major core structures (80%) shared by N-glycans. Table 1 also shows the numbers of N-glycan structures containing two major N-glycan core structures before and after applying this process. In the following prediction process, we focused on two types of glycans and used not only glycans stored in the database but also newly identified intermediate glycans.

3.2. Prediction of cancer-specific glycan structures

We performed the prediction of glycan structures using two types of gene expression data: ALL and AML dataset. To evaluate the prediction results, we collected known cancer-specific glycan epitopes from the literature [2]. Fig. 3 shows three epitopes which are known to be cancer-specific, referred to as Lewis^a, Lewis^x and sialyl Lewis^x, respectively. Glycans containing the above cancer-specific epitopes are considered to be the gold standard for the performance evaluation below. In this study we are interested in comparing the prediction performance between the previous method and our proposed method.

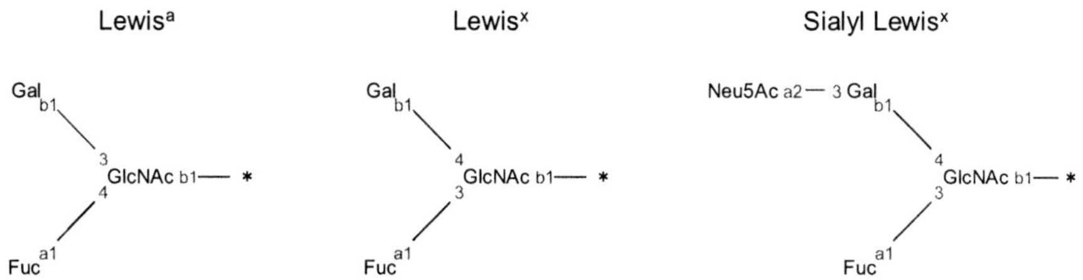

Fig. 3. Cancer-specific glycan epitopes used in this study.

First, we applied both the previous and the proposed prediction methods to ALL gene expression data. If predicted glycans contain cancer-specific substructures, they are considered to be correctly predicted cancer-specific glycans. The left and right panels in Fig. 4 show the index-plots of the sorted scores of the previous and proposed methods,

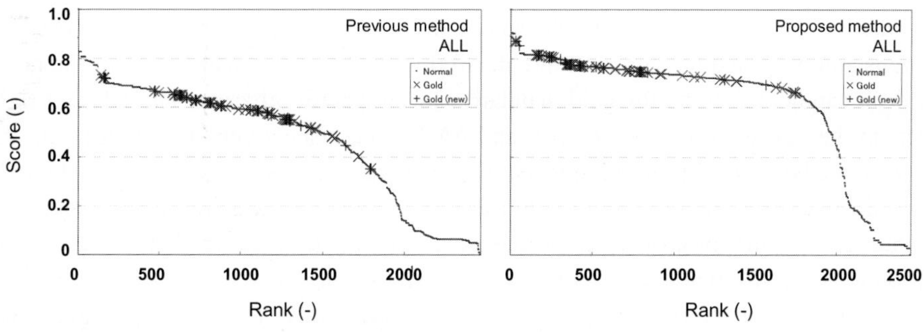

Fig. 4. Performance comparison between the previous method (left) and the proposed method (right): ALL data. Cancer-specific glycans stored in the database are indicated by red crosses, and cancer-specific glycans which are newly identified by our procedure are indicated by blue cross joints.

respectively. In each panel, cancer-specific glycans stored in the database are indicated by red crosses, and cancer-specific glycans that are newly identified by our procedure are indicated by blue cross joints. It seems that high scoring glycans tend to correspond to gold standard cancer-specific glycans. The proposed method seems to catch more information in the prediction process compared to the previous method. This suggests that the use of real-valued gene expression data is meaningful. It also seems that many newly identified glycans have high scores, which also attests to the usefulness of our data amplification procedure.

Next, we applied both prediction methods to AML gene expression data. The left and right panels in Fig. 5 show the index-plots of the sorted scores of the previous and proposed methods, respectively. As in the results for the ALL data, we observed that high scoring glycans correspond to gold standard cancer-specific glycans and many newly identified glycans have high scores. By comparison, the proposed method seems to outperform the previous method in terms of the score ranks of correctly predicted glycan structures. These results also suggest the possibility of predicting glycan structures from observed gene expression patterns in actual applications.

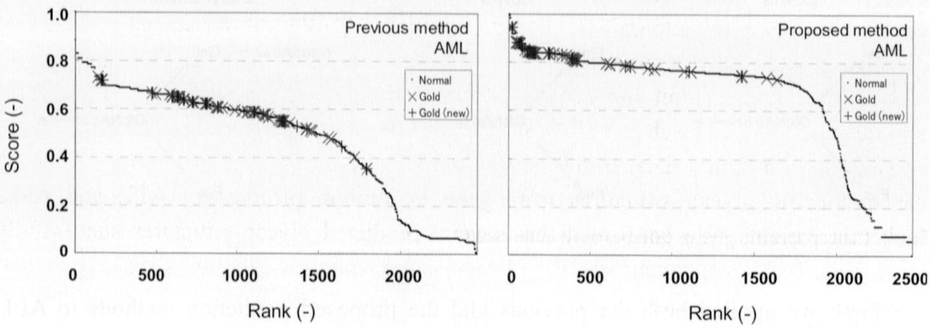

Fig. 5. Performance comparison between the previous method (left) and the proposed method (right): AML data. Cancer-specific glycans stored in the database are indicated by red crosses, and cancer-specific glycans newly identified by our procedure are indicated by blue cross joints.

Finally, we investigated the difference in prediction performance between the previous and proposed methods in more detail. Fig. 6 shows the distribution of the ranks based on candidate scores for gold standard cancer-specific glycans. The left and right panels in the figure correspond to ALL and AML data, respectively. Compared with the previous method, the proposed method seems to assign higher score ranks to gold standard cancer-specific glycans in both cases. We conducted a paired t-test to examine the statistical significance of the performance difference. It is shown that the score rank distributions are different at a statistically significant level, where the p-values in the case of ALL and AML data are 3.2e-29 and 1.6e-48, respectively. These results suggest that our proposed method significantly outperforms the previous method.

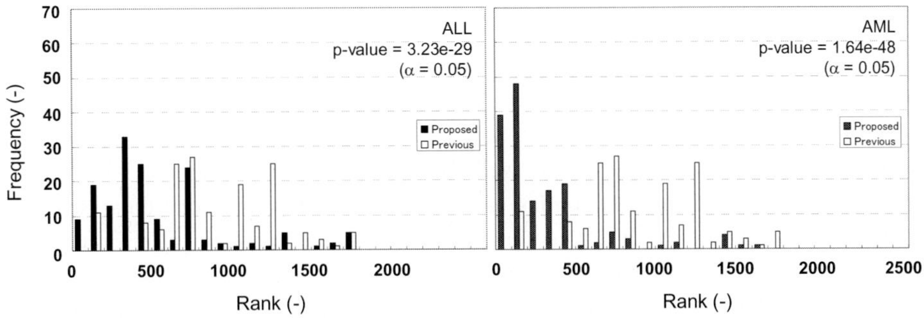

Fig. 6. Distribution of the score ranks for gold standard glycans based on the previous and proposed methods: ALL data (left) and AML data (right). Black bars and white bars correspond to the previous and proposed methods, respectively.

4. Discussion and conclusion

In this paper we developed a new method for predicting glycan structures from gene expression data. There are mainly two original themes in the proposed method. First, we proposed to increase the number of predictable glycan structure candidates by estimating missing glycans from a global glycan composite structure map, which enables us to identify new glycan structures that are not stored in the database. Secondly, we proposed a more general scoring scheme based on real-valued gene expression intensity rather than conversion into binary data. In the result we applied the proposed method to predicting cancer-specific glycan structures from gene expression profiles for ALL and AML leukemia patients. We confirmed that several predicted glycan structures successfully correspond to known cancer-specific glycan structures in the literature, and our method outperforms the previous methods at a statistically significant level.

An advantage of our method is that we can predict new glycan structures that can be synthesized theoretically according to the expression of GTs. Since the experimental determination of glycan structures is still difficult even now, computational prediction of glycan structures might contribute to obtaining new biological findings in glycobiology. In this study we focused on the prediction of cancer-specific N-glycans, but it should be pointed out that our method is applicable to any glycan structure prediction problems for other domains, for example, tissue-, organ-, organism-specific glycans. We are currently working on more comprehensive identification of such domain-specific glycans and application to not only N-glycans but also O-glycans, glycolipids, and several more.

From a technical viewpoint, our scoring scheme is based on the use of expression information about GTs. Recent biotechnology developments have also enabled us to observe the expression level of not only genes but also proteins on a large scale. It will be worth integrating both gene expression and protein expression profiles in the framework of our scoring scheme. In addition, the use of time series data for gene or protein expression might lead to further biological insights.

Acknowledgments

We would like to express our gratitude to Dr. Nelson Hayes for helpful comments and overall improvement of our manuscript. This work was supported by grants from the Ministry of Education, Culture, Sports, Science and Technology of Japan and the Japan Science and Technology Agency, as well as a bridging grant from the NIH/NIGMS Consortium for Functional Glycomics and a research fellowship for young scientists from the Japan Society for the Promotion of Science. The computational resources were provided by the Bioinformatics Center, Institute for Chemical Research, Kyoto University and the Human Genome Center, Institute of Medical Science, The University of Tokyo.

References

[1] Akama, T. O., Nakagawa, H., Sugihara, K., Narisawa, S., Ohyama, C., Nishimura, S., O'brien, D. A., Moremen, K. W., Millan, J. L., and Fukuda, M. N., Germ cell survival through carbohydrate-mediated interaction with Sertoli cells, *Science*, 295(5552):124-127, 2002.

[2] Aoki, K. F., Yamaguchi, A., Ueda, N., Akutsu, T., Mamitsuka, H., Goto, S., and Kanehisa, M., KCaM (KEGG Carbohydrate Matcher): a software tool for analyzing the structures of carbohydrate sugar chains, *Nucleic Acids Res.*, 32(Web Server issue):W267-272, 2004.

[3] Bishop, J. R. and Gagneux, P., Evolution of carbohydrate antigens--microbial forces shaping host glycomes?, *Glycobiology*, 17(5):23R-34R, 2007.

[4] Goldberg, D., Sutton-Smith, M., Paulson, J., and Dell, A., Automatic annotation of matrix-assisted laser desorption/ionization N-glycan spectra, *Proteomics*, 5(4):865-875, 2005.

[5] Golub, T. R., Slonim, D. K., Tamayo, P., Huard, C., Gaasenbeek, M., Mesirov, J. P., Coller, H., Loh, M. L., Downing, J. R., Caligiuri, M. A., Bloomfield, C. D., and Lander, E. S., Molecular classification of cancer: class discovery and class prediction by gene expression monitoring, *Science*, 286(5439):531-537, 1999.

[6] Hashimoto, K., Goto, S., Kawano, S., Aoki-Kinoshita, K. F., Ueda, N., Hamajima, M., Kawasaki, T., and Kanehisa, M., KEGG as a glycome informatics resource, *Glycobiology*, 16(5):63R-70R, 2006.

[7] Kawano, S., Hashimoto, K., Miyama, T., Goto, S., and Kanehisa, M., Prediction of glycan structures from gene expression data based on glycosyltransferase reactions, *Bioinformatics*, 21(21):3976-3982, 2005.

[8] Kim, Y. J. and Varki, A., Perspectives on the significance of altered glycosylation of glycoproteins in cancer, *Glycoconj J.*, 14(5):569-576, 1997.

[9] von der Lieth, C. W., Bohne-Lang, A., Lohmann, K. K., and Frank, M., Bioinformatics for glycomics: status, methods, requirements and perspectives, *Brief. Bioinform.*, 5(2):164-178, 2004.

COMPARISON OF SMOKING-INDUCED GENE EXPRESSION ON AFFYMETRIX EXON AND 3'-BASED EXPRESSION ARRAYS

XIAOLING ZHANG[1]
zhangxl@bu.edu

GANG LIU[2]
gliu@lung.bumc.bu.edu

MARC E LENBURG[3]
mlenburg@bu.edu

AVRUM SPIRA[2]
aspira@bu.edu

[1]*Bioinformatics Program, Boston University, 24 Cummington Street, Boston, Massachusetts 02215, USA*
[2]*The Pulmonary Center, Boston University Medical Center, 715 Albany Street, Boston, Massachusetts 02118, USA*
[3]*Department of Genetics and Genomics, Boston University, 715 Albany Street, Boston, Massachusetts 02118, USA*

Correspondence should be addressed to Avrum Spira aspira@bu.edu

Cigarette smoke is the major cause of lung cancer and chronic obstructive pulmonary disease in the United States. We have previously defined the impact of tobacco smoke on intrathoracic airway gene expression among healthy nonsmokers and smokers using standard 3'-based expression U133A arrays [12]. In this report, we compared the performance of the Affymetrix GeneChip Human Exon 1.0 ST array with the HG-U133A array for detecting smoking-related gene expression changes in large airway epithelium obtained at bronchoscopy. RNA obtained from the same bronchial airway epithelial cell samples of four current smokers and three never smokers was hybridized to both arrays. Out of 22,215 probe sets on HG-U133A, 14,741 RefSeq transcripts were mapped to 17,800 core transcripts on the Exon array and the 2 platforms were compared for this overlapping transcript set. While the reproducibility of both platforms was high, the Exon array had a slightly stronger correlation for technical replicates. A majority of the genes with the largest smoking-related fold changes were tightly correlated between platforms, but there were a number of smoking-related changes in gene expression that were detected only on the Exon arrays. Furthermore, while the HG-U133A study did not have enough power to detect any differentially expressed genes between the 4 current vs. 3 never smokers at a False Discovery Rate (FDR) < 0.05, seventy differential expressed genes were detected at FDR < 0.05 in the same set of samples using the Exon platform. These findings suggest that the all-Exon array is a more robust platform for measuring airway epithelial gene expression and can serve as an effective tool for exploring host response to and damage from cigarette smoke.

Keywords: cigarette smoke, Affymetrix exon arrays, gene expression, airway epithelium.

1. Introduction

Although cigarette smoking is well recognized as the major cause of lung cancer and chronic obstructive pulmonary disease (COPD) [7], only 10-20% of smokers actually develop these diseases [10]. It is unclear why some smokers remain healthy while others remain at high risk decades even after they have quit [5]. Unfortunately, there are currently no effective tools for identifying smokers at highest risk for developing tobacco-related lung disease.

Based on the concept that cigarette smoke creates a "field of injury" in epithelial cells throughout the respiratory tract, we have previously measured genome-wide gene expression in large airway epithelium obtained at bronchoscopy in order to gain insights into host response to and damage from smoking. Using the HG-U133A GeneChip array (Affymetrix, Santa Clara, CA), we have defined the impact of tobacco smoke on intrathoracic airway gene expression among healthy nonsmokers and smokers and have demonstrated that a subset of smoking-induced changes persist years after subjects stop smoking [12]. Further, we recently developed a profile of airway gene expression that can distinguish smokers with and without lung cancer and serve as an early diagnostic biomarker for disease [13]. The above studies, however, were limited in terms of their widespread clinical application due to the amounts and quality of RNA needed for the U133A array. That platform requires 4-8ug of total RNA as starting material (a challenge for clinical brushings and biopsy material) and does not perform well in setting of partially degraded RNA. As a result, approximately 10-20% of samples obtained at the time of bronchoscopy in our prior studies were of insufficient quality and/or quantity for microarray studies.

The recent availability of a new platform, the GeneChip Human Exon 1.0 ST array (Affymetrix, Santa Clara, CA), provides us with an opportunity to get more comprehensive and more reliable genome-wide measurements requiring only 1 μg of starting RNA. The Exon array contains >5million 25mer probes, forming 1.4 million probe sets that are together used to separately interrogate 1 million known and predicted exons [16]. This novel platform offers several key advantages over standard versions of GeneChip arrays: 1) It provides more robust measurements at the transcript level due to more probes per transcript; there are roughly 30-40 probes for each RefSeq transcript, as compared to 11 probes mostly at the 3' end in the U133 array [6]. 2) It has the potential to provide more accurate measurements in the setting of partially degraded mRNA as the probesets are distributed along the entire length of the gene (as opposed to the 3' end); 3) It has the potential to distinguish between different isoforms of a gene at the level of individual exons, and thus allows the opportunity to identify alternative splicing events that may play important role in host response to smoking.

There is very limited data comparing the robustness and reproducibility of this novel platform and the traditional U133 arrays, particularly in the setting of clinical samples. Okoniewski *et al* [8] reported that irrespective of the mapping methodology applied, Exon 1.0 ST and HG-U133 Plus2 arrays show a high degree of correspondence by hybridizing RNA samples from two human cell lines in triplicate (technical replicates) to both arrays. Gardina *et al* [6] have also demonstrated a reasonable correlation in signals for genes that are significantly differentially expressed between tissue types utilizing a subset of data (3 replicates each tissue) from a panel of 11 normal tissues that was assayed in parallel on both Exon arrays and HG-U133 Plus2 arrays. In this study, we performed a systematic comparison of gene signal estimations from the Exon 1.0 ST and the U133A arrays by hybridizing the same bronchial airway epithelial RNA obtained from smokers and nonsmokers to both arrays. Our data suggests that while both platforms show a high degree of correlation for detecting smoking-related changes, the all-exon array is a more robust platform for the genome-wide study of smoking-related gene expression changes that occur in airway epithelium.

2. Material and Methods

2.1. *Study Population and Sample Collection*

In our previous study [12], we recruited nonsmoking and smoking subjects (n = 93) to undergo fiberoptic bronchoscopy at Boston Medical Center (between November 2001 and June 2003). We obtained a sufficient quantity of good-quality RNA for U133A studies from 85 of the 93 subjects recruited into the study. 10 out of 85 samples were excluded based on a quality control filter, and 18 of the remaining 75 samples were collected from former smokers. As a result, there were 34 current smokers and 23 never smokers in our previous analysis [12]. Seven of these 57 subjects (4 smokers and 3 nonsmokers) had sufficient amount of leftover RNA (1ug) to be used in our current study comparing the U133A and the Exon arrays. Of note, there were no significant differences ($P > 0.05$) in age (mean age 40.7±12.7 vs. 39±13.3), race (3 African American (AFA), 2 Caucasian (CAU), 2 Others vs. 24 AFA, 19 CAU, 14 Others), gender (5 male, 2 female vs. 44 male, 13 female) and cumulative tobacco exposure (mean pack-years 39.4± 14 vs. 21.8 ± 20.6) between these 7 subjects and the other 50 subjects not included in this study. The Institutional Review Board at Boston Medical Center approved this study, and all participants provided written informed consent.

Bronchial airway epithelial cells were obtained from brushings of the right mainstem bronchus taken during fiberoptic bronchoscopy with an endoscopic cytobrush (Cellebrity Endoscopic Cytology Brush, Boston Scientific, Boston). After removal from the bronchoscope, the brushes were immediately placed in TRIzol (Invitrogen) and kept at -80° C until RNA isolation was performed. RNA was extracted from the brush using TRIzol reagent (Invitrogen) as per the manufacturer protocol. Integrity of the RNA was confirmed by denaturing gel electrophoresis. Epithelial cell content of representative bronchial brushing samples was quantitated by cytocentrifugation (ThermoShandon Cytospin, Pittsburgh, PA) of the cell pellet and staining with a cytokeratin antibody (Signet, Dedham MA). Using this protocol, we were able to obtain 8-15 μg of total RNA from the intrathoracic airway of these 7 subjects.

2.2. *Microarray Data Acquisition and Preprocessing*

2.2.1. *Affymetrix HG-U133A GeneChips*

6-8 μg of total RNA was processed, labeled, and hybridized to Affymetrix HG-U133A GeneChips containing 22,215 probesets (for detailed protocol, see [12]). Eight CEL files of 4 current smokers and 3 never smokers profiled on U133A arrays as part of our previous study [12] were retrieved, which includes 1 technical replicate from one of the 3 never smokers.

The Robust Multichip Average (RMA) algorithm [2] was used for background adjustment, normalization, and probe-level summarization of the microarray samples. RMA expression measures were computed using the R statistical package and the GCRMA function in the 'affy' Bioconductor package.

2.2.2. *Affymetrix GeneChip® Human Exon 1.0 ST array*

1 μg of residual RNA from bronchial airway samples for the same subjects (4 current smokers and 3 never smokers) were processed, labeled and hybridized to Affymetrix Human Exon ST 1.0 arrays as described below. Technical replicates were run on one of the current and one of the never smoker samples.

Using a random hexamer incorporating a T7 promoter, double-stranded cDNA was synthesized from 500 ng total RNA in which the majority of the ribosomal RNA had been removed using a RiboMinus Human/Mouse Transcriptome Isolation Kit (Invitrogen, Carlsbad, CA). cRNA was generated from the double-stranded cDNA template though an in-vitro transcription reaction and purified using the Affymetrix sample cleanup module. cDNA was regenerated through a random-primed reverse transcription using a dNTP mix containing dUTP. The RNA was hydrolyzed with RNase H and the cDNA purified. The cDNA was then fragmented by incubation with a mixture of UDG and APE 1 restriction endonucleases; and end-labeled via a terminal transferase reaction incorporating a biotinylated dideoxynucleotide. 5.5 μg of the fragmented, biotinylated cDNA was added to a hybridization cocktail, loaded on a Human Exon 1.0 ST GeneChip and hybridized for 16 hours at 45 ºC and 60 rpm. Following hybridization, the array was washed and stained according to the Affymetrix protocol. The stained array was scanned at 532 nm using an Affymetrix GeneChip Scanner 3000, generating CEL files for each array.

Exon-level expression values were derived from the CEL file probe-level hybridization intensities using the model-based RMA algorithm as implemented in the Affymetrix Expression Console™ software. RMA performs normalization, background correction and data summarization. Whether an exon probeset is reliably detected over background in each sample is determined from the percentile rank of the hybridization intensity of the probes relative to the hybridization intensities observed for 1000 pooled "antigenomic" probes with the same GC content as each perfect match probe (since the Exon array does not include a paired mismatch probe for each perfect match probe) [17].

Each probe's percentile rank is combined using the Fisher Test to determine an overall probability that each probeset is expressed above background. This analysis is performed using the DABG algorithm as implemented in Affymetrix Expression Console, and a p-value threshold of $p < 10^{-5}$ is used as the criterion for expression over background. Transcript-level expression values are also derived using RMA algorithm as implemented in Affymetrix Expression Console.

2.3. *Microarray Data Analysis*

Exon arrays were designed by selecting a diverse set of genomic annotations, including empirically supported and predicted transcripts. By doing so, it enables the discovery of novel transcriptional events. On the other hand, probes that may target incorrectly predicted exons (genes) could increase the exon identification errors and transcript misassignments, which could potentially decrease power to detect differential expression. As a result, in our current study, we focused our analyses on the approximately 230,000 "core" exon probesets that have been mapped to approximately 17,800 empirically supported core transcripts (RefSeq and full-length GenBank mRNAs [6]) with a high degree of confidence.

All statistical analyses below were performed with R 2.4.0 (available at http://r-project.org). The gene annotations used for each probe set were from the October 2003 NetAffx HG-U133A, and HuEx-1_0-st-transcript-annot.affy.csv.

Technical replicates were obtained from selected subjects. Pearson correlations were calculated for technical replicate samples from the same individual on both Exon 1.0 ST and U133A arrays. To compare the fold change of smoking-induced gene changes on both arrays, we considered the RefSeq mappings [3] as a nonredundant and relatively complete database of transcripts [9]. By mapping 22,215 U133A probe sets to 17,800 core transcripts on the Exon array, 14,741 RefSeq transcripts are found on both platforms. Fold changes between the log2 mean values for the smokers and nonsmoker replicates were calculated independently for Exon and U133A arrays.

Using a number of statistical methods, we also identified differentially expressed genes (DEGs) on the U133A and Exon arrays from the same subjects independently. For the U133A array, 3 common methods for identifying differentially expressed genes were applied: significance analysis of microarrays (SAM) [14], Limma [11], and Student's *t*-test. The estimated false discovery rate (FDR) for each of these analyses was calculated using the Benjamini and Hochberg approach [1] in order to correct for multiple comparisons. Due to the Exon array's unique ability to support exon-level data in which each transcript has several measurements, we were able to utilize an ANOVA model that has a term for smoking status and another that accounts for differences in exon-to-exon expression level in order to detect differential transcript-level expression on this platform. We could not apply the ANOVA model to the U133A data since only one summarization signal is available for each transcript on that platform.

For the Exon array, we leveraged the power of this platform to detect differential transcript-level expression using traditional linear model approaches to predict the observed expression level of each exon that include terms for experimental variables, terms to account for differences in exon-to-exon expression level, and a term to account for multiple measurements of transcript abundance being made from each sample. We assess the significance of the effect of experimental variables using ANOVA and correct for multiple comparisons using the Benjamini and Hochberg approach [1]. This type of modeling approach is similar to what has been proposed by others for analyzing exon array gene-expression data [15]. Additionally, we removed exon probesets that exhibited invariant expression. Measurements from exon probesets that are not significantly expressed over background were first removed by excluding probesets that are below background in > 90% of samples. Exon probesets that are expressed over background but display low variance across all samples were further removed by comparing the variance of each exon to the average variance of all exons in that transcript using a Chi-Square test.

In order to identify functional categories that were overrepresented within the genes differentially expressed between current vs. never smokers, DAVID software [4] was used to functionally classify these genes by the molecular function categories within gene ontology using total 30, 000 human genes as population background. Fisher Exact P-value was used to measure the gene-enrichment in annotation terms (for details, see [4]).

To further characterize the smoking-related genes, 2D hierarchical clustering of all never and current smokers using the differentially expressed genes was performed. Hierarchical clustering of the genes and samples was performed by using log-transformed *z*-score normalized data with a Pearson correlation (uncentered) similarity metric and average linkage clustering with CLUSTER and TREEVIEW software obtained at http://rana.lbl.gov/EisenSoftware.htm

3. Results and Discussion

3.1. *Technical Replicates comparison on the U133A and Exon arrays*

Technical replicates were run on 1 current and 1 never smoker sample. Overall transcript abundance estimates from the Exon array demonstrated a slightly higher level of reproducibility (r = 0.989) as compared to transcript abundance estimates from U133A technical replicates (r = 0.984) (see Fig. 1). However, due to the limited number of technical replicates, we were unable to get confidence intervals for the correlation coefficients in order to assess the statistical significance of this difference.

3.2. *Gene-level fold change comparison on U133A and Exon arrays*

The Exon array can serve as a gene-level expression array by summarizing multiple probes signals on different exons into an expression level of all transcripts from one gene. Since RefSeq is known to be a nonredundant and relatively complete database of high confident transcripts [9], we compared the tobacco-smoke induced gene level changes for 14,741 RefSeq transcripts found on both platforms. This approach allowed us to compare the fold change of gene level signals from the same 4 current smoker vs. 3 never smoker samples assayed on both arrays (see Fig. 2). In this scatter plot, each point corresponds to a pair of probe sets for which a successful cross-chip mapping could be found. Ideally, fold changes would be identical, and all points would fall on the major diagonal of the scatter plot. While the majority of gene-expression differences detected correlate tightly between platforms (r=.62; p-value $< 2*10^{-16}$), there were a number of smoking-related gene expression changes that were detected only on the Exon array (see colored circles along the horizontal plane in Fig. 2). This suggests that the U133A array may lack probesets to measure expression of these particular transcripts. Gardina *et al.* [6] also observed a shift of apparently low expressing genes on the U133 Plus 2 array to higher signals on the Exon array. In order to evaluate whether these results extended to the other U133A samples collected as part of our original study [12], we compared smoking-related fold changes in all 57 samples from that study to those detected on the exon array in seven subjects. These results reinforced the strong correlation between the 2 platforms (r=.4; p-value $< 2*10^{-16}$) even in the setting of unmatched samples and once again demonstrated that there are a number of smoking-related changes only detected by the exon array (data not shown).

3.3. *Gene-level differential expression changes on U133A and Exon arrays*

To examine the effect of smoking on the U133A array, three statistical methods (SAM, Limma and a two-sample Student's *t*-test) were used to test for genes differentially expressed between current ($n = 4$) and never smokers ($n = 3$). After correcting for multiple testing using the Benjamini and Hochberg FDR [1], there were no significant genes found at FDR < 0.05. We also randomly sampled, 1000 times, 4 current vs. 3 never smokers from the complete set of 57 subjects in our previous U133A study[12] and were unable to find any genes that passed the FDR cutoff in this analysis.

As mentioned in *material and methods*, we focused our analyses on the approximately 230,000 "core" exon probesets that have been mapped to approximately 17,800 core transcripts on the Exon array. With the Exon arrays, we are able to use an exon-level mixed-model ANOVA to determine whether a gene is differentially expressed at the transcript level. As a result, when comparing 4 current vs. 3 never smokers analyzed on the Exon array, we found 70 differentially expressed genes at p<0.001 (FDR<0.05), while analysis of the same samples on the U133A failed to yield any significantly differentially expressed genes. An example of how the exon-level measurements aid in the identification of differentially expressed genes is shown in Fig. 3. These results suggest that the Exon array is a more robust gene expression platform, yielding more power for detecting differential expression for a given sample size.

To further characterize the genes identified as differentially expressed on the Exon array, hierarchical clustering of all 7 subjects was performed using exon gene expression measurements for these 70 DEGs (see Fig. 4). Current and never smokers grouped with their appropriate classes according to the expression of these 70 genes.

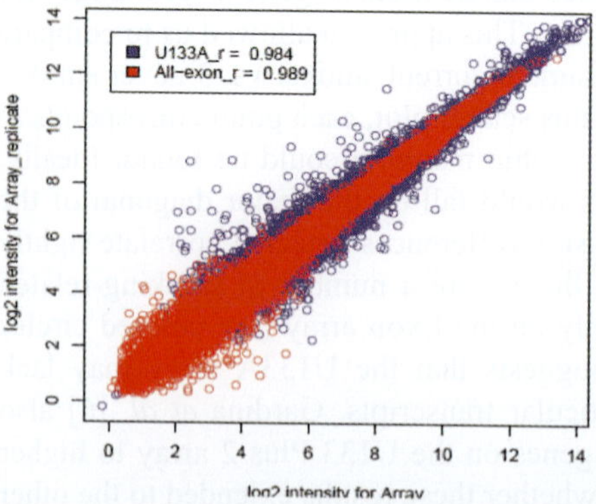

Fig. 1. Correlation between technical replicates of all-exon and U133A arrays demonstrates that all-exon arrays produce slightly more accurate estimates of transcript abundance.

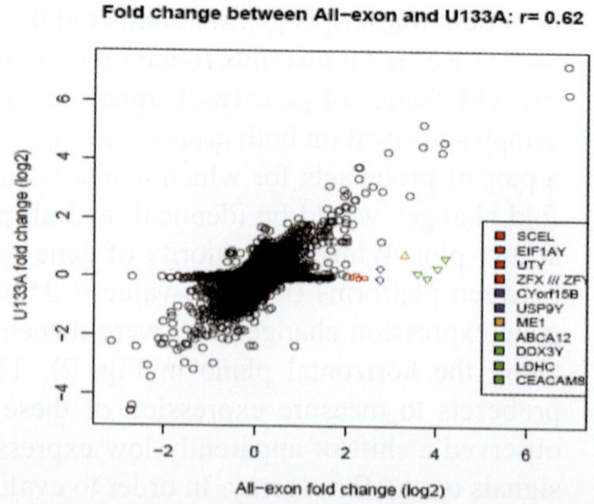

Fig. 2. Correlation between smoking induced gene-expression differences detected by all-exon (x-axis) and U133A (y-axis) arrays demonstrates that the all-exon arrays detect gene expression differences that are not detected on the U133A array. The highlighted gene expression differences (colored circles) are greater than 4 fold and detected only on the all-exon array.

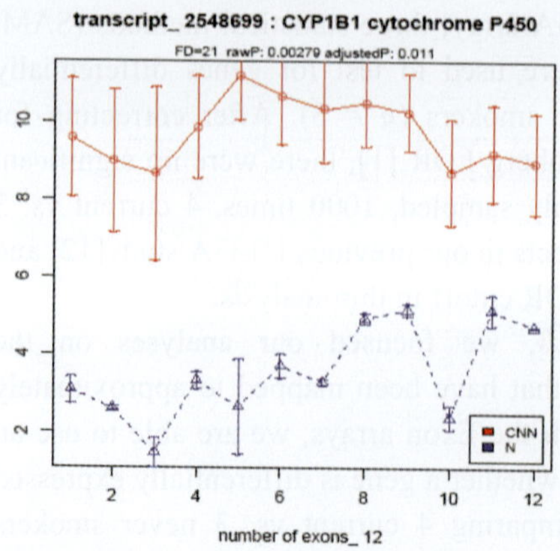

Fig. 3. Exon-level expression estimates of a gene that is detected as significantly differentially expressed on the all-exon array, but does not reach statistical significance when measured on the U133A array in the same 4 current (red: CNN) vs. 3 never smokers (blue: N).

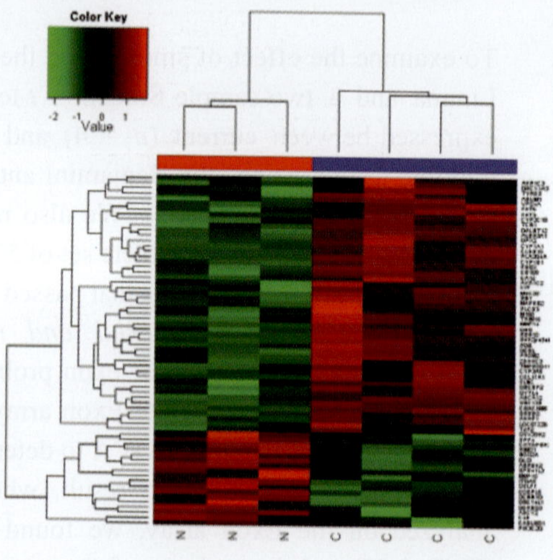

Fig. 4. Hierarchical clustering of 4 current smokers and 3 never smokers according to the expression of the 70 genes differentially expressed between current and never smokers on the Exon array. Current and never smokers are separated into their appropriate classes.

Table 1 depicts the DAVID functional annotation [4] of the 70 DEGs detected on the Exon arrays. Genes involved in oxidoreductase activity (P = 5.17E-07) and xenobiotic functions (P = 1.32E-05) are significantly enriched among the smoking-related DEG including CYP1B1, AKR1C3, PGD, NOS2A, ALDH3A1, and GPX2. We observed very similar biological findings, both at a gene level and a functional annotation level, in our previous study by analyzing a larger U133A data set (34 current vs. 23 never smokers) [12].

Table 1. Functional categories significantly enriched (by Fisher Exact test) among the genes differentially expressed between current and never smokers on the exon array with Benjamini and Hochberg adjusted p value < 0.05.

Category	Term	Count	PValue
SP_PIR_KEYWORDS	oxidoreductase	12	5.17E-07
GOTERM_MF_ALL	catalytic activity	40	1.33E-06
INTERPRO_NAME	IPR000008:C2	7	1.13E-05
KEGG_PATHWAY	HSA00980:METABOLISM OF XENOBIOTICS BY CYTOCHROME P450	7	1.32E-05
GOTERM_MF_ALL	oxidoreductase activity	13	1.84E-05
SP_PIR_KEYWORDS	metal-binding	20	2.10E-05
GOTERM_MF_ALL	transition metal ion binding	23	2.31E-05
SP_PIR_KEYWORDS	membrane	26	3.44E-05
GOTERM_MF_ALL	tetrapyrrole binding	6	3.93E-05
GOTERM_MF_ALL	heme binding	6	3.93E-05
GOTERM_MF_ALL	cation binding	27	5.04E-05
GOTERM_MF_ALL	ion binding	28	5.75E-05
GOTERM_MF_ALL	metal ion binding	28	5.75E-05
GOTERM_MF_ALL	monooxygenase activity	6	5.91E-05
SP_PIR_KEYWORDS	chromoprotein	5	6.13E-05
SMART_NAME	SM00239:C2	6	6.89E-05
GOTERM_MF_ALL	unspecific monooxygenase activity	4	8.19E-05

4. Conclusions

We compared the performance of the Affymetrix GeneChip Human Exon 1.0 ST array with the HG-U133A array for detecting smoking-related gene expression changes in airway epithelium obtained at bronchoscopy. The Exon array appears to be a reproducible platform capable of working with smaller amounts (1ug) of RNA obtained from the airway epithelium. In a gene-level fold change comparison, we found a strong correlation between the 2 platforms for smoking-related changes in gene expression, although a number of gene expression changes in airway epithelium were detected only on the Exon array. Furthermore, the exon array appears to be a more robust for detecting differentially expressed genes in the setting of a relatively small sample size. These findings suggest that the Exon array can serve as a clinically-relevant gene expression platform for measuring host response to and damage from tobacco smoke.

Acknowledgments

This work was supported by the Doris Duke Charitable Foundation (A.S.), NIH/NCI R21CA10650 (A.S.) and NIH/NCI R01CA124640 (A.S. and M.E.L)

References

[1] Benjamini, Y. and Hochberg, Y., Controlling the false discovery rate: a practical and powerful approach to multiple testing, *Journal of the Royal Statistical Society Series B*, (57):289–300, 1995.
[2] Bolstad, B.M., *et al.*, A comparison of normalization methods for high density oligonucleotide array data based on variance and bias, *Bioinformatics,* 19(2):185-193, 2003.
[3] Dai, M., *et al.*, Evolving gene/transcript definitions significantly alter the interpretation of GeneChip data, *Nucleic Acids Res.*, 33(20):e175, 2005.
[4] Dennis, G., Jr., *et al.*, DAVID: Database for Annotation, Visualization, and Integrated Discovery, *Genome Biol.*, 4(5):P3, 2003.
[5] Ebbert, J.O., *et al.*, Lung cancer risk reduction after smoking cessation: observations from a prospective cohort of women, *J. Clin. Oncol.*, 21(5):921-926, 2003.
[6] Gardina, P.J., *et al.*, Alternative splicing and differential gene expression in colon cancer detected by a whole genome exon array, *BMC Genomics*, **7**:325, 2006.
[7] Greenlee, R.T., *et al.*, Cancer statistics, 2001., *CA Cancer J. Clin.*, 51(1):15-36, 2001.
[8] Okoniewski, M.J., *et al.*, High correspondence between Affymetrix exon and standard expression arrays, *Biotechniques*, 42(2):181-185, 2007.
[9] Pruitt, K.D., Tatusova, T., and Maglott, D.R., NCBI reference sequences (RefSeq): a curated non-redundant sequence database of genomes, transcripts and proteins, *Nucleic Acids Res.*, 35(Database issue):D61-65, 2007.
[10] Shields, P.G., Molecular epidemiology of lung cancer, *Ann. Oncol.*, 10 Suppl 5:S7-11, 1999.
[11] Smyth, G.K., Linear models and empirical bayes methods for assessing differential expression in microarray experiments, *Stat. Appl. Genet. Mol. Biol.*, 3:Article3, 2004.
[12] Spira, A., *et al.*, Effects of cigarette smoke on the human airway epithelial cell transcriptome, *Proc, Natl. Acad. Sci. USA*, 101(27):10143-10148, 2004.
[13] Spira, A., *et al.*, Airway epithelial gene expression in the diagnostic evaluation of smokers with suspect lung cancer, *Nat. Med.*, 13(3):361-366, 2007.
[14] Tusher, V.G., Tibshirani, R., and Chu, G., Significance analysis of microarrays applied to the ionizing radiation response, *Proc. Natl. Acad. Sci. USA*, 98(9):5116-5121, 2001.

[15] Yoshida, R., *et al.*, A statistical framework for genome-wide discovery of biomarker splice variations with GeneChip Human Exon 1.0 ST Arrays, *Genome Informatics*, 17(1):88-99, 2006.
[16] Affymetrix. 2005. Exon probe set annotations and transcript cluster groupings. Affymetrix, Santa Clara, CA.
[17] Affymetrix_Inc. Affymetrix White Papers: http://www.affymetrix.com/support/technical/whitepapers.affx

CLUSTERING SAMPLES CHARACTERIZED BY TIME COURSE GENE EXPRESSION PROFILES USING THE MIXTURE OF STATE SPACE MODELS

OSAMU HIROSE[1] ochamu@ims.u-tokyo.ac.jp
RYO YOSHIDA[2] yoshidar@ism.ac.jp
RUI YAMAGUCHI[1] ruiy@ims.u-tokyo.ac.jp
SEIYA IMOTO[1] imoto@ims.u-tokyo.ac.jp
TOMOYUKI HIGUCHI[2] higuchi@ism.ac.jp
SATORU MIYANO[1] miyano@ims.u-tokyo.ac.jp

[1] *Human Genome Center, Institute of Medical Science, University of Tokyo, 4-6-1 Shirokanedai, Minato-ku, Tokyo, 108-8639, Japan*
[2] *Institute of Statistical Mathematics, Research Organization of Information and Systems, 4-6-7 Minami-Azabu, Minato-ku, Tokyo, 106-8569, Japan*

We propose a novel method to classify samples where each sample is characterized by a time course gene expression profile. By exploiting the mixture of state space model, the proposed method addresses the following tasks: (1) clustering samples according to temporal patterns of gene expressions, (2) automatic detection of genes that discriminate identified clusters, (3) estimation of a restricted autoregressive coefficient for each cluster. We demonstrate the proposed method along with the cluster analysis of 53 multiple sclerosis patients under recombinant interferon β therapy with the longitudinal time course expression profiles.

Keywords: clustering; mixture of state space models; multiple sclerosis; time course gene expression profile.

1. Introduction

Time course gene expression profiles enable us to understand the temporal structure of gene regulations. A number of researchers have studied the time course gene expression profiles [1, 4, 5]. One major difficulty of time course gene expression analysis comes from the fact that length of time course is usually much smaller than the dimension of data. In order to overcome the difficulties related to imbalance between dimensionality and length of data, we developed a state space model in our previous works [8, 9]. The state space models provide an approach to avoid over-parameterization of the vector autoregressive model by exploring potential sets of the co-expressed genes.

Recently, novel kinds of time series expression profiles that existing methods may fail to analyze have been appeared. Baranzini *et al.* [2] investigated the longitudinal gene expression change of multiple sclerosis (MS) patients with treatments of recombinant interferon β (rIFN β). In this data set, each MS patient is characterized by a gene expression matrix whose column vectors represent gene expression vectors for corresponding observed time points. They aimed at classifying 53 MS patients, composed of 33 good responders and 20 poor responders for the therapy of rIFN β. Hence, the problem is to classify samples, where each sample is characterized by matrix data.

It is possible to use classical clustering methods such as the hierarchical clustering and the k-means clustering, for example, by vectorizing gene expression matrices and measuring Euclid distances of all pairs of the vectorized matrices. These methods, however, often fail since they do not make use of time series of data effectively and in addition, the direct product between time and gene expands feature space and lead to increase of the imbalance between dimensionality and length of time series. Furthermore, these models lack explicit feature extraction procedure, i.e. lack of interpretablity.

In this paper, we propose a novel clustering method based on a mixture model that make use of time series of data effectively. State space models are used as component models of the mixture in order to handle high dimensional time series and to avoid the over-parameterization by considering dimension compression. The proposed method addresses the following tasks: (1) clustering samples according to temporal patterns of gene expressions, (2) automatic detection of genes that discriminate identified clusters, (3) estimation of a restricted autoregressive coefficient for each cluster. We will demonstrate the proposed method along with the cluster analysis of MS patients.

2. Methods

2.1. *Mixture of State Space Models*

Suppose that $y_n^{(l)} \in \mathbb{R}^p$ denotes a gene expression vector observed at the time point n corresponding to the lth sample among m samples, where the number of genes is p. Let us denote the set of observed time points and the time course gene expression profile of the lth sample by $\mathcal{N}_{\text{obs}}^{(l)} \subseteq \mathcal{N} = \{1, \ldots, N\}$ and $Y_N^{(l)} = \{y_n^{(l)} : n \in \mathcal{N}_{\text{obs}}^{(l)}\}$. Also, we denote an unobserved k dimensional hidden state vector by $x_n^{(l)} \in \mathbb{R}^k$. In order to avoid over-parameterization, we assume that k is much less than p. Our objective is to classify m samples into G clusters according to their time course gene expression profiles $Y_N^{(1)}, \ldots, Y_N^{(m)}$. Here, we assume that $Y_N^{(l)}$ is generated by one of state space models g:

$$y_n^{(l)} = H_g x_n^{(l)} + w_n^{(l)}, \quad n \in \mathcal{N}_{\text{obs}}^{(l)}, \tag{1}$$

$$x_n^{(l)} = F_g x_{n-1}^{(l)} + v_n^{(l)}, \quad n \in \mathcal{N}, \tag{2}$$

with probability α_g among G component models ($g = 1, \ldots, G$), starting with the initial state vector $x_0^{(l)} \sim N(\mu_{0g}, \Sigma_{0g})$. That is, $Y_N^{(l)}$ is generated by G-components mixture of the state space model. $F_g \in \mathbb{R}^k \times \mathbb{R}^k$ and $H_g \in \mathbb{R}^p \times \mathbb{R}^k$ are coefficient matrices which are often referred to as the system matrix and the observation matrix. w_n and v_n are the noise vectors which follow the normal distributions with mean zero and covariance matrices R_g and I, respectively. The lack of identifiability

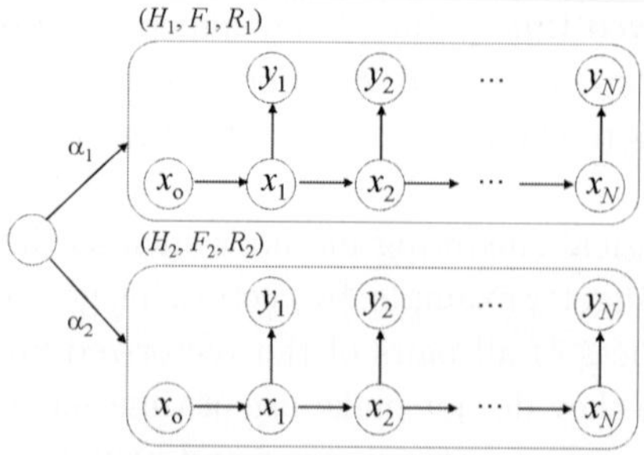

Fig. 1. A schematic expression of the two component mixture of the state space models ($G = 2$).

is a weakness of the state space model. In order to keep the uniqueness of the model, we impose parameter constraints as $H_g^T R_g^{-1} H_g = \Lambda_g$ for all g, where Λ_g is a $k \times k$ diagonal matrix. The detail of the constraints is described by our previous work [8].

2.2. *Clustering*

Let $c_g^{(l)} \in \{0, 1\}$ $(g = 1, \ldots, G)$ be an unknown class label of the lth sample which takes value one, i.e. $c_g^{(l)} = 1$, if the lth sample belongs to class g, otherwise $c_g^{(l)} = 0$. Suppose $\eta_g = (H_g, F_g, R_g, \mu_{0g})$ is the parameter set of the component state space model g and $\theta = (\alpha_1, \cdots, \alpha_G, \eta_1, \cdots, \eta_G)$ is the whole parameter set of the mixture state space model. Also, we suppose that $f_g(Y_N^{(l)}; \eta_g)$ is the gth component distribution given by

$$f_g(Y_N^{(l)}; \eta_g) = \prod_{n \in \mathcal{N}_{\text{obs}}^{(l)}} \phi(y_n^{(l)}; H_g x_{g,n|n-1}^{(l)}, H_g V_{g,n|n-1}^{(l)} H_g^T + R_g),$$

where $\phi(z; \mu, \Sigma)$ is the probability density function of a random vector z followed by the multivariate normal distribution with mean vector μ and covariance matrix Σ. $x_{g,n|n-1}^{(l)}$ and $V_{g,n|n-1}^{(l)}$ are defined by

$$x_{g,n|n-1}^{(l)} = E[x_n^{(l)} | Y_{n-1}^{(l)}, c_g^{(l)} = 1; \eta_g],$$
$$V_{g,n|n-1}^{(l)} = E[(x_n^{(l)} - x_{g,n|n-1}^{(l)})(x_n^{(l)} - x_{g,n|n-1}^{(l)})^T | Y_{n-1}^{(l)}, c_g^{(l)} = 1; \eta_g].$$

Samples can be clustered by applying the Bayes rule. That is, the sample l is assigned to gth group if the posterior probability of $c_g^{(l)} = 1$,

$$p(c_g^{(l)} = 1 | Y_N^{(l)}; \theta) = \alpha_g f_g(Y_N^{(l)}; \eta_g) / \sum_{g'=1}^{G} \alpha_{g'} f_{g'}(Y_N^{(l)}; \eta_{g'})$$

is maximum. More formally, the clustering rule is represented as follows:

$$\hat{c}_g^{(l)} = \begin{cases} 1 & \text{if } g = \text{argmax}_{g'}\, p(c_{g'}^{(l)} = 1 | Y_N^{(l)}; \theta), \\ 0 & \text{otherwise.} \end{cases}$$

Since the parameter set θ is unknown, θ will be replaced by the maximum likelihood estimator $\hat{\theta}$. The procedure of maximum likelihood estimation will be discussed in the next subsection.

2.3. *Parameter Estimation*

We use the EM algorithm for the maximum likelihood estimation. Complete data log-likelihood of the mixture state space model L_c is described as follows:

$$L_c(\theta) = \sum_{l=1}^{m}\sum_{g=1}^{G} c_g^{(l)}\{\ln \alpha_g + \ln p(Y_N^{(l)}, X_N^{(l)} | c_g^{(l)} = 1; \eta_g)\},$$

where $X_N^{(l)} = \{x_n^{(l)} : n \in \{0\} \cup \mathcal{N}\}$. Let $\rho_g^{(l)} = p(c_g^{(l)} = 1 | Y_N^{(l)}; \theta)$. Maximizing $E[L_c(\theta)|Y_N^{(1)}, \ldots, Y_N^{(m)}; \bar{\theta}]$ given an arbitrary initial parameter $\bar{\theta}$, we obtain following recursive formula for calculating the maximum likelihood estimator:

$$\alpha_g^{\text{new}} = \sum_{l=1}^{m} \rho_g^{(l)}/m,$$

$$\mu_{0g}^{\text{new}} = \sum_{l=1}^{m} \rho_g^{(l)} E[x_0^{(l)} | Y_N^{(l)}, c_g^{(l)} = 1; \bar{\eta}_g]/m,$$

$$H_g^{\text{new}} = T_{yx}(g) T_{xx}(g)^{-1},$$

$$F_g^{\text{new}} = T_{xx'}(g) T_{x'x'}(g)^{-1},$$

$$R_g^{\text{new}} = \{\sum_{l=1}^{m} \rho_g^{(l)} |\mathcal{N}_{\text{obs}}^{(l)}|\}^{-1}\{T_{yy}(g) - T_{yx}(g) T_{xx}(g)^{-1} T_{yx}(g)^T\},$$

where

$$T_{yy}(g) = \sum_{l=1}^{m} \rho_g^{(l)} \sum_{n \in \mathcal{N}_{\text{obs}}^{(l)}} E[y_n^{(l)} y_n^{(l)T} | Y_N^{(l)}, c_g = 1; \bar{\eta}_g],$$

$$T_{yx}(g) = \sum_{l=1}^{m} \rho_g^{(l)} \sum_{n \in \mathcal{N}_{\text{obs}}^{(l)}} E[y_n^{(l)} x_n^{(l)T} | Y_N^{(l)}, c_g = 1; \bar{\eta}_g],$$

$$T_{xx}(g) = \sum_{l=1}^{m} \rho_g^{(l)} \sum_{n \in \mathcal{N}_{\text{obs}}^{(l)}} E[x_n^{(l)} y_n^{(l)T} | Y_N^{(l)}, c_g = 1; \bar{\eta}_g],$$

$$T_{xx'}(g) = \sum_{l=1}^{m} \rho_g^{(l)} \sum_{n=1}^{N} E[x_n^{(l)} x_{n-1}^{(l)T} | Y_N^{(l)}, c_g = 1; \bar{\eta}_g],$$

$$T_{x'x'}(g) = \sum_{l=1}^{m} \rho_g^{(l)} \sum_{n=1}^{N} E[x_{n-1}^{(l)} x_{n-1}^{(l)T} | Y_N^{(l)}, c_g = 1; \bar{\eta}_g].$$

Expectation terms are calculated efficiently with Kalman filtering and smoothing procedure. For example, see [6, 7]. We obtain (local) maximum likelihood estimate $\hat{\theta}$ repeating the parameter updating process until a suitable convergence criterion is satisfied. Some implementation issues are described in Appendix.

2.4. *Feature Selection and Estimation of Restricted Autoregressive Coefficients*

In cluster analysis of gene expression profiles, it is of interest to detect genes that characterize identified clusters. By transforming observation equation Eq. 1 under the constraints $H_g^T R_g^{-1} H_g = \Lambda_g$, $g = 1, \dots, G$, gene expression vectors y_n can be mapped onto the state space $\mathbb{R}^k$ with the feature extraction matrix $D_g \in \mathbb{R}^k \times \mathbb{R}^k$ as follows:

$$x_n^{(l)} = D_g R_g^{-1/2}(y_n^{(l)} - w_n^{(l)}), \tag{3}$$

where $D_g = \Lambda_g H_g^T R_g^{-1/2}$ [8]. If the dimension of state k is taken to be lower than p, the dimensionality of the noise-removed gene expression vectors $R_g^{-1/2}(y_n - w_n)$ is reduced by the semi-orthogonal matrix D_g in which (i, j)th element of D_g represents the contribution of the jth gene to the ith coordinate of the state space. In practice, it is useful to extract a number of significant genes for the each state variable.
Furthermore, substituting Eq. 3 into the state space model Eq. 1 and Eq. 2, the following first-order vector autoregressive representation are obtained:

$$R_g^{-1/2}(y_n^{(l)} - w_n^{(l)}) = \Psi_g R_g^{-1/2}(y_{n-1}^{(l)} - w_{n-1}^{(l)}) + R^{-1/2} H v_n^{(l)} \tag{4}$$

where $\Psi_g = D_g^T \Lambda_g F_g D_g$ [9]. The (i, j)th element of the autoregressive coefficient matrix Ψ_g represents the influence of the jth gene on the ith gene for the component model g. After estimating parameters, we obtain the estimated Ψ_g which captures the temporal structure of the gene expressions for the identified gth cluster. Comparisons of these estimated coefficient matrices $\Psi_1, \dots, \Psi_G$ gives us an insight to understand temporal features of G groups.
Note that the degree of freedom in the autoregressive coefficient matrix Ψ_g is of order $O(p) = p(k+1) + k^2 - k(k-1)/2$. From this point of view, the state space model is considered as a parsimonious parameterization of the vector autoregressive model and provides an approach to control the model complexity by choosing dimension of state vector k.

3. Experiments

3.1. *Gene Expression Profiles of MS Patients*

The disease MS is characterized by myelin destruction and oligodendrocyte death and causes relapsing-remitting neurological disorders such as visual disorder and movement disorder. rIFN-β is routinely used for suppressing the relapse of the symptom. However, almost half of MS patients are not benefited by the therapy. Baranzini *et al.* [2] investigated long-term effects of rIFN-β on disease progression with the time course gene expression profiles of 76 genes which are mainly related to the immune system. Expression levels are measured by conducting reverse-transcription PCR at the beginning of the therapy and after 3, 6, 9, 12, 18, and 24 months. This dataset includes profiles of 53 MS patients categorized into 33 "good" responders and 20 "poor" responders according to their response levels for rIFN-β administration. A group of patients were categorized into poor responders if they suffered two or more relapses or experienced an increase of one point in the expanded disability status scale score (EDSS), a measure of progression for the MS disease, until two years after the initiation of the therapy. Good responders were defined as patients that experienced a total suppression of relapses and no increase in the EDSS.

3.2. *Results*

We applied the proposed method to the expression profiles after converting the real time set {0, 3, 6, 9, 12, 18, 24} months into $\mathcal{N}_{\rm obs} = \{1, 2, 3, 4, 5, 7, 9\}$ where the entire time points are defined by $\mathcal{N} = \{1, 2, \ldots, 7, 8, 9\}$ and $\mathcal{N}_{\rm obs}$ is the union of $\mathcal{N}_{\rm obs}^{(l)}$, $l = 1, \ldots, m$. We preset the number of clusters $G = 2$ and the dimension of states $k = 2$. Among 600 estimated parameters of the EM algorithm, which were computed by the different initial parameters, we chose the best parameter corresponding to the highest local maxima of likelihoods. Missing values included in the dataset were imputed by the EM algorithm, that is, we considered missing values as latent variables as well as the state vectors and we calculated their conditional expectations at the E-step.

The left panel of Fig. 2 shows the identified clusters of 53 MS patients. The identified two groups composed of 28 (cluster A) and 25 (cluster B) patients, respectively. The cluster A included 27 good responders and one poor responder while the cluster B included 19 poor responders and 6 good responders. The resulting two clusters A and B are likely to reflect the diagnostic categories. If we assume the cluster A corresponds to good responders and the cluster B poor responders, the total prediction accuracy is 86.8 ($= 100 \times (27 + 19)/53$)%.

We investigated the performance of the proposed method by comparing with a result of a hierarchical clustering. The result of a hierarchical clustering is illustrated in the right panel of Fig. 2. We used the complete-linkage-based hierarchical clustering by vectorizing the time course gene expression matrix for each patient, where the distance between two patients was measured based on the Pearson correlation. Patients in the good and poor responder groups are drawn by grey and white squares, respectively. Since all the patients except two belong one of clusters in case we divide patients into two clusters, the threshold value of the correlation was set to 0.80 which seems to be the best value for the classification of patients according to responder groups. As a result, the patients were divided into eight clusters except for three patients. Even if we assign the clusters to two responder groups by the majority decision, 13 patients are belonged in incorrect clusters, and thus, the proposed method seems to outperform the hierarchical clustering in performing the cluster analysis of MS patients from the dataset.

Next, we focus on the significant genes which separate the identified two clusters. Significant genes are extracted by feature extraction matrices D_1 and D_2 for the two clusters. In order to discover genes that characterize a difference between two

	Cluster A	Cluster B
Good	27	6
Poor	1	19

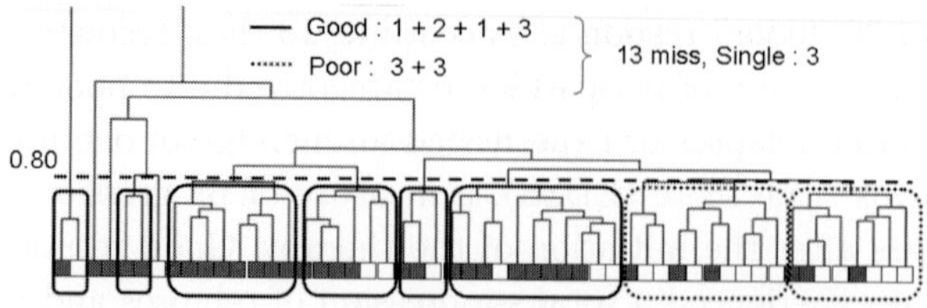

Fig. 2. The clustering result of the proposed method (left) and hierarchical clustering (right).

clusters, we computed $S = ||D_1| - |D_2||$ and selected some genes which achieved the highest score of the S. For the first coordinate of the state space, genes with the largest five score of the first row of S were TRAIL, p53, CD5, CD22 and JAK2, while CD22, CD5, TRAIL, JAK2 and p53 for the second coordinate. It was reported that TRAIL was a potential response marker for the rIFN-β treatment in MS [10]. Furthermore, Wosik *et al.* [11] reported that oligodendrocytes were protected from p53-induced cell death by blocking signaling through TRAIL receptors.

Fig. 3 display heatmaps of the estimated autoregressive coefficient matrices Ψ_1 (left) and Ψ_2 (right). Positive and negative influences are color-coded by blue and green, respectively. The estimated coefficient matrices capture clear differences in the longitudinal effect between the two clusters. For example, TRAIL has strongly negative longitudinal effects on many genes for the first cluster while does not for the second cluster. Also, p53 has large positive longitudinal effects on many genes for the first cluster, while does not for the second cluster.

4. Concluding remarks

We proposed a novel method that perform cluster analysis of samples, where each sample is characterized by a time course gene expression profile using the mixture of state space models. The proposed method were applied to the time course gene expression profiles of 53 MS patients under the therapy of rIFN-β. We succeeded in the classification of the MS patients according to the response level for rIFN-β. The proposed method exploits the mixture of the state space models which enables us to understand differences in the temporal structure of gene expressions between the identified clusters. Here, we point out a limitation in the model selection. In the context of the mixture state space models, one of the important tasks is to determinate the number of clusters and the dimension of states. For example, when we used the mixture state space model under $k = 2$ and $G = 3$, in the cluster analysis of MS patients, many of the estimated models got degenerated, that is, at least one of mixing proportions $\alpha_1, \ldots, \alpha_G$ became zero. Probably, such an unstable estimation occurred due to the imbalance between the model complexity and amount of data, i.e. over-parameterization. This indicates that applicability of the proposed method is limited in terms of the number of clusters and the dimension of state. In order to overcome the problem of over-parameterization, one possible solution is to perform Bayesian estimations, which provide an approach to control the model complexity by constructing prior distributions of parameters appropriately. It is challenging to design the plausible prior distributions and we are now investigating this problem.

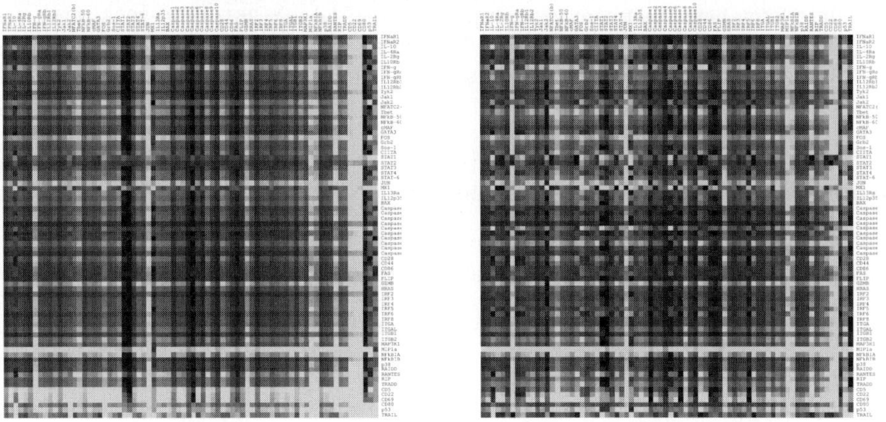

Fig. 3. Estimated autoregressive coefficient matrices Ψ_1 (left) and Ψ_2 (right).

References

[1] Akutsu, T., Kuhara, S., Maruyama, O., and Miyano, S., Identification of gene regulatory networks by strategic gene disruptions and gene overexpressions, *Proc. 9th ACM-SIAM Symp. Discrete Algorithms*, 695–702, 1998.

[2] Baranzini, S.E., Mousavi, P., Rio, J., Caillier, S.J., Stillman, A., Villoslada, P., Wyatt, M.M., Comabella, M., Greller, L.D., Somogyi, R., Montalban, X., and Ksenberg J.R., Transcription-based prediction of response to IFNb using supervised computational methods. *PLoS Biology*, 3(1):166–176, 2005.

[3] Hsiung, K.L., Kim, S.J., and Boyd, S., *Tractable approximate robust geometric programming.* Tech. Rep., Department of Electrical Engineering, Stanford University, California, 2006 2006.

[4] Imoto, S., Tamada, Y., Araki, H., Yasuda, K., Print, C.G., Charnock-Jones, S.D., Sanders, D., Savoie, C.J., Tashiro, K., Kuhara S., and Miyano, S., Computational strategy for discovering druggable gene networks from genome-wide RNA expression profiles. *Pacific Symposium on Biocomputing*, 11:559–571, 2006.

[5] Kim, S., Imoto, S., and Miyano, S., Dynamic Bayesian network and nonparametric regression for nonlinear modeling of gene networks from time series gene expression data. *Biosystems*, 75:(1-3), 57–65, 2004.

[6] Kitagawa, G. and Gersch, W., *Smoothness priors analysis of time series.* New York, Springer-Verlag, 1996.

[7] Shumway, R.H. and Stoffer, D.S., Dynamic linear models with switching. *J. American Statistical Association*, 86:763–769, 1991.

[8] Yamaguchi, R., Yoshida, R., Imoto, S., Higuchi, T., and Miyano, S., Finding module-based gene networks with state space models -Mining high-dimensional and short time-course gene expression data, *IEEE Signal Processing Magazine*, 24(1):37–46, 2007.

[9] Yoshida, R., Imoto, S., and Higuchi, T., Estimating time-dependent gene networks from time series microarray data by dynamic linear models with Markov switching. *Proc. 4th Computational Systems Bioinformatics (CSB2005)*, 289–298, 2005.

[10] Wandinger, K.P., Lunemann, J.D., Wengert, O., Bellmann-Strobl, J., Aktas, O., Weber, A., Grundstrom, E., Ehrlich, S., Wernecke, K.D., Volk, H.D., Zipp, F., TNF-related apoptosis inducing ligand (TRAIL) as a potential response marker for interferon-beta treatment in multiple sclerosis. *Lancet.* 361(9374):2036–2043, 2003.

[11] Wosik, K., Antel, J., Kuhlmann, T., Bruck, W., Massie, B., and Nalbantoglu, J., Oligodendrocyte injury in multiple sclerosis: a role for p53. *Journal of Neurochemistry*, 85:635–644, 2003.

Appendix A. Implementation Issues

We explain issues on implementations. In order to check the convergence of the EM algorithm, we need to calculate the log-likelihood of proposed model:

$$\sum_l \ln \sum_g \alpha_g f_g(Y_N^{(l)}; \bar{\eta}_g) = \sum_l \ln \sum_g \alpha_g \exp\{\ln f_g(Y_N^{(l)}; \bar{\eta}_g)\} \tag{A.1}$$

for each step. However, the component log-likelihood $\ln f_g(Y_N^{(l)}; \bar{\eta}_g)$ calculated by Kalman filtering sometimes becomes an extremely large value because of the high dimensinality of time course gene expression profiles. In such situation, $f_g(Y_N^{(l)}; \bar{\eta}_g)$ easily gets infinity from hardware limitations of floating points. Calculating formula such as (A.1) is known for log-sum-exp problem. A few methods have been proposed for approximating log-sum-exp including the method that approximates it through geometrical optimizations [3]. However, log-sum-exp formula is calculated as follows:

$$\sum_l \left\{ \ln f_{g_M^{(l)}}(Y_N^{(l)}; \bar{\eta}_g) + \ln \sum_g \alpha_g \exp\left(\ln f_g(Y_N^{(l)}; \bar{\eta}_g) - \ln f_{g_M^{(l)}}(Y_N^{(l)}; \bar{\eta}_g) \right) \right\},$$

where $g_M^{(l)} = \text{argmax}_g\, f_g(Y_N^{(l)})$ for $l = 1, \ldots, m$, since the argument of exponential is assured to be at most zero. Note that the similar calculation is needed for the posterior probabilities $p(c_g^{(l)} = 1 | Y_N^{(l)}; \bar{\theta})$.

PURE: A PUBMED ARTICLE RECOMMENDATION SYSTEM BASED ON CONTENT-BASED FILTERING

TAKASHI YONEYA[1,2] HIROSHI MAMITSUKA[1]
t-yoneya@kirin.co.jp mami@kuicr.kyoto-u.ac.jp

[1] *Bioinformatics Center, Kyoto University, Gokasho Uji, 611-0011, Japan*
[2] *Discovery Research Laboratories, Kirin Pharma Co. Ltd., 3 Miyahara, Takasaki, Gunma 370-1295, Japan*

We have developed a PubMed article recommendation system, PURE, which is based on content-based filtering. PURE has a web interface by which users can add/delete their preferred articles. Once articles are registered, PURE then performs model-based clustering of the preferred articles and recommends the highly-rated articles by the prediction using the trained model. PURE updates the PubMed articles and reports the recommendation by email on daily-base. This system will be helpful for biologists to reduce the time required for gathering information from PubMed. PURE is downloadable under GPL license, via www.bic.kyoto-u.ac.jp/pathway/mami/out/PURE.tar.gz.

Keywords: recommendation; content-based filtering; PubMed; EM algorithm.

1. Introduction

MEDLINE/PubMed [16] is one of the largest public databases on biological and medical sciences [14, 16, 17] and updated daily with thousands of new papers. Biologists devote a considerable amount of time to checking PubMed to find papers relevant to their interests. To reduce their heavy burden, we develop a system, which we call PURE (standing for a PUbmed article REcommendation system), that automatically captures the preference of a user by using this user's response to the presented papers. Using the acquired preferences, PURE then reports relevant papers with scores to this user by email. This type of system is generally called a "recommendation system" [4]. Existing methods for this system can be classified into two types: collaborative filtering [8] and content-based filtering [7]. Collaborative filtering utilizes users with similar preferences. That is, as has been done in amazon.com, item X is recommended to a user who is buying item Y, if customers who bought item Y also bought item X. Thus, the performance of collaborative filtering depends on whether there exists a user with similar taste or not. A major drawback of collaborative filtering is that this filtering requires many users to find users having similar preferences. On the other hand, content-based filtering uses the content of items which are highly rated by a user and tries to find the preference of a user more directly. That is, this filtering can be completed by a single user only, without using other users' information. The performance depends on the quality of the content of items, and if we can retrieve the item content sufficiently, we can expect that content-based filtering works well enough. PubMed provides a lot of contents of each article, divided into the title and the abstract, etc, preferably with a very small number of errors. Thus content-based filtering would be a better approach for recommending PubMed articles to a biologist than collaborating filtering, and we chose it in our system, PURE.

A user has to only register preferred articles to PURE. The preference patterns are then extracted from the registered articles by using model-based clustering, in which probabilistic parameters of a mixture model are estimated based on an EM (Expectation-Maximization) algorithm. PURE then downloads new articles from PubMed daily, which are ranked by using the likelihoods given by the trained probabilistic model to the newly downloaded articles. Finally PURE presents the highly rated articles to a user, with an interface at which a user can register preferred articles. A user can add any preferred articles at any step of the above procedure, which is iterated every time when new articles are added.

Some automatic services for recommending interesting articles are available currently, e.g. [2, 5, 9, 10, 13, 15], based on searching interesting articles by keywords. We emphasize that PURE is different from them, which present only the articles containing search keywords. PURE captures the preference pattern of the articles which are registered by a user and then recommends new articles which are the most relevant to the acquired patterns. By using this approach, PURE might find preferred papers which do not include keywords as well as those containing keywords. Another feature of PURE is that a user only has to input preferred articles by using a user-friendly and easy-to-use interface of PURE.

In experiments, we evaluated the performance of the learning method used in PURE in a two-class supervised learning manner. Experimental results indicated that PURE (or the method used in PURE) can provide a considerably favorable performance, say precision of 70% at 10% recall, even with only a very small number (e.g. 20) of training articles.

2. Implementation

Fig. 1 shows the whole scheme of the article recommendation by PURE, which has the following five steps that are iterated in this order: 1) Preferred articles are registered by a user out of their own findings or those presented by PURE. The selected articles are stored in a database of PURE. 2) A probabilistic model is trained using the stored articles to learn the preference of a user. 3) New PubMed articles are downloaded daily. 4) Prediction is done by ranking the downloaded articles using the trained model. 5) Finally the highly rated articles are presented to a user.

PURE is implemented in a client-server manner. That is, a user uses PURE at a client computer through a web interface, and the computations in the above five

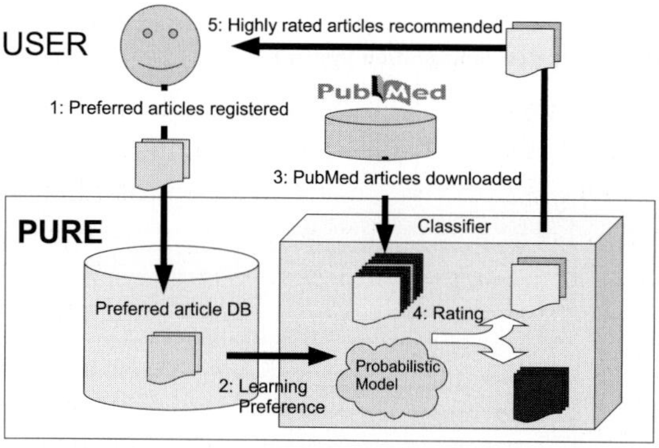

Fig. 1. Recommendation scheme of PURE.

steps are basically conducted by a central server computer. The reason why this system was adopted by PURE is the following two: 1) A large number, say normally thousands, of articles are accumulated in PubMed every day, and they must be downloaded daily by PURE. It is naturally inefficient if they are downloaded by each user, and it would be more computationally efficient that they are downloaded by one central computer only. 2) Learning from preferred documents and ranking downloaded documents are both relatively heavy loads. So it would be efficient and favorable to finish these tasks at a central computer at a convenient time, say at night, in a day.

We describe the detail of the above five steps below.

2.1. *Registration and storage of preferred articles to a PURE database*

We designed two different web interfaces for registering preferred articles, and a user can use both interfaces.

One is an interface by which a user can directly register articles (PubMed IDs). As an example of this interface, Fig. 2 shows a text box[a] at the top, in which PubMed IDs can be input by a user, and under this box, a list of articles which are already registered by this user. A user can write PubMed IDs in the box and click "submit" located at the bottom, and then a list of registered articles is shown under the input box as in Fig. 2. If a user wants to remove a registered article in the list, the checkbox in the left side of this article can be clicked and then "submit" can be clicked to delete the article.

The other is an interface by which a user can select articles out of those recommended by PURE. Fig. 3 shows an example of this interface, showing five recommended articles. A user can choose an article out of the recommended articles by

[a]For a first-time user, only this text box is displayed.

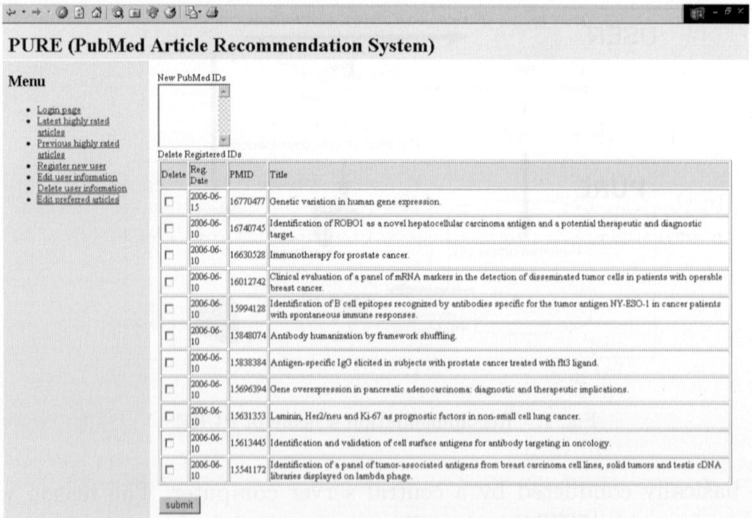

Fig. 2. A web interface for editing a list of preferred articles.

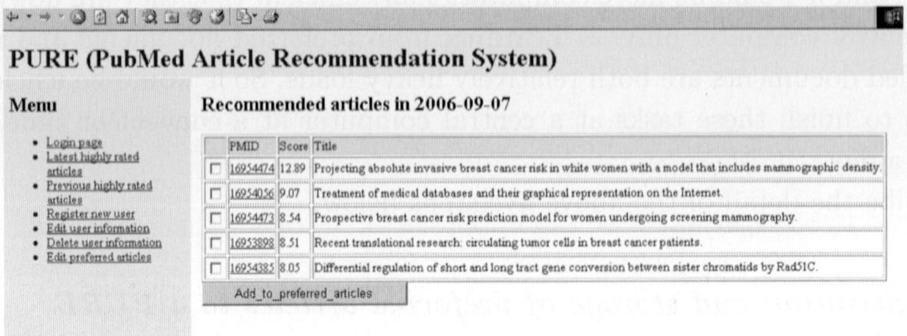

Fig. 3. A web interface for showing recommendations to a user which can be registered in this interface.

clicking the checkbox in the left side of the corresponding article, and click "submit" to register.

We note that in both interfaces, the input of PURE is PubMed ID only and no other items like keywords are required. The article which is newly registered by the first interface is downloaded from PubMed, while that by the second interface is not since it was already downloaded. The detail of the way to download articles from PubMed in PURE will be described in a later section.

2.2. *Training of a probabilistic model using stored articles to learn a user's preference*

This step is divided in the following two parts which are performed in this order.

2.2.1. *Selecting Words and Assigning Initial Scores*

We treat a PubMed article as a set of words when it is rated or used for training a model. In doing so, we first generate a set of stop words to be removed from articles. The stop words can be classified into two types: those which are generated from PubMed articles and not.

Stop words in the first type are generated in the following manner, when PURE is installed on a computer. We first randomly download a large number (e.g. 10,000) of articles from PubMed and compute the *df* (document frequency) and the *tf-idf* (term frequency - inverse document frequency) [3] of each word. We note that for a word, the document frequency is defined as the number of documents having this word, and the term frequency is defined as the number of appearances of this word. Then a word with a high *df* or with a low *tf-idf* is a stop word.

Stop words in the second type are pre-defined in PURE, and they belong to at least one of the following three categories: 1) A word of less than three letters, e.g. I, IV. 2) A word of no alphabets, e.g. 10%, 10.3. 3) A word appearing in Journal of Business Research from Jan. 2005 to Apr. 2006, since they must be unrelated to biological and medical sciences.

Out of each downloaded article, we delete all these stop words to obtain a set of words for this article, and then for each word of the word set, the *tf* (term frequency) is computed to be assigned as the initial word score.

2.2.2. *Learning a Probabilistic Model*

The selected words and initial scores in preferred articles are used to learn a probabilistic model, which is used to give a score to a new article. We use the preferred articles only for learning the preference, meaning that unsupervised learning, more precisely soft clustering of preferred documents (and words), is conducted. In this learning, we note that a user only has to show preferred articles. That is, the rating of each article is not required. This is the advantage of this method in terms of easy handling of preferred articles. We describe the detail of our probabilistic model and the learning process below.

We denote d as an article, z as a latent variable corresponding to a cluster, s as a *field*, e.g. the title or the abstract, and w as a word, and $n_{s,d}(w)$ as the count of word w occurring in field s of document d. We model the probabilistic structure of an article (a set of words) by using a finite mixture model [6] as follows:

$$p(d) = \sum_{z} p(d, z) = \sum_{z} p(z) \prod_{s \in d} \prod_{w \in s} p_s(w|z), \tag{1}$$

We then train probability parameters $p(z)$ and $p_s(w|z)$ from preferred articles. This is a so-called mixture model which is often used for clustering. An important feature which makes this model different from a normal mixture model is that we distinguish the fields, meaning that preferences in each field can be captured independently. In fact, we used two fields, the abstract and the title, independently. The probability parameters are trained by an EM (Expectation-Maximization) algorithm [1], which repeats the following E-and M-steps alternately until some stopping condition is satisfied.

E-step:

$$p(z|d) = \frac{p(d,z)}{\sum_z p(d,z)}$$

M-step:

$$\hat{p}(z) = \frac{\sum_d p(z|d)}{\sum_{z,d} p(z|d)}$$

$$\hat{p}_s(w|z) = \frac{\sum_d n_{s,d}(w)p(z|d)}{\sum_w \sum_d n_{s,d}(w)p(z|d)}$$

This parameter estimation is carried out by a cron script, every time when a table of preferred articles is modified.

2.3. *Daily download of PubMed articles*

New articles of PubMed are retrieved every day in a MEDLINE format through the "Entrez Data (EDAT)" field by a cron-scheduled script, which is written by using an eFetch script [11] as a reference tool. Daily download of PURE is performed in the night time of Eastern Standard Time, i.e. the standard time at NCBI, because a large number of downloads must be done at night, as notified in [12]. Articles, downloaded from PubMed, are stored in a MySQL table with additional information like a user ID and the registered date, etc.

2.4. *Rating articles*

All downloaded articles are ranked by the trained probabilistic model. For each article, words for rating are extracted, and then the likelihood that an article d is preferred can be computed by using Eq. (1). However, as is easily expected from Eq. (1), the larger the number of words in an article, the less the probability (or likelihood) of the article. To correct this bias, we use the Z-score for each article. That is, we gathered a large number of articles and grouped them into those having the same number of words. We then computed the mean μ and the standard deviation σ of $p(d)$ in each group. Given a new article d, we counted the number of words of d and computed the Z-score of d using the μ and the σ of the group containing articles with the same number of words as that of d, as follows: $Z = \frac{p(d)-\mu}{\sigma}$. Thus, downloaded papers are ranked by using their Z-scores.

2.5. *Presentation of highly rated articles*

A pre-defined number of highly rated articles can be recommended articles, which are stored in a MySQL table of PURE, and all other downloaded articles are discarded. The recommended articles are presented to each user by the following two

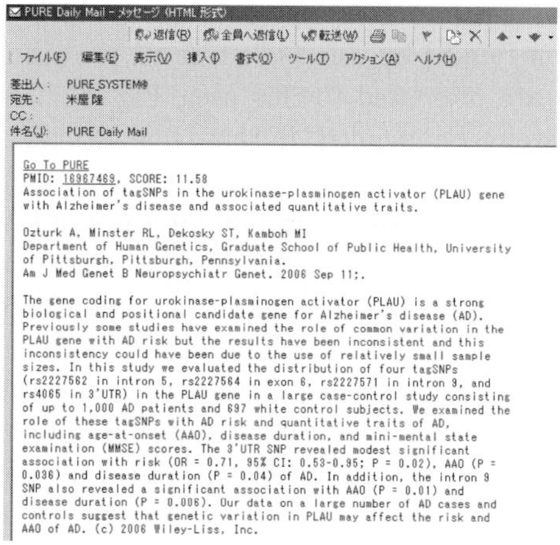

Fig. 4. A notification email of recommended articles.

ways. One is through a web interface, and Fig. 3 shows an example. As in this figure, the title of a recommended article is presented with its score (Z-score), and the whole abstract of an article can be displayed by clicking the corresponding PubMed ID[b]. If a user finds an article to be added to this user's preferred article list, this user only has to click the "check box" of this article and then click the "submit" button. The time period to keep the recommended articles in PURE can be set by a user, and expired articles are removed automatically. The other way of presenting recommended articles is using an email notification service, and Fig. 4 is an example of a notification email. A user can click "Go To PURE" at the top of this email to login to PURE and can see a web page showing a list of recommended articles, which is in Fig. 3.

3. Results and Discussion

3.1. *Recommendation of articles relevant to cancer diagnostics*

PURE is a web-based software with a user-friendly interface. A user can easily check the articles recommended already as well as those currently, and maintain preferred articles through the easy-to-use web-interface of PURE.

We show an actual result obtained by using PURE. We here assume that a user's interests are in diagnostics and target discovery for cancer therapy. Based on these interests, PubMed IDs of this user's preferred papers are registered to PURE. Fig. 2 shows a resultant interface at which eleven preferred papers are already shown. You can see that they are actually relevant with the above topic of this user's interests. The probabilistic model in PURE is trained by using the eleven registered articles to

[b]Previously recommended articles can be found just by clicking "Previously highly rated articles" in the menu as shown in the left column of Fig. 3.

capture the patterns of this user's interests. Articles, which are newly downloaded from PubMed, are then ranked by using the trained probabilistic model, and only highly ranked articles are presented to this user. Fig. 3 shows an actual interface by which five articles are recommended. In these five articles, the top article, which has a significantly high score, is an article on risk prediction of breast cancer, which we believe, is well suited to this user's original interests of cancer therapy.

You might think that this function of PURE is very similar to a function of PubMed which given an article, tries to find related articles. However, we emphasize that they are totally different from each other, because this function of PubMed is to find articles related with only one article, whereas the input of PURE is not only one article but multiple articles. PURE captures the pattern of interests from the input articles and presents articles relevant to this pattern. We note that any number of inputs can be allowed and the performance of PURE will be improved more by inputting a larger number of inputs.

3.2. *Evaluation of the performance of PURE with various training sets*

We then measured the performance of PURE in terms of how many examples we need to obtain satisfactory recommendations, using the probabilistic model in PURE. In this experiment, we used a two-class supervised learning manner to check the performance of PURE. The performance is measured by the precision at a low recall, exactly 10% recall, because only a small number of highly rated articles are recommended by PURE. Precision and recall are defined as $Precision = \frac{TP}{TP+FP}$ and $Recall = \frac{TP}{TP+FN}$, respectively, where TP, FP and FN represent true positives, false positives and false negatives, respectively.

3.2.1. *Data*

Each dataset in this experiment was obtained in the following manner. We first retrieved articles from PubMed by inputting a key word pattern. These articles are positive examples, which were divided into two groups: 500 for test (rating) and the remaining examples for training. Then for test, 4,500 negative examples were randomly collected so that the test set contains 10% (500) of positive articles and 90% (4,500) of negatives[c]. We repeated the division for positives ten times randomly and averaged the results over the ten runs.

We conducted the above experiment five times using five different keyword patterns. Table 1 shows the five keyword patterns we used and the number of articles retrieved from PubMed.

[c]The reason why we used this setting is that a biologist must be interested in probably only 10% or less articles out of all articles.

Table 1. Five datasets used in our experiments.

Name	Keyword pattern	# articles
marker	cancer AND marker AND gene AND expression	1593
structure	protein AND compound AND structure	1360
rheumatoid	(autoimmune OR rheumatoid) AND drug	5618
cancer	cancer AND compound	5167
solubility	(compound OR small molecule OR drug) AND solubility	3729

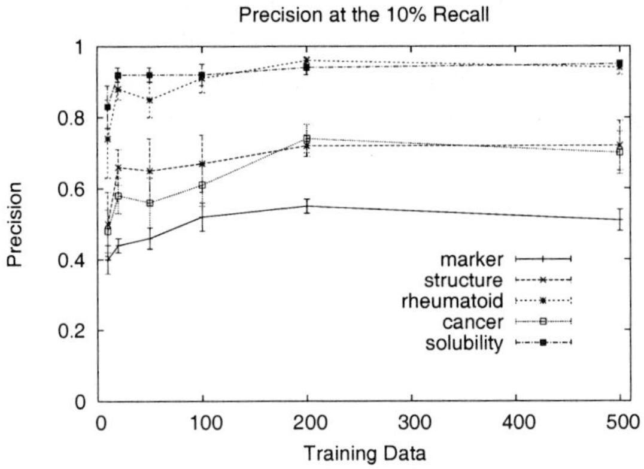

Fig. 5. Precisions at 10% recall with various training datasets

Table 2. Average precisions at 10% recall with changing the size of training articles.

Training Articles	10	20	50	100	200	500
Average Precision (%)	59.0	69.6	68.7	72.6	78.2	76.4

3.2.2. *Results*

Fig. 5 shows the precision at 10% recall for five datasets, which was obtained by increasing the number of training (positive) articles. Although the precision values at the same number of training articles were different depending on keyword patterns, we can see that the precisions were saturated at a relatively small number, e.g. twenty, of training articles. This result implies that nearly the best performance will be obtained simply by inputting a relatively small number of articles as preferred articles. Table 2 shows the average precision at 10% recall over the five datasets, when the number of training articles varied. This result indicates that our method can achieve an accuracy of roughly 70% when the top 10% articles are recommended, even if the number of training articles is only twenty. From these results, we can say that PURE must be useful for automatically finding articles which are relevant to the interests of a biologist.

4. Conclusions

We developed a PubMed article recommendation system, PURE, based on a content-based filtering. The results obtained by our various experiments imply that this system is useful in automatically finding articles which are relevant to a user's interest. A key feature of this system is an easy handling of preferred articles. That is, a user only has to input preferred articles into the system, which then captures the preference of this user from the inputs. Then by using the captured preferences, the system recommends the articles, which are, as shown in our experiment, highly ranked among the new articles downloaded from PubMed daily. Thus, this system will be helpful for finding preferred articles without using any other information such as keywords. PURE is downloadable under GPL license, via www.bic.kyoto-u.ac.jp/pathway/mami/out/PURE.tar.gz.

5. Acknowledgments

The authors would like to thank Ichigaku Takigawa, Raymond Wan, Shanfeng Zhu and Motoki Shiga of Kyoto University and Takayuki Onuma and Reina Nishida of Kirin Pharma for fruitful discussions and valuable comments.

References

[1] Dempster, A., Laird, N., and Rubin, E., Maximum likelihood from incomplete data via the EM algorithm (with discussion), *J. Royal Stat. Soc. B.*, 39:1–38, 1997.
[2] Hokamp, K. and Wolfe, K.H., PubCrawler: keeping up comfortably with PubMed and GenBank, *Nucl. Acids Res.*, 32(Web Server issue):W16–W19, 2004.
[3] Joachims, T., A Probabilistic Analysis of the Rocchio Algorithm with TFIDF for Text Categorization, *ICML-97*, 143–151, 1997.
[4] Leavitt, N., Recommendation technology: Will it boost e-commerce?, *IEEE Computer*, 39:13–16, 2006.
[5] Marchin, M., Kelly, P.T., and Fang, J., Tracker: continuous HMMER and BLAST searching, *Bioinformatics*, 21:388–389, 2005.
[6] McLachlan, G. and Peel, D., Finite Mixture Models, Wiley, 2000.
[7] Mooney, R. and Roy, L., Content-based book recommending using learning for text categorization, *ACM DL-2000*, 195–204, 2000.
[8] Resnick, P. and Varian, H., Recommender systems, *Communications of the ACM*, 40:56–58, 1997.
[9] Shultz, M. and De Groote, S.L., MEDLINE SDI services: how do they compare?, *J. Med. Libr. Assoc.*, 91:460–467, 2003.
[10] `http://biomail.sourceforge.net/biomail/`
[11] `http://eutils.ncbi.nlm.nih.gov/entrez/query/static/efetchseq_help.html`
[12] `http://eutils.ncbi.nlm.nih.gov/entrez/query/static/eutils_help.html`
[13] `http://www.amedeo.com/`
[14] `http://www.info.scopus.com/`
[15] `http://www.leaddiscovery.co.uk/PubMed-dailyupdates.html`
[16] `http://www.ncbi.nlm.nih.gov/entrez/`
[17] `http://www.sciencedirect.com/`

PERFORMANCE IMPROVEMENT IN PROTEIN *N*-MYRISTOYL CLASSIFICATION BY BONSAI WITH INSIGNIFICANT INDEXING SYMBOL

MANABU SUGII[1]
manabu@yamaguchi-u.ac.jp

RYO OKADA[2]
r-okada@hcu.co.jp

HIROSHI MATSUNO[3]
matsuno@sci.yamaguchi-u.ac.jp

SATORU MIYANO[4]
miyano@ims.u-tokyo.ac.jp

[1] *Media and Information Technology Center, Organization for Academic Information, Yamaguchi University, 1677-1 Yoshida, Yamaguchi 753-8511, Japan*
[2] *Network Solution Group, Hitachi Chugoku Solutions, Ltd., 11-10 motomachi, Hiroshima 730-0011, Japan*
[3] *Graduate School of Science and Engineering, Yamaguchi University, 1677-1 Yoshida, Yamaguchi 753-8511, Japan*
[4] *Human Genome Center, University of Tokyo, Tokyo 108-8639, Japan.*

Many *N*-myristoylated proteins play key roles in regulating cellular structure and function. In the previous study, we have applied the machine learning system BONSAI to predict patterns based on which positive and negative examples could be classified. Although BONSAI has helped establish 2 interesting rules regarding the requirements for *N*-myristoylation, the accuracy rates of these rules are not satisfactory. This paper suggests an enhancement of BONSAI by introducing an "insignificant indexing symbol" and demonstrates the efficiency of this enhancement by showing an improvement in the accuracy rates. We further examine the performance of this enhanced BONSAI by comparing the results of classification obtained the proposed method and an existing public method for the same sets of positive and negative examples.

Keywords: *N*-myristoylation; machine learning; alphabet indexing; protein classification.

1. Introduction

Protein *N*-myristoylation is a lipid modification of proteins, and many *N*-myristoylated proteins play key roles in regulating cellular structure and function such as the BH3-interacting domain death agonist (BID) which is involved in apoptosis that occurs via the alpha subunit of a G-protein localized on the cell membrane. *N*-myristoylated proteins have a specific sequence at the N-terminus called the *N*-myristoylation signal sequence, and this sequence is probably composed of 6 to 9 amino acids (up to 17) [1].

In order to determine the N-terminal sequence requirements for protein *N*-myristoylation, the amino acid sequences of *N*-myristoylated proteins have been examined [2, 3]. Most of the methods used by researchers predict the patterns for *N*-myristoylation based on the data obtained through biological experimentations. However, the information on the amino acid sequences is very vast, and *N*-myristoylation is not based on one simple rule but many specific rules. Hence, computational techniques are essential for predicting the rules from a huge amount of data on the sequence required for *N*-myristoylation.

The machine learning system BONSAI is a system for knowledge acquisition based on the theory of Probably Approximately Correct Learning (PAC learnability) and uses the method of local search [4] [5]. By using BONSAI, we carried out the computational experiment to establish new rules characterizing the difference between positive examples and negative examples of *N*-myristoylation sequences, and established the following 2 types of new rules: one, a rule that supports the existing *N*-myristoylation rule; the other, a rule that has not been discovered thus far [6]. Thus, the usefulness of BONSAI has been proved for the characterization of *N*-myristoylation sequences. However, the accuracy rates obtained using BONSAI are not sufficiently satisfactory for application in searching for new *N*-myristoylation sequences from real data. In addition, the difficulties of using our BONSAI-based method remained, specifically in terms of the complex decision trees and long processing time involved in obtaining rules.

In order to resolve with these, this paper introduces a modified BONSAI system called "BONSAI with insignificant indexing symbol." *Insignificant indexing symbol* is a special indexing symbol to which the sysytem assigns letters that do not concern with the rules of classification as either positive or negative. The results of the computational experiments show that this introduction improves the accuracy of decision trees particularly for positive examples, i.e., for sequences known to be *N*-myristoylated. In addition, this introduction allows BONSAI to generate decision trees that have smaller depth and fewer numbers of nodes than the decision trees produced by the original BONSAI [6]. We further report the results of a comparison between the proposed method and an existing public method, demonstrating that our method performs better than the existing method with respect to the accuracy of extraction of both positive and negative examples despite shorter extraction time.

2. Protein N-Myristoylation

Protein *N*-myristoylation is the lipid modification of proteins in which the 14-carbon saturated fatty acid binds covalently to the N-terminus of viral and eukaryotic proteins. Approximately 0.5% of human proteins are estimated to be *N*-myristoylated [1]. Protein *N*-myristoylation is a cotranslational protein modification catalyzed by 2 enzymes, namely, methionine aminopeptidase and *N*-myristoyltransferase (NMT). It is estimated that for undergoing *N*-myristoylation, a protein must at least have a Met-Gly sequence on its N-terminus. The initial Met is removed cotranslationally by the Met aminopeptidase, and then the myristic acid is linked to the next Gly via an amide bond through catalysis by NMT. NMT catalyzes the transfer of myristic acid from myristoyl-CoA to the N-terminus Gly residue of the substrate protein (Fig. 1). Most of myristoylated proteins are involved in physiological activities such as cell

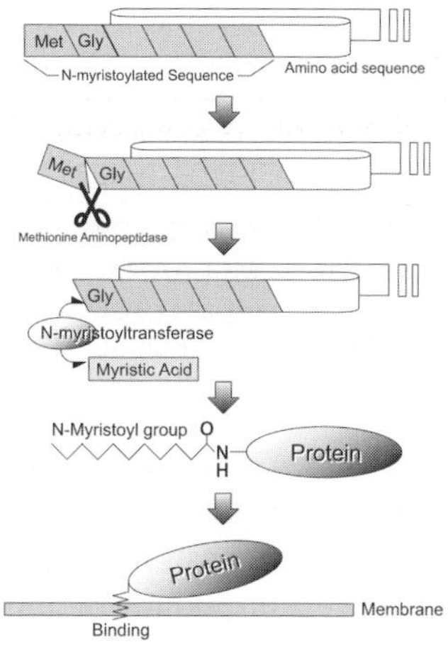

Fig. 1. Protein *N*-myristoylation is the lipid modification of proteins in which the 14-carbon saturated fatty acid binds covalently to the N-terminus. The initial Met is removed by methionine aminopeptidase. Gly is required at position 2 from the N-terminus for the formation of a bond with myristic acid through catalysis by *N*-myristoyltransferase.

signaling and exerting specific functions through binding with organelle membranes. It is known that the membrane binding mediated by myristoylation is controlled in various manners and plays a crucial role in the functional regulation mechanisms of proteins in cell signaling pathways and virus growth [7]. For example, the HIV-1 Gag protein is transferred to the plasma membrane via an *N*-myristoyl group and is involved in the formation and release of virus particles. Additionally, it is known that the apoptosis-inducing factor BID is digested by protease and that the new N-terminus of the digested peptide is also myristoylated [8].

N-myristoylated proteins have a specific sequence at the N-terminus called an *N*-myristoylation signal sequence. This sequence is probably composed typically of 6 to 9 amino acids, but this number can be as high as 17 [1]. The effect of the amino acid sequence on *N*-myristoylation depends on the distance and position from N-terminus; with the increase in the distance, this effect decreases. Table 1 shows examples of the N-terminus sequences in myristoylated proteins. Amino acids are usually denoted by 1-letter or 3-letter codes.

Biologists have revealed that the combination of amino acid residues at positions 3 and 6 constitutes a major determinant for the susceptibility to protein *N*-myristoylation. As shown in Fig. 1, when Ser is located at position 6, 11 amino acid residues (Gly, Ala, Ser, Cys, Thr, Val, Asn, Leu, Ile, Gln, His) may be located at position 3 to direct efficient protein *N*-myristoylation [2] [3]. Most of these 11 amino acids satisfy a rule that the radius of gyration of the residue is smaller than 1.80Å. In fact, other amino acids that have a radius of gyration larger than 1.80Åcannot be present at position 3. In addition to the restriction of the radius of gyration of the amino acid residues, it has been also revealed that the presence of negatively

Table 1. The sequences at the N-terminus of *N*-myristoyl proteins.

Protein	Amino acid sequence
GAG SIVM1	MGARNSVLSGKKADE
GAG MPMV	MGQELSQHERYVEQL
KCRF STRPU	MGCAASSQQTTATGG
Q26368	MGCNTSQELKTKDGA
GBAZ HUMAN	MGCRQSSEEKEAARR

charged residues (Asp and Glu) and a Pro residue at this position completely inhibited *N*-myristoylation. On the other hand, when Ala is located at position 6, 5 kinds of amino acid residues can occupy position 3 for *N*-myristoylation. When Thr or Phe is located at position 6, only 2 or 3 kinds of amino acid residues can occupy position 3 for *N*-myristoylation. In addition, some amino acid residues at position 7 dictate the amino acid requirement at position 3 for *N*-myristoylation. For example, although the presence of Ser at position 6 does not basically allow Lys to occupy position 3, the presence of Lys at position 7 alters to the requirement for amino acid residue at position 3; Lys can be present at position 3 [3].

3. BONSAI with Insignificant Indexing Symbol

BONSAI is a machine learning system for knowledge acquisition from positive and negative examples of strings (Fig. 2) [5]. A hypothesis generated by this system is presented using 2 kinds of classification of symbols called an alphabet indexing and a decision tree that classifies the given examples as either positives or negatives. *Alphabet indexing* (indexing, in short) is the transformation of symbols to reduce the number of letters assigned to positive and negative examples, without omitting important information in the original data. In the case of amino acid residues,

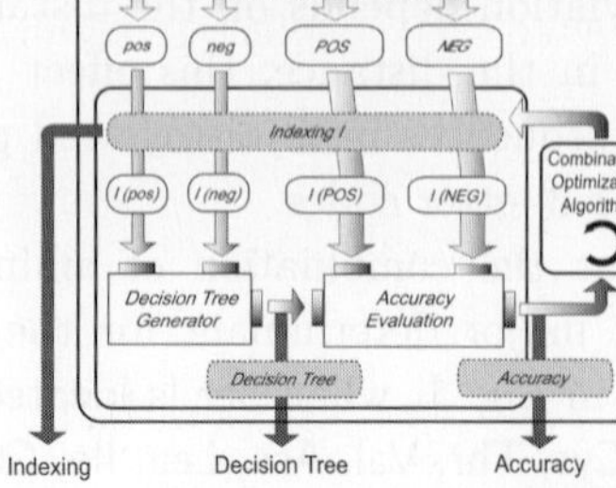

Fig. 2. For the *Positive Examples* and *Negative Examples* inputted, BONSAI computes indexings, decision trees, and accuracy. From the positive and negative examples randomly selected and transformed by indexing function I, *Decision Tree Generator* constructs decision trees. *Accuracy Evaluation* is used to search for a better indexing. With *Combinatorial Optimization Algorithm*, these are repeated until a locally optimal indexing and a decision tree are found.

alphabet indexing can be regarded as a classification of 20 kinds of amino acid residues to a few categories. Indexing contributes not only quicken the computations involved in finding rules but also to simplify expression patterns assigned at the nodes of decision trees.

3.1. *Decision Tree for N-Myristoyl Sequences Generated using Original BONSAI*

Fig. 3 shows the result of BONSAI for some positive and negative examples of *N*-myristoylation. By analyzing binary patterns shown in the table in the Fig. 3, we found a rule that classifies the given positive and negative examples [6]. However, the accuracy of this rule is not high-61.1% for positive examples and 92.0% for negative examples.

A discriminative indexing pattern found by the original BONSAI is assigned at each node of the decision tree in Fig. 3. This decision tree classifies the given sequences by sequentially performing "OR operation" over the discriminative indexing patterns. This decision tree similar to a decision list has a large depth and small width, because the original BONSAI can only find such a decision tree if only poor rules exist naturally in positive and negative examples. According to the widely believed principle that "a smaller decision tree indicates essential knowledge," a tree with such a structure is not desirable.

3.2. *Introducing Insignificant Indexing Symbol*

This paper introduces a new concept of "insignificant indexing symbol" in BONSAI. *Insignificant indexing symbol* is a special indexing symbol to which the sysytem assigns letters that do not concern with the rules of classification as either positive or negative.

Insignificant indexing symbol can be realized by a simple modification to BONSAI as shown below.

(1) Choose 1 symbol from all indexing symbols as an insignificant indexing symbol, and

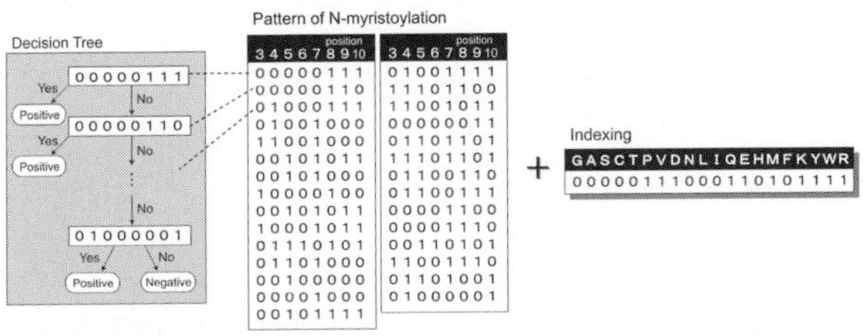

Fig. 3. Generation of decision tree and indexing for given positive and negative examples for *N*-myristoylation sequences using BONSAI [6].

(2) When evaluating a decision tree during the computation using BONSAI, the chosen insignificant indexing symbol is dealt with as "wildcard," that is, any single character can be matched at the locations of the insignificant indexing symbol.

In the following, we use sequential numbers i.e., 0, 1, 2,... for indexing symbols and assign a symbol 0 to function as the insignificant indexing symbol. Further, BONSAI_{iis} (BONSAI with the insignificant indexing symbol) is also used Fig. 4 shows an example of the BONSAI_{iis} process. The letters S, C, N, Q, H, M, Y, W and R are assigned as insignificant indexing symbols This implies that a more accurate decision tree can be obtained unless these letters are used in decision trees. In other words, if some of these letters are important for classifying positive and negative examples, these should be assigned to either of indexing symbol 1 or 2 in the case of Fig. 4.

4. Verification of the Effect of a Modified Algorithm for BONSAI

We have examined the performance of BONSAI_{iis} with the same positive and negative examples as the experiment in section 3. The positive examples include 78 myristoylated amino acid sequences, and the negative examples include 800 amino acid sequences randomly selected from the human protein database. The indexing size was set to 3, and the length of the input sequences is 9 or 8, which excludes the first amino acid Met or the both of the first and second amino acid from the N-terminus.

Fig. 5 shows accuracy rates of classifications using 10 decision trees obtained from the original BONSAI and BONSAI_{iis}. We used 10 decision trees because BONSAI generally creates different trees with the same input data. The vertical axis indicates the accuracy rate of classification, where the white bar represents positive examples and the black bar, negative examples.

Fig. 5(a) clearly indicates that decision trees obtained using BONSAI can classify the example data more accurately than those obtained using the original BONSAI.

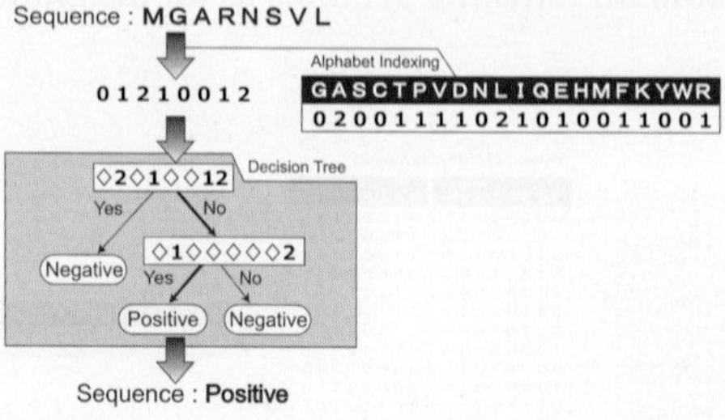

Fig. 4. Example of the BONSAI_{iis} process. A given sequence MGARNSVL is converted to the sequence 01210012 according to the alphabet indexing table. The decision tree classifies this converted sequence to Positive. In order to clearly express that this symbol 0 works as a wildcard, the symbol ◇ is used instead of the previous symbol in the decision tree.

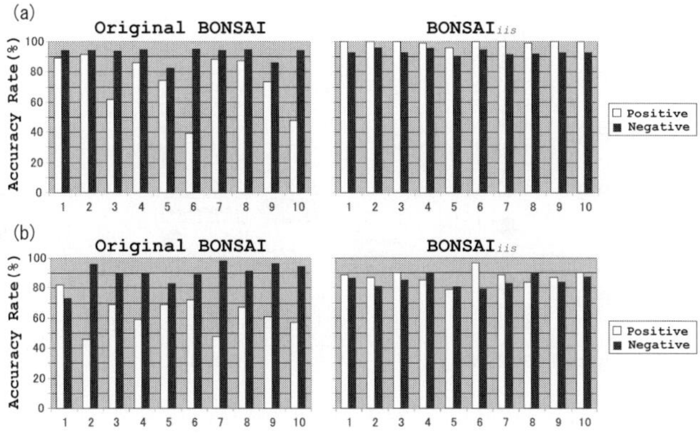

Fig. 5. Accuracy rates of classifications using 10 decision trees obtained from the original BONSAI (left) and BONSAI_{iis} (right) with 9 amino acids sequences eliminating the first Met (a) and with 8 amino acids sequences eliminating the first Met and second Gly (b). The vertical axis indicates the accuracy rate of classification, the white bar indicates positive examples; the black bar, negative examples.

The accuracy rate of BONSAI_{iis} is 96.3%, which is superior to 83.1% of original BONSAI. Hence, the decision trees obtained using BONSAI_{iis} can more accurately provide a signal sequence required for *N*-myristoylation. A comparison of the results obtained from the original BONSAI and BONSAI_{iis}, shows that BONSAI_{iis} shows more stable performance than the original BONSAI. Fluctuation in the accuracy rates of the original BONSAI depends on the structure of a decision tree, as shown in Fig. 3. This structure of the decision tree obtained using BONSAI_{iis} also contributes to finding decision trees with higher accuracy rates.

BONSAI_{iis} found a more accurate decision tree than the original BONSAI. However, BONSAI_{iis} attempts to search for an *N*-myristoylation signal whether the second position of the pattern in the decision tree is occupied by Gly. This is not desirable for our research to find a new rule for N-myristoylation. Thus, we further examined the performance of the 2 BONSAI systems with the 8 amino acid sequences eliminating the first Met and second Gly to find new rules while excluding those already established for known myristoylation signals.

The result is shown in Fig. 5(b). Both the BONSAI systems show low accuracy rates compared to above experiments; this is because of the lack of the second Gly that is indispensable for *N*-myristoylation. But BONSAI_{iis} has an advantage in that it provides accurate classification, with an accuracy rate of 86.1%, while the accuracy rate of the original BONSAI is 76.6%. Moreover, BONSAI_{iis} maintains stable and high performance and the smaller difference in the accuracy rates between positives and negatives.

Fig. 6 shows 2 typical decision trees-one is chosen from the decision trees of the original BONSAI and the other, from those of BONSAI_{iis}. The average depth indicated in this figure for each of 10 decision trees obtained by the original BONSAI or

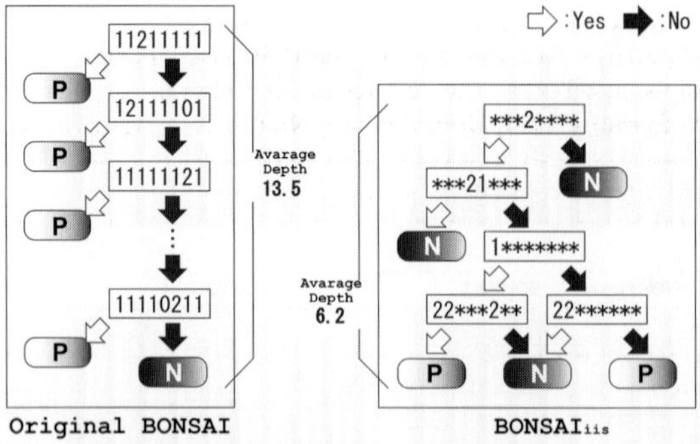

Fig. 6. Typical decision trees from the original BONSAI and BONSAI$_{iis}$. The left tree has been selected from decision trees obtained by the original BONSAI and the right one obtained by BONSAI$_{iis}$.

BONSAI$_{iis}$. We can see that the tree depth is large for a decision tree obtained by the original BONSAI, while it is small for a decision tree obtained by BONSAI$_{iis}$. At the same time, the widths of these trees are different. The decision tree obtained by BONSAI$_{iis}$ is more desirable since this decision tree is more compact than the tree from the original BONSAI. In addition, this decision tree provides a more clear representation of rules classifying positive and negative examples.

5. Comparison with Results on an Existing Website for Predicting *N*-Myristoylation

Currently, a website predicts whether a given sequence will be *N*-myristoylated or not [9]. The prediction function on this website comprises terms evaluating amino acid type preferences at sequences that are close to the N-terminus as well as terms indicate deviations from the pattern of the physical properties of amino acid side-chains encoded in a multi-residue correlation within the motif sequence [10]. The underlying biological facts for determining the scores of the prediction function are described in the paper [1]. We have compared the method used in that paper [10] with our method by using the same amino acid sequence set for both of these methods.

Table 2 shows the result of performance comparison of these 2 methods. Seventy-eight and 88 sequences were selected as positive and negative examples, respectively. Positive examples were the same sequences as those used in section 4, while negative examples were sequences presented in the literature [2, 3] as sequences that are not *N*-myristoylated sequences.

The classification results for the sequences used in our method (BONSAI$_{iis}$) are expressed as probabilities ranging from 0% to 100%. On the other hand, the classification results on the website [9] (NMT) are expressed as RELIABLE, TWILIGHT ZONE, and NO, which indicate that *N*-myristoylation of a given sequence will occur, can not be judged, and will not occur, respectively. Hence, we have derived relation-

Table 2. Performance comparison between the proposed method (BONSAI$_{iis}$) and the method used in the website [9] (NMT). Symbols used for the classification results are as follows: P=*N*-myristoylated, U=unknown, N=not *N*-myristoylated.

(a) 78 *N*-myristoylated sequences

	P	U	N	accuracy
BONSAI$_{iis}$	74	1	3	94.9%
NMT	72	6	0	92.3%

(b) 88 not *N*-myristoylated sequences

	P	U	N	accuracy
BONSAI$_{iis}$	9	6	73	83.0%
NMT	6	13	69	74.0%

ships for these 2 different expressions as follows: RELIABLE = $55\% \leq p$, TWILIGHT ZONE = $45\% \leq p < 55\%$, and NO = $p < 45\%$ for probability p provided by BONSAI$_{iis}$. From Table 2, we can see that

- the 2 methods express almost the same accuracy rates for positive examples (*N*-myristoylated sequences), but BONSAI$_{iis}$ expresses a higher accuracy rate than NMT for negative sequences, and
- the number of false positives in case of NMT is less than that obtained by BONSAI$_{iis}$ for both positive and negative examples.

Based on these results, we cannot emphasize that our BONSAI$_{iis}$ method is superior to the method used in NMT in terms of the accuracy rate. However, our method offers a great advantage over the NMT method with respect to computational time owing to the structure of the decision tree used in our method for classification rules. False positives in the NMT method were less in number than in BONSAI$_{iis}$, because the algorithm used in the NMT method [10] is more complex than decision trees. The number of false positives and negatives will increase as an error when only poor rules exist naturally in examples because BONSAI creates decision trees using the local optimum solution by the local search method. Thus, if BONSAI$_{iis}$ uses other system such as the database used in the NMT method, the accuracy rate of classification will be higher and the number of false positives will be reduced. BONSAI$_{iis}$ can find desirable rules in addition to reducing the process time without requiring complex algorithms.

6. Conclusion

In the previous paper [6], we modified BONSAI in order that it enables the assigning of positions of amino acids from the N-terminus. When sequences that occupy the low selective positions for amino acid were given, the modified BONSAI in [6] have produced decision trees with large depths, similar to a decision list, such as the tree

in Fig. 3, in which all conditions for classifying positive and negative examples are reflected as node labels. We reported 2 types of new rules in the previous paper [6]; however, it is difficult to interpret rules for *N*-myristoylated sequences with a decision tree having such a large depth.
Futher, taking into account the fact that *N*-myristoylated sequences have the low selective positions for amino acid, we have further modified BONSAI by introducing a new concept called an "insignificant indexing symbol." The insignificant indexing symbol will be assigned to amino acid symbols unimportant for *N*-myristoylation. This introduction allows BONSAI to distinguish letters that do not concern with the rules of classification as either positive and negative examples. We have not yet found a new rule regarding the node patterns in decision trees obtained using BONSAI_{iis}, although several known biological rules were confirmed.
However BONSAI is based on the theory that PAC learnability can search for the local optimum solution using local search, the local optimum solution found using BONSAI does not necessarily represent the global optimum solution. Other learning systems based on other algorithms such as support vector machine may be expected to improve the accuracy rate of classification and to find new rules for *N*-myristoylation.

References

[1] Maurer-Stroh, S., Eisenhaber, B., and Eisenhaber, F., N-terminal *N*-myristoylation of proteins: refinement of the sequence motif and its taxon-specific differences, *J. Mol. Biol.*, 317, 523-540, 2002.

[2] Utsumi, T., Sato, M., Nakano, K., Takemura, D., Iwata, H., and Ishisaka, R., Amino acid residue penultimate to amino-terminal Gly residue strongly affects two cotranslational protein modifications, *N*-myristoylation and *N*-acetylation, *J. Biol. Chem.*, 276, 10505-10513, 2001.

[3] Utsumi, T., Nakano, K., Funakoshi, T., Kayano, Y., Nakao, S., Sakurai, N., Iwata, H., and Ishisaka, R., Vertical-scanning mutagenesis of amino acid in a model *N*-myristoylation motif reveals the major amino-terminal sequence requirements for protein *N*-myristoylation, *Eur. J. Mol. Biochem.*, 271, 863–874, 2004.

[4] Valiant, L. G., A theory of the learnable, *CACM* 27(11), 1134–1142, 1984.

[5] Shimozono, S., Shinohara, A., Shinohara, T., Miyano, S., Kuhara, S., Arikawa, S., Knowledge acquisition from amino acid sequences by machine learning system BONSAI, *Trans. Inform. Process. Soc. Japan,* 35, 2009-2018, 1994.

[6] Okada, R., Sugii, M., Matsuno, H., and Miyano, S., Machine learning prediction of amino acid patterns in protein N-myristoylation, *Pattern Recognition in Bioinformatics (LNBI)*, 4146, 4-14, 2006.

[7] Farazi, T.A., Waksman, G., and Gordon, L.I., The biology and enzymology of protein *N*-myristoylation, *J. Biol. Chem.*, 276, 39501-39504, 2001.

[8] Zha, J., Weiler. S., Oh, K.J., Wei, M.C., Korsmeyer, S.J., Posttranslational *N*-myristoylation of BID as a molecular switch for targeting mitochondria and apoptosis, *Science*, 290, 1761-1765, 2000.

[9] `http://mendel.imp.ac.at/myristate/`

[10] Maurer-Stroh, S., Eisenhaber, B., and Eisenhaber, F., N-terminal *N*-myristoylation of proteins: prediction of substrate proteins from amino acid sequence, *J. Mol. Biol.*, 317, 541-557, 2002.

IDENTIFICATION OF DIVERSE CARBON UTILIZATION PATHWAYS IN *SHEWANELLA ONEIDENSIS* MR-1 VIA EXPRESSION PROFILING

MICHAEL E. DRISCOLL[1]
mdriscol@bu.edu

MARGIE F. ROMINE[2]
margie.romine@pnl.gov

FRANK S. JUHN[3]
frnkjn@bu.edu

MARGRETHE H. SERRES[4]
mserres@mbl.edu

LEE ANNE MCCUE[2]
leeann.mccue@pnl.gov

ALEX S. BELIAEV[2]
alex.beliaev@pnl.gov

JAMES K. FREDRICKSON[2]
jim.fredrickson@pnl.gov

TIMOTHY S. GARDNER[1,3]
tgardner@bu.edu

[1] *Bioinformatics Program, Boston University, St. Boston, MA 02215, USA*
[2] *Biological Sciences Division, Pacific Northwest National Laboratory, Richland, WA 99352, USA*
[3] *Department of Biomedical Engineering, Boston University, Boston, MA 02215, USA*
[4] *Josephine Bay Paul Center , Marine Biological Laboratory, Woods Hole, MA, USA*

To identify pathways of carbon utilization in the metal-reducing marine bacterium *Shewanella oneidensis* MR-1, we assayed the expression of cells grown with various carbon sources using a high-density oligonucleotide Affymetrix microarray. Our expression profiles reveal genes and regulatory mechanisms which govern the sensing, import, and utilization of the nucleoside inosine, the chitin monomer N-acetylglucosamine, and a casein-derived mixture of amino acids. Our analysis suggests a prominent role for the pentose-phosphate and Entner-Doudoroff pathways in energy metabolism, and regulatory coupling between carbon catabolism and electron acceptor pathways. In sum, these results indicate that *S. oneidensis* possesses a broader capacity for carbon utilization than previously reported, a view with implications for optimizing its role in microbial fuel cell and bioremediative applications.

Keywords: *Shewanella*; microarray; carbon metabolism.

1. Introduction

Shewanella oneidensis MR-1 is an environmentally ubiquitous, metabolically versatile gamma-proteobacteria with a broad capacity for the respiration of metals. *Shewanella*'s ability to shuttle electrons onto metals – including arsenic and uranium – has made it a model organism for use in microbial fuel cells and environmental remediation applications [10, 17, 23]. In contrast to its well-known utilization of electron acceptors, *Shewanella* has been considered to possess a relatively narrow capacity for utilizing electron donors, preferring simple carbon compounds such as formate, lactate, and other fermentative end products [24]. However, comparative sequence analysis, phenotypic assays, and recent experimental work have suggested that more complex carbon compounds may also drive respiration in this organism [29, 33, 36].

To investigate the pathways involved in the electron donor metabolism of *S. oneidensis*, we measured cell growth and performed whole-genome expression profiling using five substrates as both carbon source and electron donor: a casein-derived mixture of amino acids, the nucleoside inosine, the amino sugar N-acetylglucosamine, and the carboxylic acids pyruvate and lactate. These substrates were chosen because they support robust growth of *S. oneidensis* in the laboratory and are known to exist in the natural environments that *Shewanella* species occupy. N-acetylglucosamine is a monomer of chitin, which is among the largest stores of dissolved organic carbon in the oceans [27]. Free amino acids and inosine are major byproducts in the post-mortem breakdown of marine vertebrate tissue [7, 30]. Finally, lactate is a fermentative end product common to the natural sediments from which *Shewanella* has been isolated, and pyruvate is chemically similar to lactate.

Here we describe expression of chemotactic, transport, and catabolic genes involved in the utilization of these carbon sources, show evidence of transcription factors regulating these genes, and identify the likely metabolic pathways by which these sources are coupled to energy production in the cell.

2. Materials and Methods

2.1. *Experimental Design*

Our experimental design consisted of a two-way layout testing five carbon sources with varying salt and H2O2 levels in a defined minimal medium (Fig. 1C). For each of the carbon sources investigated, three samples were collected. The salt and H2O2 conditions were chosen at levels which did not significantly diminish growth (data not shown), but for which we might be able to detect changes via expression signatures. This two-way layout allowed for the multiple observations of experimental factors without the explicit use of technical replicates.

2.2. *Cell Culture & RNA Isolation*

For RNA samples, *S. oneidensis* MR-1 cells (ATCC strain 700550) were grown overnight to high density in Luria Broth, then washed and diluted to a starting inocolum (OD600 0.05) in a condition-specific minimal media. Cells were then grown for an additional thirty hours at 30 C to achieve sufficient biomass and RNA was extracted. To minimize the effects of metabolic waste products, cells were centrifuged at 1350g for 10 minutes at 4 C, washed, and resuspended in fresh condition-specific media after 12 and 24 hours of growth. Due to differences in growth rates on the carbon sources used, final biomass yields ranged widely (OD600s: inosine 0.22 +/-0.05; N-acetylglucosamine 0.55 +/- 0.02; Casamino 1.95 +/-0.05; lactate 0.38 +/-0.28; pyruvate 0.41 +/-0.15).

The media used for each of the conditions studied was composed of one of five carbon sources (7.5 mM inosine, 7.5 mM N-acetylglucosamine, 0.2% Casamino acids (Difco Laboratories, P/N 223120), 15 mM lactate, or 15 mM pyruvate) crossed with a salt or peroxide treatment (200mM NaCl or 10uM H2O2), in a background of modified M1 media (3 mM PIPES, 28 mM NH4Cl, 1.3 mM KCl, 4.4mM NaH2PO4-H2O), with micromolar supplements of amino acids, mineral, and vitamins (Table S2). Aeration was achieved via shaking at 220 RPM, and while tests with the oxygen-sensitive dye resazurin indicated presence of oxygen in all wells, the low surface area to volume ratio of our microplates (Whatman, USA, PN 7701-6102) suggests that our cultures were oxygen-limited.

RNA extraction was performed by suspending cells in RNAProtect, followed by isolation using RNEasy spin columns (Qiagen Inc, Valencia, CA), which included an on-column DNAse treatment step to reduce genomic contamination. RNA yields were estimated from UV 260nm/280nm absorbance ratios and normalized to 10ug for all samples. Reverse transcription of RNA to cDNA, cDNA fragmentation and labeling, hybridization to Affymetrix microarrays followed the GeneChip Expression Analysis Technical Manual (Affymetrix, Santa Clara, CA). All cell culturing, RNA isolation, and array hybridization steps were performed in parallel, and scanning of hybridized microarrays was performed on the same scanner.

2.3. *Microarray Design & Expression Data Analysis*

The custom microarray we developed for this study is a high-density array manufactured by Affymetrix (Santa Clara, CA) containing 103,797 pairs of perfect-match and single-mismatch 25-mer oligonucleotide probes with a feature size of 11um [16]. provided our chip design file (CDF), which describes array coordinates for probes and their genome targets, in the Supplementary Material.

Probe-level intensity values from Affymetrix CEL files were processed using the RMA algorithm to compute gene expression values [2]. For a given condition, a gene's mean expression was compared against a background mean of all other conditions. Using this difference of means, along with a pooled estimate of variance, the significance of up or down regulation (expressed as P values) was computed according to a two-sample T test statistic. This was repeated for all conditions.

2.4. *Motif Alignment*

Motif discovery was performed by aligning upstream regions of selected genes using the MEME algorithm as implemented by the Microbes Online web site (http://www.microbesonline.org). Visualization of the identified motifs was performed using WebLogo (http://weblogo.berkeley.edu).

3. Results

3.1. *Use of Pentose Phosphate and Entner-Doudoroff Pathways for Growth on Inosine*

The nucleosides adenosine, inosine, and uridine have been identified as the top three most "electrogenic" carbon sources for *S. oneidensis*, on the basis of NADH-

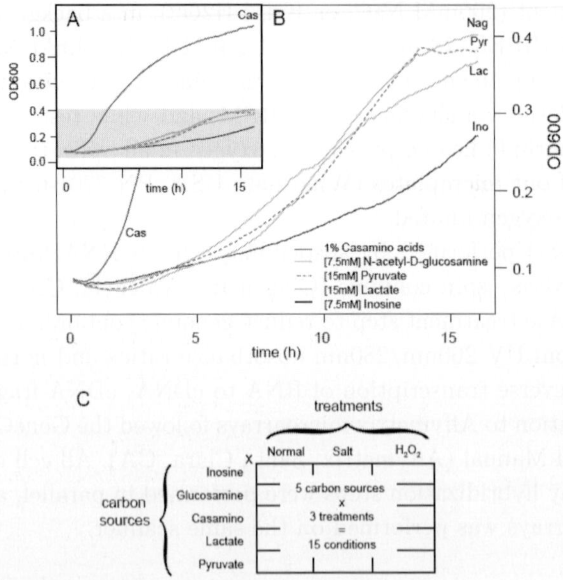

Fig. 1. Sixteen-hour growth profiles for carbon sources, shown at two scales. (A) Growth of Casamino acids (1% or 10 mg/mL) is highlighted in red, showing its rapid growth rate as compared with other carbon sources studied. These slower curves are expanded in panel (B) to resolve the growth rates of N-acetylglucosamine, pyruvate, lactate, and inosine, at concentrations as labeled. Cells were inoculated from overnight LB cultures into a minimal modified M1 media with carbon sources as described in Materials and Methods, and grown in 24-well plates in an aerated, incubated plate reader (Biotek P/N SIAFRTD) at 30 C, with optical density readings (OD600) taken at six-minute intervals. (C) Illustration of the two-way layout experimental design that was used for the fifteen microarrays used in this study.

sensitive dyes which detect electron transport [33]. In this assay, a carbon source is considered more "electrogenic" if its utilization results in more electrons being routed down its electron transport chain.

Previous work has also demonstrated growth on inosine and ribose in other *Shewanella* species [15, 24, 37]. Our growth profiles confirm that inosine supports growth as the sole carbon source in *S. oneidensis*, with a growth rate approximately half that of lactate, the fourth most electrogenic carbon source, at comparable molarities (Fig. 1).

Based on our expression profiles, *S. oneidensis* appears to drive energy production from inosine in several steps: sensing and transport into the cell, release of its ribose base, conversion of ribose to hexose via the pentose phosphate pathway, and finally, cleavage by the Entner-Doudoroff enzymes to generate a triose phosphate and pyruvate for central metabolism (Fig. 2).

The use of the pentose phosphate cycle for non-oxidative conversion of ribose to hexose is a common strategy in environmental microorganisms [3]. The activity of pentose-phosphate and Entner-Doudoroff pathway enzymes in *S. oneidensis* extracts has previously been demonstrated [22], and recent 13C labeling experiments have established significant fluxes through these pathways during growth on lactate [31], both under aerobic conditions. Our microarray profiles show high constitutive expression of pentose-phosphate pathway enzymes during all growth conditions we assayed (Fig. 2), though no significant differential expression during inosine growth.

In the Entner-Doudoroff pathway, we observed significant upregulation of both the *eda* and *edd* genes, part of the *zwf-pgl-edd-eda* operon (SO2489-SO2486) encoding enzymes for conversion of glucose-6-phosphate to pyruvate and glyceraldehyde-3-phosphate. These products form precursors that can flow into central energy metabolism or be used a substrates for direct generation of NADH via dehydrogenases.

In addition to the activation of these catabolic pathways, growth with inosine also results in differential regulation of several genes involved in nucleotide biosynthesis. For example, the *nrdAB* locus, encoding an aerobic ribonucleoside-diphosphate reductase, which converts nucleotides to deoxynucleotides, showed significant upregulation, reflecting greater flux through this conversion pathway (Table S1). Conversely, six of the ten genes annotated as being in the de novo purine synthesis pathway were significantly repressed (Fig. 2B).

In *E.coli* the purine synthesis pathway is controlled by the transcriptional regulator PurR [20], for which no homolog exists in *S. oneidensis* or in any of the twelve sequenced *Shewanella* genomes currently available (http://img.jgi.doe.gov). Here, as with several of the other pathways discussed, *cis*-regulatory sequences of the top differentially expressed genes were aligned, but no significant motifs were detected. The coordinated repression of these six genes, which are spread across five distinct loci on the genome, could be effected through an as yet unidentified transcription factor, or via a small RNA mechanism (antisense or riboswitch).

3.2. *Growth on N-acetylglucosamine and Chitin-Related Pathways*

In the natural aquatic environments where N-acetylglucosamine is found, marine bacteria are thought to coordinate several activities to facilitate its catabolism: chemotaxis and adhesion to chitin, breakdown into N-acetylglucosamine, transport into the cell, and finally conversion to fructose [14]. Our expression profiles of *S. oneidensis* grown with N-acetylglucosamine indicate several transcriptional programs which activate these cellular functions.

The most significant transcriptional response to growth on N-acetylglucosamine in *S. oneidensis* occurs in the neighborhood of the *nag* operon, encompassing a set of eleven adjacent genes SO3514 thru SO3503 (Fig. 3B). Recent analysis has

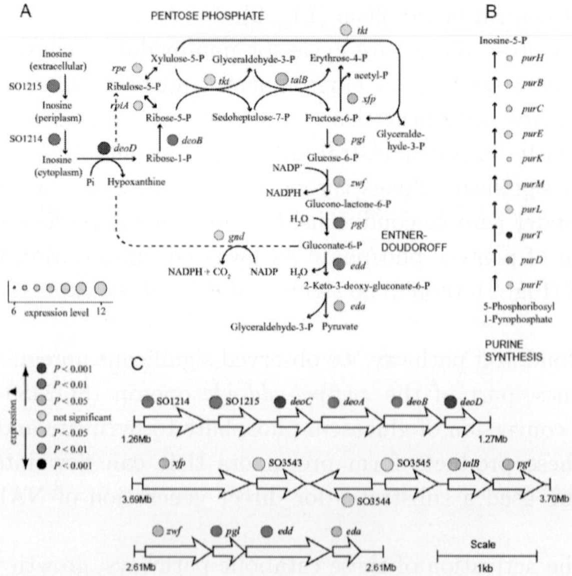

Fig. 2. Expression of enzymes involved in pathways related to inosine metabolism. Red and blue colors indicate up and downregulation, respectively, during growth with inosine; circle circumference represents absolute levels of expression. (A) A liberated ribose moiety from inosine is thought to be converted to fructose-6-phosphate by the non-oxidative branch of the pentose phosphate cycle, and fed into the Entner-Doudoroff pathway. (B) Repression of genes involved in the first steps of purine synthesis, for which inosine-5-phosphate is a key intermediate; for brevity, not all metabolites are shown. (C) Expression of three key gene neighborhoods in their genomic contexts. Refer to Table S1 for full gene descriptions, Enzyme Commission (EC) numbers, and detailed data for these genes.

suggested that this cluster contains the major components of *S. oneidensis* MR-1's chitin metabolism: specifically, these genes encode two permeases specific for chitin oligosaccharides, a chemotactic-response protein, and enzymes involved in converting N-acetylglucosamine into fructose [36]. The extraordinarily high level of expression detected for SO3514 (Fig. 3A and Table S1), the TonB-dependent outer membrane transporter, indicates the highly specific regulation and functional importance of its encoded protein.

The *nag* locus does not, however, contain any genes which govern adhesion and attachment of cells to chitin surfaces, which has been described as a central part of the chitin utilization program in *Vibrio cholerae* [19]. *S. oneidensis* appears to modulate such a response via a locus of five genes (SO0854 thru SO0850) that are coordinately upregulated during growth on N-acetylglucosamine (Fig. 3). Existing annotations for this locus indicate the presence of type IV pili biogenesis domains in SO0853, SO0852, and SO0851, and the presence of multiple repeats characteristic

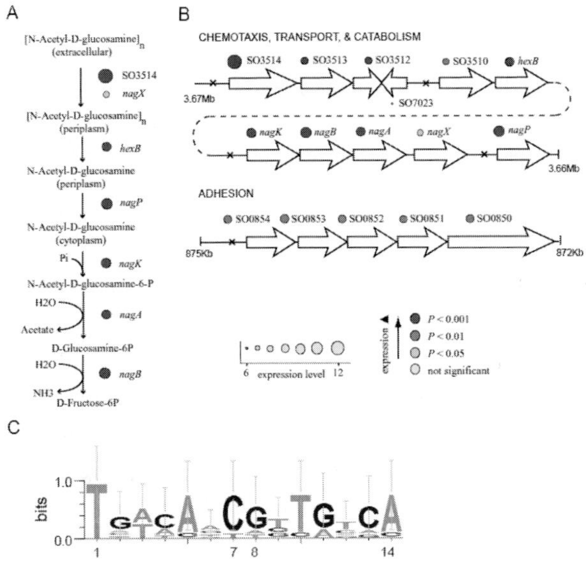

Fig. 3. Expression of enzymes involved in pathways related to N-acetylglucosmaine metabolism. Red and blue colors indicate up and downregulation, respectively, during growth with inosine; circle circumference represents absolute levels of expression. (A) Expression of chemotaxis, transport, catabolic, and adhesion pathways for N-acetylglucosamine. (B) Genomic context of the *nag* operon (top) and the cluster of type IV pili proteins, SO0854-SO0850 (bottom) implicated for a role for attachment to chitin. Black crosses indicate presence of the conserved binding motif described in (C). (C) The 14-base long *cis*-regulatory motif which was found by aligning promoters from the top 15 N-acetylglucosamine upregulated genes, consisting of two palindromic seven-base halves.

of adhesins in SO0850. We therefore suggest that these genes are involved in the assembly and extension of type IV pili for adhesion onto chitin and chitin-derived oligosaccharides.

To investigate whether these two gene neighborhoods are modulated by a common transcriptional regulator we aligned the promoter sequences of the top 15 most significantly overexpressed genes, which included all but one of the 11 nag neighborhood of genes and three of the five type IV pili genes (all P values < 0.01). This alignment revealed a statistically significant 14-base long motif (P value < 0.001) present at six distinct promoter sites, consistent with a motif previously inferred by [36] (Fig. 3C).

Coordinated changes in gene expression in response to external stimuli are typically mediated by two-component systems in bacteria [13], and it is possible that this detected motif may be a target of such a response regulator. Based on its proximity to the *nag* operon, and its homology with a LacI family of regulators, the gene

SO3516 was previously denoted as *nagR* and proposed to encode for a repressor of N-acetylglucosamine metabolic genes [36]. However, though this gene was expressed above background, it did not show a significant change in expression during growth with N-acetylglucosamine (Table S3), suggesting that if SO3516 is a regulator of the *nag* operon, its activation may be mediated non-transcriptionally via allosteric binding of substrate.

3.3. *Biosynthetic and Degradative Pathways Induced by Growth with Amino Acids*

Growth of *S. oneidensis* using amino acids as the sole carbon source has been previously reported [28], and a recently published sequence analysis described several likely pathways for their utilization [29]. Our growth profiles reveal growth rates exceeding all that of the other four carbon sources used in this study (Fig. 1).

Amino acid metabolism is tightly controlled in the cell, and the regulation of transcription represents just one of several modes of control. During growth with amino acids, over 5% of the genome (238 genes) were significantly changed (P value $<$ 0.05).

Here we focus on utilization of the amino acid valine which, by way of having distinct pathways for its degradation and synthesis, offers a clear example of the inverse regulation of these processes. Pathways for valine metabolism share enzymes with those for the other branched-chain amino acids isoleucine and leucine. With respect to valine, we observed three trends: the differential regulation of genes involved in passive and active transport, upregulation of degradative enzymes, and repression of synthesis enzymes.

For transport, two APC Superfamily amino acid transporters (SO0313, SO4565), and an AzlCD-like branched chain amino acid transporter (SO1759, SO1760) were found to be upregulated during growth with amino acids. In addition, we observed repression of an operon annotated to encode an ABC arginine transporter complex (SO1042-SO1044), as well as downregulation of two cation:alanine/glycine transporters of the AGCS family (SO3063, SO3541). The repression of these latter two active transport mechanisms suggests that when concentrations of amino acids are high outside the cell, passive channels may be the primary mode of uptake.

Enzymes involved in the degradation of valine were strongly induced in the presence of amino acids (Fig. 4). The first step of valine catalysis, the removal of the amino group from the amino acids, appears to be effected by a leucine dehydrogenase (SO2638) rather than an aminotransferase as previously predicted [29]. Leucine dehydrogenase is also known to act on isoleucine and valine. New assignments were also made to intermediary steps in the isoleucine and valine degradation pathways based on significant increases in the expression of adjacently located genes (SO1677-SO1683).

The synthesis pathway of valine comprises several multi-component enzymes which catalyze the steps of its formation from pyruvate. As indicated in Fig. 4, these en-

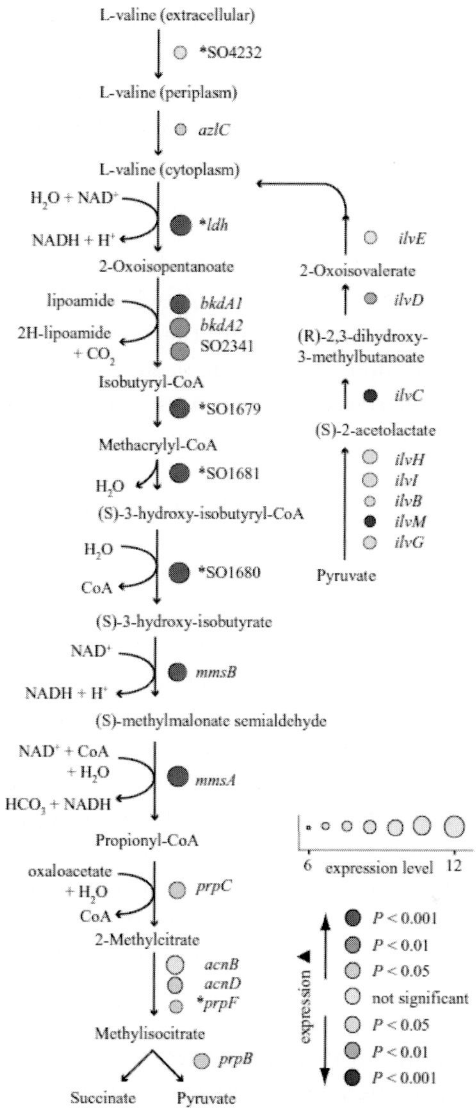

Fig. 4. Differential induction of degradative enzymes and concomitant repression of the biosynthetic pathway for L-valine during growth with an amino acid mixture. Red and blue colors indicate up and downregulation, respectively; circle circumference represents absolute levels of expression. Refer to Table S1 for full gene descriptions, Enzyme Commission (EC) numbers, and detailed data for these genes.

zymes are significantly repressed during growth with amino acids. As with degradative pathways described above, the enzymes of the valine synthesis pathways overlap considerably with those of isoleucine and valine. Thus the observed repression of these enzymes may be due, in sum or in part, to the presence of any of the branched chain amino acids in our amino acid mixture.

4. Discussion

Shewanella's metabolic versatility reflects its diverse environmental ecology. In recent years, *Shewanella* species have been isolated from fresh water and marine sediments around the world, surface waters of the Sargasso sea, hydrothermal vents of the deep Pacific, mollusks and spoiling fish [9, 24, 25, 34, 37]. Reflecting its aquatic lifestyle, many of the preferred electron donors and acceptors for *S. oneidensis* are organic breakdown products of marine tissues and minerals abundant in marine water and sediments.

To successfully compete across these different niches and in shifting nutritive environments, *Shewanella* species must be catabolic and respiratory generalists. For example, *S. oneidensis*' capacity to utilize carbon sources such as inosine and N-acetylglucosamine as electron donors may be an infrequently utilized but beneficial trait, which confers a competitive advantage over other niche organisms.

The pathways and regulation of carbon metabolism in environmental organisms such as *S. oneidensis* differ in several ways from the canonical models derived from *E.coli* and *B. subtilis*. The pentose phosphate and Entner-Doudoroff pathways, for example, appear to play a key role in sugar catabolism, as suggested here and in a recent study of seven phylogenetically diverse bacteria [8]. In addition, the absence of homologs for several transcription factors governing central intermediary metabolism – such the purine biosynthesis repressor PurR – imply that alternative regulatory mechanisms have yet to be discovered. Additional isotopic labeling experiments and further whole-genome expression assays will be necessary to close these and other gaps in our understanding of the metabolic and regulatory networks of *S. oneidensis*.

The interaction of carbon catabolism and respiratory pathways may also impact the efficency and rate of energy conservation in cells. Utilization of some carbon sources can yield more ATP per substrate molecule, but result in slower growth rates [26]. This may partially explain why inosine, the most electrogenic carbon source known for *S. oneidensis*, has the slowest growth kinetics of the carbon sources we tested. How carbon sources regulate the bioenergetic strategies cells use, and how these strategies influence an organism's competitive fitness, are questions that merit further study.

While terminal electron acceptor pathways have been a dominant focus of study to date in *S. oneidensis*, an understanding of carbon metabolism is equally important in defining its ecophysiology. Moreover, the existence of regulatory coupling between catabolic and respiratory pathways suggests novel approaches for optimization of *Shewanella* and other metal-reducing bacteria in microbial fuel cells and bioremedial applications.

References

[1] Abreu-Goodger, C. and Merino, E., Ribex: a web server for locating riboswitches and other conserved bacterial regulatory elements, *Nucleic Acids Res.*, 33(Web Server issue):W690–W692, 2005.

[2] Bolstad, B. M., Irizarry, R.A., Astrand, M., and Speed, T. P., A comparison of normalization methods for high density oligonucleotide array data based on variance and bias, *Bioinformatics* , 19(2):185–193, 2003.

[3] Buckel, B., *Biology of the Prokaryotes*, ch. 12, pp.278–326. Blackwell Science, 1999.

[4] Charlier, D. and Glansdorff, N., *EcoSal - Escherichia coli and Salmonella: cellular and molecular biology*, ASM Press, 2004.

[5] Daraselia, N., Dernovoy, D., Tian, Y., Borodovsky, M., Tatusov, R., and Tatusova, T., Reannotation of *Shewanella oneidensis* genome, *OMICS*, 7(2):171–175, 2003.

[6] Fisher. R.A., *Statistical methods for research workers*, Oliver and Boyd, 1932.

[7] Fraser, O.P. and Sumar. S., Compositional changes and spoilage in fish (part ii) microbiological induced deterioration, *Nutrition & Food Science*, 6:325–329, 1998.

[8] Fuhrer, T., Fischer, E., and Sauer, U., Experimental identification and quantification of glucose metabolism in seven bacterial species, *J. Bacteriol*, 187(5):1581–1590, 2005.

[9] Gao, W., Liu, Y., Giometti, C.S., Tollaksen, S.L., Khare, T., Wu, L., Klingeman, D.M., Fields, M.W., and Zhou, J., Knock-out of SO1377 gene, which encodes the member of a conserved hypothetical bacterial protein family COG2268, results in alteration of iron metabolism, increased spontaneous mutation and hydrogen peroxide sensitivity in *Shewanella oneidensis* MR-1, *BMC Genomics*, 7(1):76, 2006.

[10] Gralnick, J.A. and Hau, H.H., Ecology and Biotechnology of the Genus *Shewanella*, *Annu. Rev. Microbiol*, 2006.

[11] Grimek, T.L. and Escalante-Semerena, J.C., The *acn*D genes of *Shewenella oneidensis* and *Vibrio cholerae* encode a new Fe/S-dependent 2-methylcitrate dehydratase enzyme that requires *prp*F function *in vivo*, *J. Bacteriol*, 186(2):454–462, 2004.

[12] Heidelberg, J.F.,*et al.*, Genome sequence of the dissimilatory metal ion-reducing bacterium *Shewanella oneidensis*, *Nat. Biotechnol.*, 20(11):1118–1123, 2002.

[13] Hoch, J.A., Two-component and phosphorelay signal transduction, *Curr. Opin. Microbiol*, 3(2):165–170, 2000.

[14] Keyhani, N.O. and Roseman, S., Physiological aspects of chitin catabolism in marine bacteria, *Biochim Biophys Acta.*, 1473(1):108–122, 1999.

[15] Khashe, S. and Janda, J.M., Biochemical and pathogenic properties of Shewanella alga and *Shewanella putrefaciens*, *J. Clin. Microbiol.*, 36(3):783–787, 1998.

[16] Lipshutz, R.J., Fodor, S.P., Gingeras, T.R., and Lockhart, D.J., High density synthetic oligonucleotide arrays, *Nat. Genet.*, 21(1 Suppl):20–24, 1999.

[17] Lovley, D.R., Bug juice: harvesting electricity with microorganisms, *Nat. Rev. Microbiol.*, 4(7):497–508, 2006.

[18] Mass, E., Vanderpool, C.K.,and Gottesman, S., Effect of *ryh*B small RNA on global iron use in *Escherichia coli*, *J. Bacteriol*, 187(20):6962–6971, 2005.

[19] Meibom, K.L., Li, X.B., Nielsen, A.T., Wu, C.Y., Roseman, S., and Schoolnik, G.K., The *Vibrio cholerae* chitin utilization program, *Proc. Natl. Acad. Sci. USA*, 101(8):2524–2529, 2004.

[20] Meng, L.M., Kilstrup, M., and Nygaard, P., Autoregulation of *pur*R repressor synthesis and involvement of *pur*R in the regulation of *pur*B, *pur*C, *pur*L, *pur*MN and *gua*BA expression in *Escherichia coli*, *Eur. J. Biochem.*, 187(2):373–379, 1990.

[21] Myers, J.M. and Myers, C.R., Role for outer membrane cytochromes OmcA and

OmcB of *Shewanella putrefaciens* MR-1 in reduction of manganese dioxide, *Appl. Environ. Microbiol.*, 67(1):260–269, 2001.

[22] Nealson, K.H. and Saffarini, D., Iron and manganese in anaerobic respiration: environmental significance, physiology, regulation, *Annu. Rev. Microbiol.*, 48:311–343, 1994.

[23] Nealson, K.H., Harnessing microbial appetites for remediation, *Nat. Biotechnol.*, 21(3):243–244, 2003.

[24] Nealson, K.H. and Scott, J., *The Prokaryotes: An Evolving Electronic Resource for the Microbial Community*, Springer-Verlag, 2004.

[25] Onishchenko, O.M. and Kiprianova, E.A., *Shewanella* genus bacteria isolated from the Black Sea water and molluscs, *Mikrobiol. Z.*, 68(2):12–21, 2006.

[26] Pfeiffer, T., Schuster, S., and Bonhoeffer, S., Cooperation and competition in the evolution of ATP-producing pathways, *Science*, 292(5516):504–507, 2001.

[27] Riemann, L. and Azam, F., Widespread N-acetyl-D-glucosamine uptake among pelagic marine bacteria and its ecological implications, *Appl. Environ. Microbiol.*, 68(11):5554–5562, 2002.

[28] Ring, E., Stenberg, E., and Strm, A.R., Amino acid and lactate catabolism in trimethylamine oxide respiration of *Alteromonas Putrefaciens* NCMB 1735, *Appl. Environ. Microbiol.*, 47(5):1084–1089, 1984.

[29] Serres, M.H. and Riley, M., Genomic Analysis of Carbon Source Metabolism of *Shewanella oneidensis* MR-1: Predictions versus Experiments, *J. Bacteriol.*, 188(13):4601–4609, 2006.

[30] Surette, M.E., Gill, T.A., and LeBlanc, P.J., Biochemical basis of postmortem nucleotide catabolism in Cod (*Gadus morhua*) and its relationship to spoilage, *J. Agric. Food Chem.*, 36:19–22, 1988.

[31] Tang, Y.J., Hwang, J.S., Wemmer, D.E., and Keasling, J.D., *Shewanella oneidensis* MR-1 fluxome under various oxygen conditions, *Appl. Environ. Microbiol.*, 73(3):718–729, 2007.

[32] Tapparel, C., Monod, A., and Kelley, W.L., The DNA-binding domain of the *Escherichia coli* CpxR two-component response regulator is constitutively active and cannot be fully attenuated by fused adjacent heterologous regulatory domains, *Microbiology*, 152(Pt 2):431–441, 2006.

[33] Tiedje, J., Personal communication.

[34] Venter, J.C., *et al.*, Environmental genome shotgun sequencing of the Sargasso Sea, *Science*, 304(5667):66–74, 2004.

[35] Wan, X.F., *et al.*, Transcriptomic and proteomic characterization of the Fur modulon in the metal-reducing bacterium *Shewanella oneidensis*, *J. Bacteriol.*, 186(24):8385–8400, 2004.

[36] Yang, C., *et al.*, Comparative genomics and experimental characterization of Nacetylglucosamine utilization pathway of *Shewanella oneidensis*, *J. Biol. Chem.*, 281(40):29872–29885, 2006.

[37] Zhao, J.S., Manno, D., Leggiadro, C., O'Neil, D., and Hawari, J., *Shewanella halifaxensis* sp. nov. , a novel obligately respiratory and denitrifying psychrophile, *Int. J. Syst. Evol. Microbiol.*, 56(Pt 1):205–212, 2006.

ANALYSIS OF COMMON SUBSTRUCTURES OF METABOLIC COMPOUNDS WITHIN THE DIFFERENT ORGANISM GROUPS

AI MUTO
muto@kuicr.kyoto-u.ac.jp

MASAHIRO HATTORI
hattori@kuicr.kyoto-u.ac.jp

MINORU KANEHISA
kanehisa@kuicr.kyoto-u.ac.jp

Bioinformatics Center, Institute for Chemical Research, Kyoto University, Gokasho, Uji, Kyoto 611-0011, Japan

With the increase in available post-genomic data and metabolic pathway information, we have been focusing on revealing the biological meaning of higher phenomena such as relationships of metabolic systems in different organisms. Metabolism plays an essential role in all cellular organisms, e.g. energy transportation, signal transduction and structural formation of cell components. The metabolic pathway of each organism has a different landscape from all others because of the different sets of enzymes encoded in the genome. The organisms that are incapable of producing their own essential chemical compounds should acquire them in some way from other organisms that can produce them. For example, several vitamins are required by animals to survive. In this manner we can assume that the different availabilities of metabolites may influence the relationship between organisms in nature. In this study, we focus on the differences in available metabolites among organisms. First, we divided 239 species with complete genomes into 9 organism groups in accordance with phylogeny and averaged out the annotation quality and the phylogenetic sparsity. Then, we calculated the commonly used chemical compounds between organism groups and the uniquely used chemical compounds in an organism group. The total number of metabolites we consider in this study is 1,074, which is about one-third of all metabolites that appear in the KEGG metabolic pathways. Finally we show the differences and the similarities between organism groups on every metabolic pathway map, illustrating the commonly observed substructures within the uniquely used metabolites. These results will help us to better comprehend the architecture of metabolic pathways and the relationships between organisms.

Keywords: metabolism, metabolite, organism group, KEGG.

1. Introduction

Using increasingly advanced experimental techniques available to analyze cellular functions, a massive amount of valuable data for biological functions is being assembled through transcriptome [13], proteome [8] and metabolome analyses [11]. The complex network information like metabolic systems and protein regulation under cellular processes has also been compiled in KEGG PATHWAY database [9]. Using such genome scale data we can now investigate the complexity of biological systems. In particular, metabolic systems are essential for all cellular organisms, and most enzymatic reactions and metabolic compounds that comprise their metabolisms play an important role such as energy transportation, signal transduction and structural transformation of chemical compounds to construct cellular components. Every organism should produce or degrade chemical compounds through their metabolic activities to survive in the biosphere. This means that metabolic systems contribute to the maintenance of life for each organism and it is very worthwhile to understand them.

In the course of considering such a biological system, we have analyzed the chemical aspects of metabolic pathways to better elucidate the biological meaning of metabolism so far [6, 10]. Here, we are focusing on the differences in available metabolites among organisms. The metabolic pathway of each organism has a different look from all others because of the different sets of enzymes which are encoded in each genome. While a fraction of the available metabolites frequently overlaps between organisms, other compounds rarely overlap because of differences in their available enzymes. Therefore, the organisms which never metabolize their own essential chemical compounds should acquire them from other organisms that can produce them in order to survive. For example, animals should acquire several vitamins by predation activity from other organisms such as plants. In another case, we can discover that the metabolic difference is directly linked to a phenotypic difference because of different usage of metabolites. For instance, some prokaryotes can utilize hydrogen sulfide (H_2S) as nutrient, but it is well-known that this chemical compound is poisonous to animals in the vapor state [4]. In this manner we can assume that the different availabilities of metabolites may influence the relationship between organisms in nature.

To examine such a correlation we can use organism-specific metabolism information from the KEGG database, which is compiled from genome information, i.e., the set of enzymes encoded in each genome. However, levels of gene annotation in each organism vary widely, causing problems when iterating over organism pairs. In order to avoid such difficulties, we first divided 239 species with complete genomes into 9 organism groups in accordance with phylogeny. Then, we performed a comprehensive analysis on metabolite usage among different organism groups. For each pair of organism groups, we determined those chemical compounds that were common to both and those that were specific to each member of the pair. After that, we verified the differences and similarities between groups on every metabolic pathway map. Finally, we identified specific common substructures within the uniquely used compounds by surveying structural differences among metabolites. This information on preferred substructures will help us to know the utility and the limitations of each metabolic pathway in terms of chemical structure; thus, we will be able to understand the structural design of metabolic pathways and the relationship between organisms.

2. Results

2.1. *Extraction of the commonly possessed metabolites*

In the KEGG database, information on metabolic pathways has been manually curated and stored in the PATHWAY database on the basis of genome information, IUBMB enzyme information, textbook information and so forth. KEGG also provides the chemical information for metabolites in the KEGG LIGAND database. Using these databases, we can easily elucidate which components of metabolic pathways appear within each organism group. In this section, we first extracted the whole set of metabolic compounds on all metabolic pathways for each species, using the KGML information obtained from KEGG PATHWAY database. The number of metabolic compounds we used is 3,499.

After that, we identified 1,074 commonly used metabolites among all species in each organism group, as illustrated in Fig. 1. The results are shown in the "Total" column in Table 1. Here the commonly used compounds are defined as metabolites that over 80% of the species in each group possess in each metabolic system. We determined that the total number of identified compounds is about one-third of the metabolites that appear in metabolic pathway maps in KEGG. The organism group which has the highest number of metabolites is Fungi with 743 compounds, four times that of Spirochete, which has the fewest metabolites. The pair of groups sharing the highest number of common compounds is Fungi and gamma-Proteobacteria with 500 compounds in common.

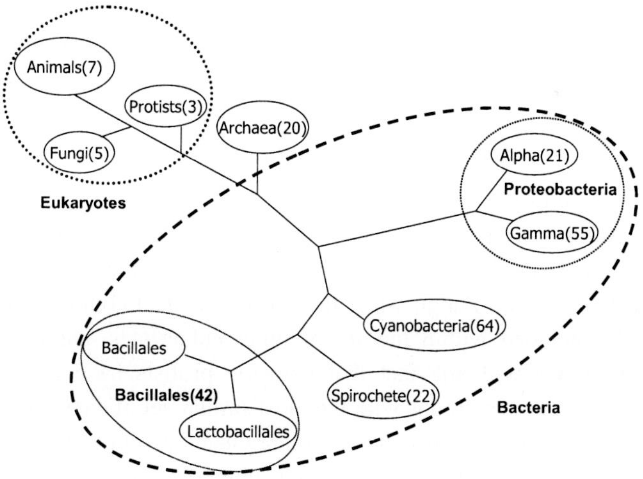

Fig. 1. Organism groups based on phylogeny
We obtained all phylogenetic information from the categorization used in KEGG database (http://www.genome.jp/kegg/catalog/org_list.html). The numbers in parentheses represent the numbers of organisms contained in each group.

The distribution of compounds in each organism group in each pathway category is shown in Table 2. Because some metabolites appear in more than one pathway, the sum of metabolites is not equal to the "Total" in Table 1.

Table 1. The number of commonly used compounds between organism groups
Each number represents the number of commonly observed chemical compounds between two organism groups. The "Total" in the last column means the unique number of compounds obtained from one organism group. Some labels are abbreviated as follows: alpha-Proteobacteria as Alpha, gamma-Proteobacteria as Gamma, and Cyanobacteria as Cyano.

	Animals	Fungi	Protists	Archaea	Alpha	Gamma	Bacillales	Spirochete	Cyano	Total
Animals	–	403	166	217	226	320	263	128	178	579
Fungi	403	–	214	354	316	500	404	177	237	743
Protists	166	214	–	143	149	171	179	118	150	220
Archaea	217	354	143	–	230	341	267	147	187	387
Alpha	226	316	149	230	–	351	283	146	207	408
Gamma	320	500	171	341	351	–	398	178	248	662
Bacillales	263	404	179	267	283	398	–	181	268	468
Spirochete	128	177	118	147	146	178	181	–	181	182
Cyano	178	237	150	187	207	248	268	181	–	268

Table 2. Distribution of compounds of each organism group in each pathway category
Each number represents the number of chemical compounds of each organism group that are observed in each pathway category. The pathway categories are obtained from the categorization used in the KEGG database (http://www.genome.jp/kegg/pathway.html).

	Animals	Fungi	Protists	Archaea	Alpha	Gamma	Bacillales	Spirochete	Cyano
Carbohydrate Metabolism	115	133	47	50	59	102	94	33	55
Energy Metabolism	41	60	23	35	35	55	35	14	20
Lipid Metabolism	122	126	31	55	54	80	62	8	15
Nucleotide Metabolism	66	86	36	62	58	79	74	54	59
Amino Acid Metabolism	143	226	51	127	108	185	145	45	78
Metabolism of Other Amino Acids	39	45	16	11	21	48	30	6	18
Glycan Biosynthesis and Metabolism	8	12	8	6	27	41	24	0	20
Biosynthesis of PK and NRP*	0	0	0	0	0	19	0	0	0
Metabolism of Cofactors and Vitamins	36	93	11	46	73	111	57	20	32
Biosynthesis of Secondary Metabolites	23	64	11	42	6	59	27	0	5
Xenobiotics Biodegradation and Metabolism	80	25	0	8	34	14	12	0	0

*PK and NRP: Polyketides and Nonribosomal Peptides

2.2. *Extraction of the different usage of metabolites*

In order to identify the different availability of chemical compounds within each organism group, we extracted the set of chemical compounds unique to one organism group, as shown in Table 3. In eukaryotes, the set of chemical compounds overlap greatly between Animals and Fungi, while the set of chemical compounds of Protists is almost completely included within the set of other eukaryotes, namely Animals and Fungi. This corresponds with the fact that parasitic protists have reduced metabolic systems. The pair of organism groups with the most similar set of chemical compounds is Fungi and gamma-Proteobacteria despite their more distant evolutionary relationship.

As for the "Specific" compounds that are uniquely observed in each organism group, many chemical compounds are extracted from eukaryotes. In contrast, bacteria have a very small number of intrinsic metabolites. Most of the frequently observed chemical compounds that are specifically possessed by eukaryotes are found to be lipid-related metabolites whose chemical structures are relatively large (data not shown), indicating the existence of eukaryote-specific lipid-metabolisms.

Table 3. The number of uniquely used compounds in each organism group

Each number represents the number of uniquely used chemical compounds by the organism group in the row against the group in the column. The number in "Specific" is the number of specifically observed compounds in each organism group listed in each row.

	Animals	Fungi	Protists	Archaea	Alpha	Gamma	Bacillales	Spirochete	Cyano	Specific
Animals	–	176	413	362	353	259	316	451	401	111
Fungi	340	–	529	389	427	243	339	566	506	87
Protists	54	6	–	77	71	49	41	102	70	24
Archaea	170	33	244	–	157	46	120	240	200	6
Alpha	182	92	259	178	–	57	125	262	201	1
Gamma	342	162	491	321	311	–	264	484	414	1
Bacillales	205	64	289	201	185	70	–	287	200	3
Spirochete	54	5	64	35	36	4	1	–	1	0
Cyano	90	31	118	81	61	20	0	87	–	0

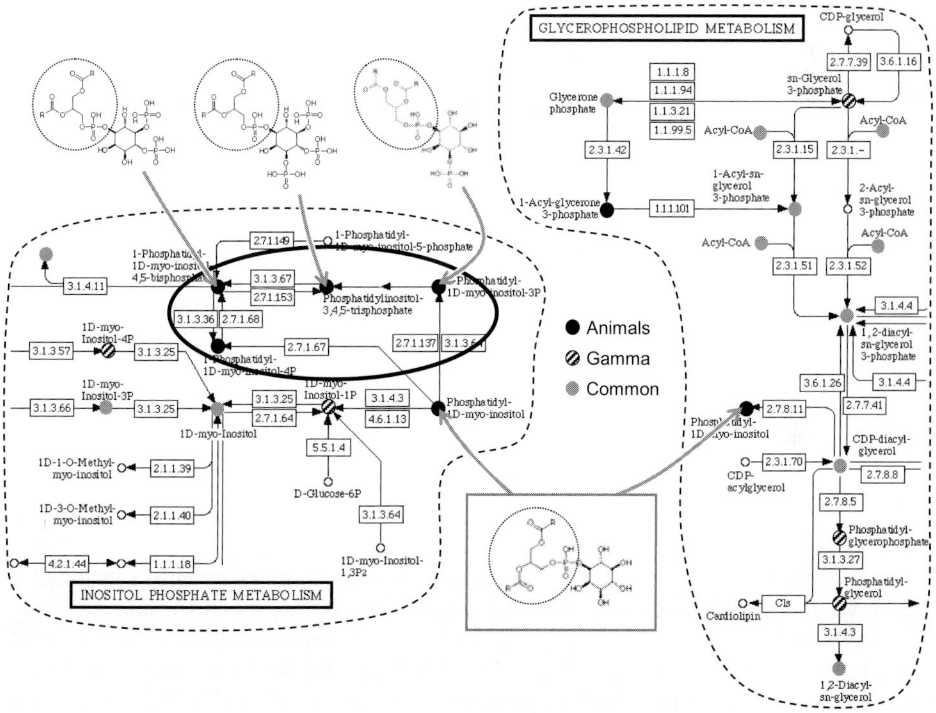

Fig. 2. An example of the set of structurally conserved compounds

The figure shows the crossover region between two metabolic pathways, the "Inositol phosphate metabolism" and the "Glycerophospholipid metabolism", represented by a dotted line. Here, Glycerophospholipid metabolism is a subcategory of Lipid metabolism and Inositol phosphate metabolism is that of Carbohydrate metabolism. These two pathways are connected by phosphatidyl-1D-myo-inositol, indicated in a square and arrows. The set of specifically used chemical compounds by Animals are also emphasized by a bold oval, indicating the observed common substructures on each chemical structure.

2.3. *Common substructures of organism-specific metabolites*

In order to know the biological meaning of organism-specific metabolic pathways, we demonstrated the set of differently used chemical compounds between two organism groups on metabolic pathway maps.

The phosphatidyl-1D-myo-inositol is one of the products of Glycerophospholipid metabolism and is specifically observed within three eukaryotes; Animals, Fungi, and Protists. It is also utilized in the Inositol phosphate metabolism (Fig. 2). In the subsequent path in the latter metabolism, we found that there was a common substructure containing a phosphatidyl group within the block of eukaryote-specific compounds, i.e., phosphatidyl-1D-myo-inositol derivatives. In contrast, the other derivatives, which have no phosphatidyl groups, have also been extracted from other organism groups, indicating that there may be other pathways for non-eukaryotic species.

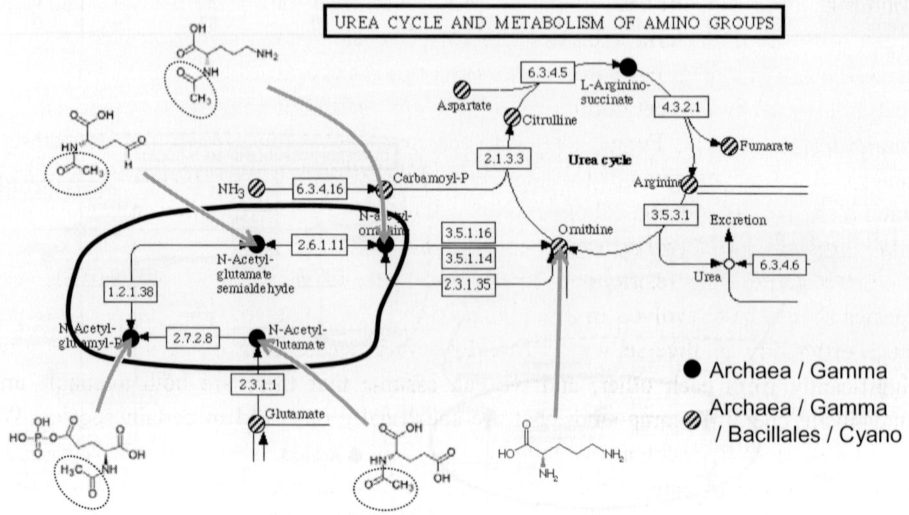

Fig. 3. Another example of the set of structurally conserved compounds

The figure shows a part of the metabolic pathway, "Urea Cycle and Metabolism of Amino Groups". The pathway is a part of Amino Acid metabolism. The set of specifically used chemical compounds by Archaea and gamma-Proteobacteria are surrounded by a bold line, and all acetyl groups are indicated by a dotted line.

In another case of the Urea cycle and metabolism of amino groups, illustrated in Fig. 3, we found a block of uniquely-used compounds. Within the block four glutamate derivatives comprise a series of paths. All of these compounds contain an acetyl group. The gamma-Proteobacteria, Fungi, alpha-Proteobacteria and Archaea possess all of these four chemical compounds, while Cyanobacteria, Protists, Bacillales and Spirochete have none of the four compounds. In this sub-pathway there are 2 terminals; N-acetylglutamate and N-acetylornithine. Animals have N-acetyl-ornithine; however, they do not have the other three compounds. The existence of a common substructure within the block suggests that the enzyme that adds an acetyl group to a precursor makes not only its product but also the following pathway components. In addition, the alpha-

Proteobacteria do not contain other components of a urea cycle, and are able to fill the gap of the pathway in Cyanobacteria or Bacillales.

3. Discussion

We demonstrated that there were specific common substructures within the uniquely used compounds on metabolic pathways. In this paper, two common substructures are closely depicted; one is the phosphatidyl group of phosphatidyl-1D-myo-inositol in the Inositol phosphate metabolism and the other is acetyl group of the N-acetylglutamate in the Urea cycle and metabolism of amino groups. In the former case, the organism groups are separated into two classes according to the possession of phosphatidyl derivatives. The first class contains three organism groups of eukaryotes, Animals, Fungi, and Protists, which possess all of phosphatidyl derivatives. The other organism groups, Bacteria and Archaea, comprise the second class and have no such derivatives. In the latter case, the organism groups are characterized by possession of N-acetylglutamate or N-acetylornithine, which are incorporated in the metabolisms of gamma-Proteobacteria, Fungi, alpha-Protebacteria and Archaea. On the other hand, neither chemical compound was found to be utilized by Cyanobacteria, Protists, Bacillales and Spirochete. These results suggest that there are metabolite-mediated relationships between two different organism groups.

Because of the restrictions of biological resources, metabolic systems of each species should have evolved in a way that utilizes their available metabolites more and more efficiently in diverse ways. Therefore, the metabolisms of each species differ significantly from each other, and we can assume that there are both available and unavailable chemical compounds that are specifically observed in certain species. We can also suppose that such a restriction of chemical compounds may cause the symbiosis or the predator-prey interaction between species. In this study, we found that several pairs of organism groups are highly correlated in terms of utilizing metabolites and the specific substructures are conserved within the uniquely used compounds. In particular, the lipid-related metabolites are specifically observed in eukaryotic groups. It corresponds to the experimental fact that there are some eukaryote-specific pathways of lipid biosynthesis. Thus, those findings may also help us to better understand the relationships between organisms. Of course, the annotation qualities of complete genomes are now completed in each species and the information on metabolisms has not been fulfilled. However, this type of research should be more important in this post-genomic era in order to understand the biological meaning of metabolites and metabolisms; hence we believe our current study may also contribute to solving such a grand challenge.

4. Materials and Methods

4.1. *Datasets on metabolic pathways and metabolites*

We obtained the information on enzymatic reactions and their reactants belonging to each organism from the KEGG PATHWAY database (version 42.0 + update 2007/04/20), which is comprised of a series of XML files written in KGML (KEGG Markup Language) at the KEGG FTP site; ftp://ftp.genome.jp/pub/kegg/xml/. We also obtained the chemical compound structures from the COMPOUND section of the KEGG LIGAND database (version 42.0 + update 2007/04/20), which contains information on 12,338 chemical compounds with 2D graph representations of structures in MDL/MOL format. However, many of the chemical compounds in COMPOUND are not always assigned to metabolic pathways. In this study, we used only the 3,499 metabolites that appear on metabolic pathway maps.

4.2. *Phylogenetic groups of organisms and their specific metabolites*

All 239 species that have complete genomes have been collected in the KEGG GENES database; however we decided to first classify all organisms into several organism groups and treat each group as one collective organism because of the limitations of available gene annotation data. When compiling the set of metabolites that each organism group possesses we estimate that a chemical compound should be included when over 80% of the genomes in the group possess a gene for the relevant enzyme. Also, we estimated that the chemical compound should be specific to the organism group when less than 20% of the genomes in other groups possess it. Here, the total number of organism groups is set to 9, and each group is identified using the following labels; Animals, Fungi, Protists, Archaea, alpha-Proteobacteria, gamma-Proteobacteria, Bacillales+Lactobacillales, Spirochete, Cyanobacteria, as illustrated in Fig. 1. These categorizations come originally from the definition of species categorization used in KEGG.

4.3. *Commonly used or uniquely used compounds*

After obtaining the organism group-specific metabolites, we performed a correlation to check the commonly used chemical compounds between two organism groups as well as the metabolites that appear only within one of the groups. This information is represented on each pathway map, by coloring the commonly used compounds in green and the uniquely possessed compounds in blue or yellow, to easily visualize their distribution on pathway maps. Then we carefully checked all the colored pathways for interesting regions and elucidated the common substructures specifically observed within those regions of pathway maps, using SIMCOMP. SIMCOMP can compute the pairwise atom alignments between two chemical compound structures. Here, we calculated the alignments for all possible pairs among the concerned set of chemical compounds and then extracted the commonly aligned area as the common substructures.

Acknowledgments

This work was supported by the 21st Century COE Program "Genome Science" and a grant-in-aid for scientific research on the priority area "Comprehensive Genomics", both from the Ministry of Education, Culture, Sports, Science and Technology of Japan, and by the Institute for Bioinformatics Research and Development of the Japan Science and Technology Agency. The computational resources were provided by the Bioinformatics Center, Institute for Chemical Research, Kyoto University.

References

[1] Aguilar, D., Aviles, F.X., Querol, E., Sternberg, and M.J., Analysis of phenetic trees based on metabolic capabilities across the three domains of life, *J. Mol. Biol.*, 340(3):491-512, 2004.

[2] Brooksbank, C., Cameron, G., and Thornton, J., The European Bioinformatics Institute's data resources: towards systems biology, *Nucleic Acids Res.*, 33(Database issue):D46-53, 2005.

[3] Feldman, H.J., Dumontier, M., Ling, S., Haider, N., and Hogue, C.W., CO: A chemical ontology for identification of functional groups and semantic comparison of small molecules, *FEBS Lett.*, 579(21):4685-4691, 2005.

[4] Friedrich, C.G., Physiology and genetics of sulfur-oxidizing bacteria, *Adv. Microb. Physiol.*, 39:235-289, 1998.

[5] Goto, S., Nishioka, T., and Kanehisa, M., LIGAND: chemical database for enzyme reactions, *Bioinformatics*, 14(7):591-599, 1998.

[6] Hattori, M., Okuno, Y., Goto, S., and Kanehisa, M., Development of a chemical structure comparison method for integrated analysis of chemical and genomic information in the metabolic pathways, *J. Am. Chem. Soc.*, 125:11853-11865, 2003.

[7] Hattori, M., Okuno, Y., Goto, S., and Kanehisa, M., Heuristics for chemical compound matching, *Genome Inform.*, 14:144-153, 2003.

[8] Humphery-Smith, I. and Blackstock, W., Proteome analysis: genomics via the output rather than the input code, *J. Protein. Chem.*, 16(5):537-544, 1997.

[9] Kanehisa, M. and Goto, S., KEGG: Kyoto encyclopedia of genes and genomes, *Nucleic Acids Res.*, 28(1):27-30, 2000.

[10] Kotera, M., Okuno, Y., Hattori, M., Goto, S., and Kanehisa, M., Computational assignment of the EC numbers for genomic-scale analysis of enzymatic reactions, *J. Am. Chem. Soc.*, 126(50):16487-16498, 2004.

[11] Oliver, S.G., Winson, M.K., Kell, D.B., and Baganz, F., Systematic functional analysis of the yeast genome, *Trends Biotechnol.*, 16(9):373-378, 1998.

[12] Rison, S.C. and Thornton, J.M., Pathway evolution, structurally speaking, *Curr. Opin. Struct. Biol.*, 12(3):374-382, 2002.

[13] Subramanian, A., Tamayo, P., Mootha, V.K., Mukherjee, S., Ebert, B.L., Gillette, M.A., Paulovich, A., Pomeroy, S.L., Golub, T.R., Lander, E.S., and Mesirov, J.P., Gene set enrichment analysis: a knowledge-based approach for interpreting genome-wide expression profiles, *Proc. Natl. Acad. Sci. USA*, 102(43):15545-15550, 2005.

PRUNING GENOME-SCALE METABOLIC MODELS TO CONSISTENT *AD FUNCTIONEM* NETWORKS

SABRINA HOFFMANN ANDREAS HOPPE
sabrina.hoffmann@charite.de andreas.hoppe@charite.de

HERMANN-GEORG HOLZHÜTTER
hergo@charite.de

Institute of Biochemistry, Medical Faculty of the Humboldt University, Charité, Monbijoustr. 2, 10117 Berlin, Germany

Metabolic networks represent a set of reactions and associated metabolites that may occur in a given cell or tissue. They are frequently reconstructed from pure genomic data without thorough biochemical validation. Such genome-scale metabolic networks may thus either lack relevant or contain non-existent reactions and metabolites. Filling gaps and removing falsely predicted reactions can be a cumbersome procedure. On the other hand, using the network to build mathematical models addressing a specific problem (e.g. analyzing changes in the level of cellular ATP at substrate depletion) it may turn out that the network comprises more reactions and metabolites than actually needed or, on the contrary, that essential reactions are missing. Therefore, we propose a method to prune the whole network to a smaller sub-network which contains no dead ends and blocked reactions, i.e reactions that may neither proceed in forward nor backward direction. Inspection of this reduced network reveals its actual functional capabilities in terms of producible metabolites. We apply our method to a genome-scale metabolic network of *E. coli*. Depending on the choice of the exchangeable metabolites, composition of the external medium, and type of thermodynamic constraints we obtain different reduced network variants that may serve as a basis for flux balance models.

Keywords: FBA; flux balance analysis; iJR904 *Escherichia coli*; genome-scale metabolic model; consistency; minimal flux mode; MinMode; flux minimization.

1. Introduction

The process of metabolic network reconstruction from genomic information has significantly advanced and the number and size of such published networks increased considerably during the last few years [2, 4, 5, 16, 17, 19]. While traditional metabolic modelling implied a step by step construction of a specific pathway [18, 22], the so-called genome-scale metabolic models are reconstructed by a "top-down" procedure. Such genomic reconstructions exploit sequence homologies to genes of enzymes and membrane transporters already known in other cell types. Organism–specific collections of automatically assigned biochemical reactions on this level can be downloaded from resources such as KEGG [26] or BioCyc [24]. For a detailed review on the reconstruction of microbial genome-scale metabolic models the reader is referred to Francke *et al.* [6]. As opposed to the traditional bottom-up models the purpose of genome-scale models is not the exploration of a specific question but rather to collect all available biochemical reactions of a certain cell type. Obviously, for some parts of the network the available knowledge is fragmentary. Assignment of genes to enzymes can be wrong, the expression of certain genes can be inhibited due to DNA methylation, enzymes may catalyze different chemical reactions than already reported for other cell types, or important metabolic genes may have escaped identification because of too large dissimilarities of sequences, to name a few reasons for errors in genome-wide reconstructed metabolic networks. Despite of these errors genome-scale metabolic models, especially the manually curated ones, contain a complete and correct description of many cellular functions which may serve as model objectives. Here, we propose an approach to identify these complete parts and to elucidate the functional capabilities of a given genome-scale metabolic model. To this end we apply a two-step strategy which consists of pruning the network of all blocked reactions, i.e. reactions that may neither proceed in forward nor backward direction, followed by the determination of its synthesizing capacity given by the set of metabolites for which a net synthesis is possible. We demonstrated this procedure with a genome-scale metabolic network of *E. coli*. Furthermore, we imposed different constraints on the composition of the external medium, exchangeable metabolites, and flux directionalities and found remarkable differences in the size and synthesizing capacities of the resulting metabolic models.

2. System and Methods

Definition of the FBA model. Because of sparsely known enzyme-kinetic details, constraint-based modeling currently represents the method of choice to analyze large-scale metabolic models. Basically, this method requires only knowledge of the network topology, which can be described mathematically formalized and compact by the stoichiometric matrix S, an $m \times n$ matrix where m corresponds to the number of metabolites (rows), n to the number of reactions (columns) or fluxes. Its positive or negative elements S_{ij} specify the amount of metabolite M_i formed or consumed in reaction j, respectively. Neglecting the spatial distribution, the time-dependent change of the metabolite concentrations is determined by the kinetic equation system:

$$\frac{d[M_i]}{dt} = \sum_j S_{ij} v_j - b_i, \tag{1}$$

where v_j refers to the flux rate of the j-th reaction, $[M_i]$ denotes the concentration of metabolite M_i, and b_i refers to the unspecific metabolic use of metabolite M_i not covered by the chemical reactions in the considered network. A common assumption of constraint-based modeling approaches is the so-called flux balance constraint that assumes a steady-state behavior, i.e. the metabolite concentrations remain constant over time:

$$\sum_j S_{ij} v_j - b_i = 0 \quad \forall i \in \mathbb{N}_m \tag{2}$$

No metabolic use ($b_i = 0$) will be referred to as strict steady state condition for metabolite M_i. In biological systems, this strict assumption is valid if the characteristic time constant for changes of the metabolic output is much larger than the time constant of metabolic conversions. As this condition is actually not met even under conditions of cellular growth (because the amount of all metabolites has to be increased) several studies relaxed the flux balance constraint (Eq. 2) by allowing accumulation of all metabolic species, i.e. assuming $b_i \geq 0$ [12, 13, 15]. Although quite unusual, variable b_i might also take negative values and represent a metabolic repository from which metabolites may be fed into the system. Both, the availability of a repository and further metabolic use imply an exchange of the metabolite over systems boundary. This is actually true in both directions for metabolites whose concentrations are large enough to remain constant despite consuming and producing fluxes, e.g. compounds of the external media. Therfore, for these compounds variable b_i is unconstraint and no flux balance constraint is applied to these so called external metabolites.

Furthermore a flux-balance model has to specify the metabolic output of the network, i.e. a set of metabolites delivered by the network either as cellular building blocks for macro-molecules, degradation products (e.g. of toxic compounds), or exported material that represent metabolic functions. As these metabolites are required by the cell, their metabolic use has to be greater than zero, e.g. $b_i = 1$. Fig. 1 summarizes the necessary definitions of a flux balance model. In addition, we consider thermodynamic information to constrain the reversibility of reaction directions as described in the next subsection.

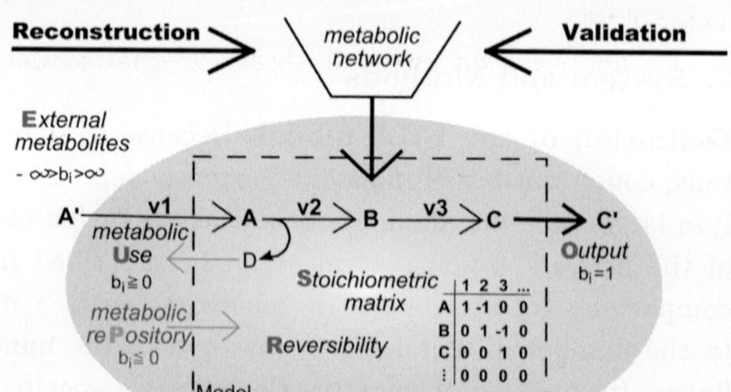

Fig. 1. Characteristics of a flux balance model

The definition of an appropriate objective function, i.e. the measure which is assumed as optimized in cellular system, is most crucial for constraint-based optimization analysis. Possible functions include the maximization of biomass production and minimizing the sum of fluxes [10].

Determination of blocked reactions and non-producible metabolites. The method used to determine blocked reactions and non-producible metabolites is based on the concept of minimal flux modes [9]. These modes are minimal flux distributions required either for the net synthesis of a single metabolite, so called metabolite minimal modes (**MetabMinMode**), or to maintain an unit flux through a given single reaction (either in forward or backward direction) as reaction minimal modes (**ReactMinMode**). The advantage of this approach lies in its simplicity and extensibility. Here, we applied additional constraints and focussed on model inconsistencies instead of predicting physiological flux distributions. If for a given metabolite the MetabMinMode does not exist the metabolite is called non-producible. If for a given reaction the ReactMinMode does neither exist in forward nor backward direction the reaction is called blocked. It has to be noted that non-existence of a minimal flux distribution excludes the existence of any flux distribution, our criteria for the determination of blocked reactions and producible metabolites are thus sufficiently general.

Specification of metabolic use, external metabolites and flux directions. All models used in this study consider only extracellularly located metabolites as external. For the remaining set of internal metabolites two different assumptions on metabolic use are investigated: In the first situation, denoted by (U_{bio}), metabolic use is assumed $(b_i \geq 0)$ only for metabolites serving as biomass precursors according to Reed *et al.* [19]. All other metabolites are strictly balanced with zero metabolic use $(b_i = 0)$. In the second situation (U_{all}), metabolic use is allowed $(b_i \geq 0)$ for all metabolites.

In our analysis, we included three external media differing in available carbon sources. Environment E_{gluc} stands for a glucose minimal media (composition and concentrations taken from Henry *et al.* [8]). A slightly richer medium E_{rich} contains the following carbon sources in addition to glucose used in E_{gluc}: ac, akg, lac-D, lac-L, glyc, mal-L, pyr and succ, each at 0.02M. E_{cplx} represents a complex medium that includes the carbon sources of E_{rich} plus a larger number of other exchangeable species as defined by Reed *et al.* [19]. All three media contain unlimited oxygen, phosphate, sulfate, nitrogen, potassium, iron, and sodium. The concentrations of the external metabolites are given in the supplementary material.

Three different variants of restrictions on flux directions are considered. In the first variant, denoted with R_{all}, no restrictions are imposed on the directions, i.e. all fluxes may proceed in both forward and backward direction. In the second variant, denoted with R_{irr}, fixed heuristic reversibility constraints as proposed in the original iJR904 model [19] are used. In addition, as proposed by Reed *et al.* [20], for 17 reactions[a] forward and backward directions are constrained to prevent thermodynamically infeasible cycles. In the third variant, denoted with R_{tr}, no a priori

[a]Reactions: VPAMT, ALARi, LCADi, ACCOAL, GALUi, ADK, CYTDt2, ABUTt2, GLUt4, INSt2, ADNt2, PROt4, SERt4, THMDt2, THRt4, URAt2, URIt2

assumptions on the reversibility of reactions are made. Instead, MetabMinModes and ReactMinModes are calculated with the constraint of thermodynamic realizability (TR) [11], i.e. metabolite concentrations are restricted to physiologically feasible ranges and must be determined such that the changes of Gibb's free energies $(\Delta G_r)_j$ are consistent with flux directions. The Gibb's free energy depends on the standard Gibb's free energy $\left(\Delta G_r^0\right)_j$ and active concentrations $[M_i]$ of the reactants by the formula [1]:

$$(\Delta G_r)_j = \left(\Delta G_r^0\right)_j + RTln \prod_{i=1}^{M} [M_i]^{S_{ij}}, \tag{3}$$

where R is the gas constant, T the temperature, and S_{ij} the stoichiometric coefficient of metabolite M_i in the j-the reaction. The linear problem to be minimized reads as follows:

$$\textbf{minimize} \quad \sum_{j=1}^{N} v_j$$

$$\textbf{subject to} \quad v_j \geq 0 \quad \forall j \in \mathbb{N}_n, \quad \sum_{j=1}^{N} S_{ij} v_j - b_i = 0 \quad \forall i \in \mathbb{N}_m$$

if the kth *MetabMinMode*: $b_k = 1$	**if** the rth *ReactMinMode*: $v_r = 1$
if U_{bio}: $b_i = 0 \quad \forall i \notin M_{biomass}$ $b_i \geq 0 \quad \forall i \in M_{biomass}$	**if** U_{all}: $b_i \geq 0 \quad \forall i \in \mathbb{N}_m$
if R_{tr}: $0 \leq v_j + \alpha d_j \leq \alpha, \quad \forall j \in \mathbb{N}_n$ $0 \leq -(\Delta G_r)_j + \alpha d_j \leq \alpha, \quad \forall j \in \mathbb{N}_n$ $(\Delta G_r)_j = \left(\Delta G_r^0\right)_j + \sum_{i=1}^{n} c_i^{S_{ij}}$ $c_i^{\max} \leq c_i \leq c_i^{\min}, \quad \forall i \in \mathbb{N}_m$	**if** R_{irr}: $v_j = 0 \quad \forall j \in R_{irr}$

where n is the number of reactions and m the number of metabolites; for any $1 \leq j \leq n$, d_j is a binary variable; $c_i = RT\ln([M_i])$ is a coefficient calculated from the active concentration of metabolite M_i, $c_i^{\max}$ and $c_i^{\min}$ are minimal and maximal values related to a realistic concentration range of metabolite M_i; α is set to a positive number which is larger than any possible flux value and larger than any possible Gibb's energy value, and it can easily be shown that the constraints $0 \leq v_j + \alpha d_j \leq \alpha$ and $0 \leq -(\Delta G_r)_j + \alpha d_j \leq \alpha$ are equivalent to $v_j \neq 0 \rightarrow \text{sgn}(v_j) = -\text{sgn}\left((\Delta G_r)_j\right)$. Physiological concentration ranges were available for 22 internal metabolites (given in Kümmel *et al.* [14]) and 10 external metabolites (given in Henry *et al.* [8]). For the other metabolites generic concentration bounds were used based on typical cellular concentration ranges reported in the literature: 5 μM–2 mM. Standard Gibb's free energies computed by Henry *et al.* [8] are used.

Definitions:	
Metabolic network	Set of metabolites and reactions combined by the stoichiometric matrix
Flux-balance model	A metabolic network combined with definitions of exchangeable metabolites, metabolic output, declaration of irreversible reactions, and further constraints that are optimized towards a flux objective to predict steady-state flux distributions
Metabolic output	A metabolite whose production is an essential metabolic function.
Non-producible metabolite	A metabolite for which net production is not possible by any flux distribution with respect to given model assumptions — determined by the computation of a MetabMinMode
Blocked reaction	Reaction that may neither proceed in forward nor in backward direction with respect to given model assumptions — determined by the computation of a ReactMinMode

3. Results

We based our studies on the genome-scale metabolic network iJR904 of *E. coli* [19] which already has been analyzed in several studies [7, 20, 23, 25]. The network encompasses 931 reactions and 618 metabolites. The problem is that this network contains a large number of blocked reactions and non-producible metabolites. For example, 408 blocked reactions were reported for a flux model of a previous network (iJE660a) when biomass production is maximized under aerobic growth conditions in a glucose minimal medium [3]. The fact that such a large part of the network is comprised of disabled reactions hampers an unbiased statistical analysis of flux distributions, e.g. when analyzing the impact of enzyme mutants. Therefore, we pruned the network to its consistent core by eliminating all blocked reactions based on the calculation of ReactMinModes for all reactions of the network (see subsection 2). In total, we constructed 18 sub-networks from the original iJR904 network using different constraints on exchangeable metabolites, environment (medium composition) and flux directions explained in subsection 2 and summarized in Fig. 3. The large impact of the various constraints on the number of blocked reactions is depicted in Table 1. The minimal number of blocked reactions is 20 if metabolic use

Table 1. Numbers of blocked reactions for different environments, exchangeable metabolites, and reversibility constraints.

	R_{all}		R_{irr}		R_{tr}	
	U_{bio}	U_{all}	U_{bio}	U_{all}	U_{bio}	U_{all}
E_{cplx}	141	**20**	185	75	160	70
E_{rich}	155	49	365	251	324	221
E_{gluc}	159	52	**377**	265	326	224
E_{zero}	564	564	931	931	931	931

is assumed for all metabolites (case U_{all}), the direction of fluxes is not restricted (R_{all}), and the cells grow in a complex medium (E_{cplx}). In contrast, 377 reactions are blocked if metabolic use is restricted to precursor metabolites for biomass production (case U_{bio},R_{irr},E_{gluc}).

Removal of blocked reactions does not affect reactions involved in internally balanced reactions. Internal cycles that are thermodynamically infeasible are excluded automatically if the calculation of the ReactMinModes is performed under the TR constraint and the principle of micro-reversibility is taken into account [11]. For other types of thermodynamic constraints on flux directions, the internally balanced cycles can be identified by ReactMinModes that remain if the concentration of all external substrates is put to zero (case E_{zero}, see last row in Table 1). Intriguingly, for the fully reversible network there are $367 = 931 - 564$ reactions belonging to internal cycles.

Blocked reactions were removed from the original model by canceling the respective column in the stoichiometric matrix S. The synthesizing capacity of these reduced networks is given by the total number of producible metabolites, i.e. metabolites for which a MetabMinMode can be found (see subsection 2). Fig. 2 illustrates the

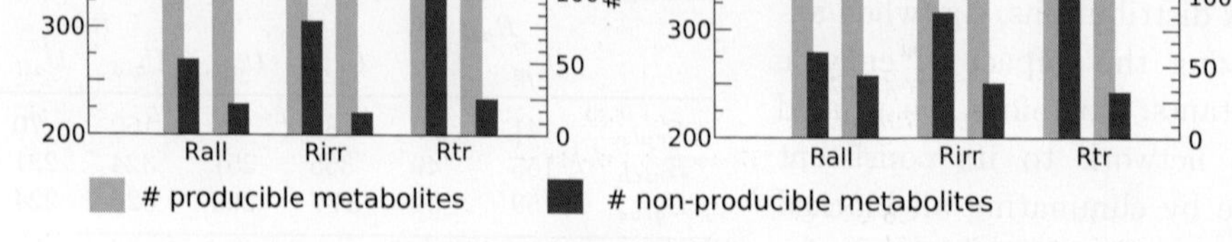

Fig. 2. Producible and non-producible metabolites for complex metabolic input (Ecplx)

synthesizing capacities of the original, complete network (cpl) and the various reduced networks (red) for cells growing in a complex medium (E_{cplx}). The weaker the applied constraints, the higher the number of producible metabolites. In U_{all} there is no difference of producible metabolites between the reduced and the original model, whereas for U_{bio} the number of producible metabolites is lower in the reduced model. This decrease results from the strict flux balance constraint for non-biomass producing metabolites assumed for U_{bio}.

In contrast to the calculation of the ReactMinModes, for the calculation of a MetabMinMode, metabolic use is assumed for the corresponding metabolite to drain it out of the network. Therefore, blocked reactions always include non-producible metabolites although reactions including non-producible metabolites are not necessarily blocked. If synthesis of a metabolite is not possible itself, subsequent reactions may only proceed if a reaction exists that regenerates the consumed non-producible metabolite. If this is the case, a so-called regeneration cycle exists and the reactions of which may carry a non-zero flux and thus will not be identified as blocked. For example, the reduction of ATP (`atp`) to its desoxy form (`datp`) is driven by an oxidation of the cofactor thioredoxin (`trdrd/trdox`):

```
RNTR1:        atp + trdrd --> datp + h2o + trdox
TRDR:         h + nadph + trdox --> nadp + trdrd
```

The oxidized as well as the reduced form of thioredoxin are non-producible. However, there exists a NADPH (`nadph`) dependent thioredoxin reductase that catalyzes the regeneration of thioredoxin (`TRDR`) and thus enables a flux through `RNTR1`. Therefore, after elimination of blocked reactions the reduced model only comprises non-producible metabolites involved in such regenerating cycles that may contribute to the synthesis of other metabolites while lacking an own *de novo* synthesis. The

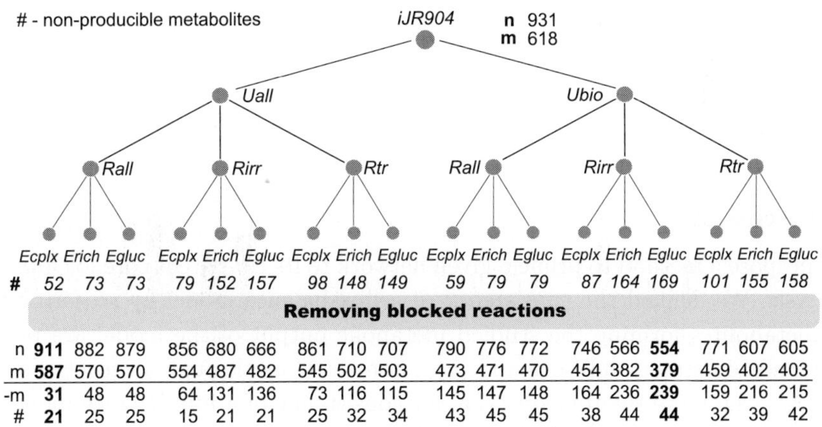

Fig. 3. Overview over the obtained results with n and m denoting the total number of reactions and metabolites, respectively, $-m$ stands for the number of metabolites that have been removed from the original model and # indicates the number of non-producible metabolites.

number of metabolites removed from the original network and the remaining number of non-producible metabolites for the 18 considered different combinations of constraints among others are depicted in Fig. 3. It has to be noted that the 49 biomass relevant metabolites are producible by all reduced network variants. However, in none of the reduced networks all metabolites are producible. For the variant with assumed metabolic use of all metabolites (U_{all}), complex substrate composition (E_{cplx}) and heuristic *a priori* reversibility constraints (R_{irr}) in total 64 non-producible metabolites have been removed and only 15 of the remaining 554 metabolites are non-producible. As reasoned above, these 15 metabolites must be involved in regenerating cycles, which according to the applied reversibility constraints, are thermodynamically feasible. Determination of moiety conservation relations [21] of the reduced stoichiometric matrix allows to group these 15 non-producible metabolites into three sets containing: 1) the acetyl carrier protein (`ACP`) and metabolites coupled to this protein[b], 2) the reduced and oxidized form of thioredoxin (`trdrd` and `trdox`) and 3) the loaded and unloaded form of L-Glutamyl-tRNA (`trnaglu` and `glutrna`). Hence, there are actually only three metabolites (for example `ACP`, `trdrd`, and `trnaglu`) which, if producible, would render all metabolites of the model producible. However, without any further modifications the cellular functionality, in terms of metabolic output by this model variant, can be described by a combination of the 554 producible metabolites.

The model above lacks 33 metabolites present in the respective model without constraints on reversibility ($E_{cplx}, U_{all}, R_{all}$). To determine which metabolites could additionally be produced if one reversibility constraint is relaxed, we investigated the synthesizing capacity of the completely reversible model, modified by dropping the reversibility for reaction r_i (E_{cplx}, U_{all}, $R_{all} \setminus \{i\}$). Iteration over all reactions assumed to be irreversible in R_{irr} showed 23 reactions whose irreversibility blocks the synthesis of exactly one metabolite. We modified the model to allow these 23 reactions to proceed in both directions by removing their reversibility constraint. After this modification 9 additional metabolites can be produced by the model.

4. Discussion

We propose a method to prune a given network to its consistent core for flux-balance analysis. We defined the consistency of a flux-balance model by requiring that (i) all metabolites forming the studied metabolic output of the model are producible under steady-state conditions and (ii) a non-zero flux through all reactions is possible. An even more rigorous criterion for consistency of a flux-balance models based on genome-scale networks had been proposed by Kumar *et al.* [15] demanding producibility of all metabolites in the cytosol, as well as producibility and degradability of all metabolites in intracellular compartments other than the cytosol. However,

[b] `acACP, actACP, palmACP, myrsACP, hdeACP, malACP, ddcaACP, tdeACP, octeACP` and `3hrmsACP`

the idea behind this criterion is that during growth the molar amount of every metabolite increases as the cellular volume increases and the concentrations remain constant (dilution effect). This criterion is useful if the knowledge presented by genome-scale metabolic networks is considered to be complete. In the context of a specific question, e.g. biomass production, producibility of a relevant set of metabolites is sufficient.

By removing reactions and non-producible metabolites according to the specific constraints formulated in the corresponding flux-balance model we reduced the original metabolic network of *E. coli* to a core network capable of providing a metabolic output in accordance with the defined exchange processes in the original network. This is demonstrated by the fact that irrespective of the applied constraints the reduced network still allows for the synthesis of the 49 metabolites required for biomass production — the functional output whose investigation motivated the compilation of the iJR904 network. However, metabolic functions other than biomass production may be of interest as well. The full spectrum of possible metabolic functions requiring a net synthesis of output metabolites can be directly inferred from the synthesizing capacity of the pruned network. As demonstrated for the *E. coli* network this synthesizing capacity depends strongly upon the constraints placed to the substrate composition of the extracellular space, directionality of the fluxes, and exchangeability of metabolites.

Our calculations have shown that under certain constraints only three metabolite sets of the *E. coli* network remain for which producibility has to be somehow enabled in order to render all metabolites producible. These are the acetyl carrier protein, thioredoxin (in reduced or oxidized form) and L-Glutamyl tRNA (loaded or unloaded with the amino acid). Obviously, these are compounds that do not represent metabolites in a strict sense and thus should not be included into the metabolic network. This issue shows a weak point in the definition of metabolic networks, a definition which usually is confined to chemical compounds having a lower molecular weight of several hundreds of Dalton. On the other hand, gene regulatory or signaling networks are composed of interacting proteins and nucleic acids also without explicitly taken into account the synthesis and degradation of these macromolecules. The reconstruction of cellular reactions networks modelling the turnover of high-molecular weight compounds as proteins, nucleic acids and phospholipids is currently "no man's land". To overcome this situation is certainly a great challenge for future work in computational systems biology.

References

[1] Atkins, P. W. and De Paula, J., *Atkins' Physical Chemistry*,Oxford University Press, 1995.

[2] Becker, S. A. and Palsson, B. O., Genome-scale reconstruction of the metabolic network in *Staphylococcus aureus* N315: an initial draft to the two-dimensional annotation, *BMC Microbiol*, 5(1):8, 2005.

[3] Burgard, A.-P., Nikolaev, E.-V., Schilling, C.-H., and Maranas, C.-D., Flux coupling

analysis of genome-scale metabolic network reconstructions, *Genome Res.*, 14(2):301–312, 2004.

[4] Duarte, N. C., Becker, S. A., Jamshidi, N., Thiele, I., Mo, M. L., Vo, T. D., Srivas, R., and Palsson, B. O., Global reconstruction of the human metabolic network based on genomic and bibliomic data, *Proc. Natl. Acad. Sci. USA*, 104(6):1777–1782, 2007.

[5] Feist, A. M., Scholten, J. C., Palsson, B. O., Brockman, F. J., and Ideker, T., Modeling methanogenesis with a genome-scale metabolic reconstruction of *Methanosarcina barkeri*, *Mol. Syst. Biol.*, 2:2006.0004, 2006.

[6] Francke, C., Siezen, R. J., and Teusink, B., Reconstructing the metabolic network of a bacterium from its genome, *Trends Microbiol.*, 13(11):550–558, 2005.

[7] Henry, C. S., Janokowski, M. D., Broadbelt, L. J., and Hatzimanikatis, V., Genome-scale thermodynamic analysis of *Escherichia coli* metabolism, *Biophys J.*, 90(4):1453–1461, 2006.

[8] Henry, C. S., Broadbelt, L. J., and Hatzimanikatis, V., Thermodynamics-based metabolic flux analysis, *Biophys J.*, 92(5):1792–1805, 2007.

[9] Hoffmann, S., Hoppe, A., and Holzhütter, H.-G., Composition of Metabolic Flux Distributions by Functionally Interpretable Minimal Flux Modes (MinModes), *Genome Informatics*, 17(1):195–207, 2006.

[10] Holzhütter, H.-G., The principle of flux minimization and its application to estimate stationary fluxes in metabolic networks, *Eur. J. Biochem.*, 271(14):2905–2922, 2004.

[11] Hoppe, A., Hoffmann, S., and Holzhütter, H.-G., Including metabolite concentrations into flux balance analysis: Thermodynamic realizability as a constraint on flux distributions in metabolic network, *BMS Sys. Biol.*, 1:23, 2007.

[12] Imielinski, M., Belta, C., Halasz, A., and Rubin, H., Investigating metabolite essentiality through genome-scale analysis of *Escherichia coli* production capabilities, *Bioinformatics*, 21(9):2008–2016, 2005.

[13] Imielinski, M., Belta, C., Rubin, H., and Halasz, A., Systematic analysis of conservation relations in *Escherichia coli* genome-scale metabolic network reveals novel growth media, *Biophys J.*, 90(8):2659–2672, 2006.

[14] Kümmel, A., Panke, S., and Heinemann, M., Systematic assignment of thermodynamic constraints in metabolic network models, *BMC Bioinformatics*, 7:512, 2006.

[15] Kumar, V. S., Madhukar, S. D., and Maranas, C. D., Optimization based automated curation of metabolic reconstructions, *BMC Bioinformatics*, 2007.

[16] Notebaart, R. A., van Enckevort, F. H., Francke, C., Siezen, R. J., and Teusink, B., Accelerating the reconstruction of genome-scale metabolic networks, *BMC Bioinformatics*, 7:296, 2006.

[17] Poolman, M. G., Bonde, B. K., Gevorgyan, A., Patel, H. H., and Fell, D. A., Challenges to be faced in the reconstruction of metabolic networks from public databases, *Syst Biol (Stevenage)*, 153(5):379–384, 2006.

[18] Rapoport, T. A., Heinrich, R., and Rapoport, S. M., The regulatory principles of glycolysis in erythrocytes in vivo and in vitro. A minimal comprehensive model describing steady states, quasi-steady states and time-dependent processes, *Biochem J.*, 154(2):449–469, 1974.

[19] Reed, J. L., Vo, T. D., Schilling, C. H., and Palsson, B. O., An expanded genome-scale model of *Escherichia coli* K-12 (iJR904 GSM/GPR), *Genome Biol.*, 4(9):R54, 2003.

[20] Reed, J. L. and Palsson, B. O., Genome-Scale in silico models of *E. coli* have multiple equivalent phenotypic states: Assessment of correlated reaction subsets that comprise network states, *Genome Res.*, 14:1797–1805, 2007.

[21] Sauro, H. M. and Ingalls, B., Conservation analysis in biochemical networks: computational issues for software writers, *Biophys. Chem.*, 109(1):1–15, 2004.

[22] Schuster, R. and Holzhütter, H. G., Use of mathematical models for predicting the metabolic effect of large-scale enzyme activity alterations. Application to enzyme deficiencies of red blood cells, *Eur. J. Biochem.*, 229(2):403–418, 1995.

[23] Wang, Q., Chen, X., Yang, Y., and Zhao, X., Genome-scale in silico aided metabolic analysis and flux comparisons of *Escherichia coli* to improve succinate production, *Appl. Microbiol Biotechnol.*, 73:887–894, 2006.

[24] `http://www.biocyc.org/`

[25] `http://gcrg.ucsd.edu/organisms/ecoli/ecoli_others.html`

[26] `http://www.genome.jp/kegg/`

METABOLIC SYNERGY: INCREASING BIOSYNTHETIC CAPABILITIES BY NETWORK COOPERATION

NILS CHRISTIAN[1]
nils.christian@physik.fu-berlin.de

THOMAS HANDORF[1]
thomas.handorf@physik.hu-berlin.de

OLIVER EBENHÖH[2]
ebenhoeh@mpimp-golm.mpg.de

[1] *Institute for Biology, Humboldt University Berlin, Germany*
[2] *Max Planck Institute for Molecular Plant Physiology, Potsdam-Golm, Germany*

Cooperation between organisms of different species is a widely observed phenomenon in biology, ranging from large scale systems such as whole ecosystems to more direct interactions like symbiotic relationships. In the present work, we explore inter-species cooperations on the level of metabolic networks.
For our analysis, we extract 447 organism specific metabolic networks from the KEGG database [7] and assess their biosynthetic capabilities by applying the method of network expansion [5]. We simulate the cooperation of two organisms by unifying their metabolic networks and introduce a measure, the gain Γ, quantifying the amount by which the biosynthetic capability of an organism is enhanced due to the cooperation with another species. For all theoretically possible pairs of organisms, this synergetic effect is determined and we systematically analyze its dependency on the dissimilarities of the interacting partners. We describe these dissimilarities by two different distance measures, where one is based on structural, the other on evolutionary differences.
With the presented method, we provide a conceptional framework to study the metabolic effects resulting from an interaction of different species. We outline possible enhancements of our analysis: by defining more realistic interacting networks and applying alternative structural investigation methods, our concept can be used to study specific symbiotic and parasitic relationships and may help to understand the global interplay of metabolic pathways over the boundary of organism specific systems.

Keywords: metabolism; scope; KEGG; symbiosis; systems biology; synergy.

1. Introduction

For a few years, the number of fully sequenced and annotated genomes is increasing with an amazing speed and, considering the number of ongoing sequencing projects, is likely to increase even faster in the near future. Using homology matching methods and a tedious manual curation, for a substantial number of organisms largely complete metabolic networks have been characterized. With the emergence of comprehensive metabolic databases such as KEGG [7] or MetaCyc [9], such networks have become readily accessible. Existing methods to analyze large scale metabolic networks include elementary flux modes [11, 12], the closely related concept of extreme fluxes [10], flux balance analysis [8] as well as graph theoretical approaches [6, 14].

All these methods have in common that they can be performed even without the specific knowledge of the kinetic properties of the enzymes catalyzing the biochemical reactions.

So far, large scale metabolic network analyses focused to a large extent on single organism networks, see e. g. [13, 16]. In their natural habitats, however, all species are in constant interaction with organisms belonging to other species. On a population level, the interaction of different species is mathematically described in the research field of population dynamics, ranging from simple predator prey models to very complex ecosystem models, for an overview see e. g. [3]. However, the interaction does not only take place on the level of populations, but also on the level of single individuals by the exchange of metabolites. Examples are given by a predator that consumes and digests its prey, or, more directly, by an intracellular symbiont living inside a host cell and exchanging intermediates by specific transporters. Inspired by these facts, we have developed a conceptional framework to study such interactions on the metabolic level, and to quantify the benefit for each of the organisms.

We retrieve 447 organism specific networks from the KEGG database and determine their capability to incorporate glucose as a sole carbon source into their metabolism. To quantify this capability, we apply the concept of scopes [5], where a scope characterizes the biosynthetic capability of a network when it is provided with certain external resources. To determine how a cooperation of two organism's metabolic networks may enhance the biosynthetic capabilities of each other, we construct unified networks for all possible pairs of organisms. We introduce a measure, called the gain, quantifying the increase in biosynthetic capabilities, by comparing the performance of the unified network with those of the single organism networks. We investigate how the gain correlates with the dissimilarity of the networks, for which we provide two measures, one based purely on structural properties and the other exclusively on phylogenetic information.

The introduced methodology as well as the results from the systematic interaction analysis provides a basis for the investigation of specific biological examples of parasitic or symbiotic behavior. We expect that the biological significance of such investigations may be considerably enhanced by refining the models for interacting networks as well as by applying other network analysis techniques.

2. Concepts

2.1. *Biosynthetic capabilities*

The metabolic network of a particular organism, denoted O, is defined by a specific set of biochemical reactions. We evaluate the biosynthetic capability of a network using the method of network expansion [5]. Starting from a set of compounds, called the seed and denoted S, a series of expanding networks is constructed in an iterative manner. In each step, those reactions from O are added to the network which use exclusively those metabolites as substrates which occur either in the seed or as products of reactions included in earlier steps. The iteration stops if no new reactions can be included. The set of compounds within the final network is called the scope of the seed, denoted $\Sigma^{O}(S)$, and by construction comprises all those compounds which can in principle be produced under the condition that only the seed compounds are available. Often, we are concerned not with the exact composition of a scope but rather with the number of metabolites it contains. We denote the scope size by $|\Sigma^{O}(S)|$. The scope is in general a useful measure to relate structural and functional network properties and is used in this work to characterize the biosynthetic capabilities of a metabolic network.

In cellular metabolism, there exist a small number of key metabolites, the cofactors, which occur in many reactions and mostly perform one particular function. For example, the most common usage of ATP is the transferral of a phosphate group to another molecule yielding ADP and, due to the free energy change of the hydrolyzation, drive reactions that would otherwise be thermodynamically unfeasible. Similarly, NADH is involved in a large number of redox reactions in which it functions as an electron donor yielding the oxidized form, NAD^{+}. In the process of network expansion, a reaction involving a cofactor may only be used if the cofactor has already been synthesized from the seed compounds by reactions incorporated into the network in previous steps. Under most physiological conditions, however, a cell maintains a substantial level of such cofactors and therefore it is unrealistic to assume that they have to be manufactured de novo. Throughout this work, we use a modified form of the expansion algorithm, which allows that cofactor functions can be performed even if the cofactors have not yet been synthesized. The inclusion of cofactor functionalities in the algorithm was introduced in [5] and [4].

Some reactions are considered as irreversible because under physiological conditions they can only proceed in one direction. However, in principle every biochemical reaction may be reversed and the actual direction depends strongly on the present state of a cell as well as on the considered cell or tissue type. In our analysis, we have considered all reactions as reversible.

Clearly, the scope strongly depends on the network composition as well as on the available seed compounds. Naturally, the choice of seed compounds is crucial for the biological interpretation of the biosynthetic capability. One important function of metabolism is to incorporate external carbon sources and convert them into organic compounds used by other processes. It is therefore of interest to study how different carbon sources may be incorporated into cellular metabolism. In this work, we focus on glucose, which is a central metabolite in the energy metabolism ubiquitous throughout all domains of life. To assess this capacity, we identify all non carbon containing compounds occurring in at least one network and include them, together with glucose, in the seed.

2.2. *Metabolic Synergy*

In biological environments, no species lives completely isolated. Rather, metabolites are exchanged between different species by a variety of mechanisms.

The aim of this work is to study how metabolic networks may mutually benefit from each other by sharing their metabolic resources. For this, we investigate pairs of organisms and assume the simplest possible metabolic interaction, namely that the two organisms may exchange all intermediate metabolites. Such a scenario is simply described by a metabolic network which is the union of the two single species networks. Let O_1 and O_2 denote the networks of the single organisms, then the unified (interacting) network is written as $O_1 \cup O_2$. For a pair of organisms, we determine the metabolic capabilities of the single organisms, $\Sigma^{O_1}(S)$ and $\Sigma^{O_2}(S)$, as well as the metabolic capability for the unified network, $\Sigma^{O_1 \cup O_2}(S)$. Clearly,

$$\Sigma^{O_1 \cup O_2}(S) \supseteq \Sigma^{O_1}(S) \cup \Sigma^{O_2}(S), \tag{1}$$

where equality signifies that there is no increase in the metabolic capabilities as a result of network interaction. To quantify the positive synergetic effect resulting from sharing the metabolic resources of two networks, we introduce the gain Γ as the increase in size of the scope of the unified network over the union of the scopes of the single networks,

$$\Gamma(O_1, O_2; S) = \left|\Sigma^{O_1 \cup O_2}(S)\right| - \left|\Sigma^{O_1}(S) \cup \Sigma^{O_2}(S)\right|. \tag{2}$$

Thus, the gain equals to the number of metabolites which can be produced from the interacting network, but not from either of the single networks.
While the gain Γ describes the synergetic effect for a pair of organisms, it is also of relevance to study how each of the partners benefits from the interaction. For this, we introduce the quantities

$$\Gamma_{1/2}(O_1, O_2; S) = \left|\Sigma^{O_1 \cup O_2}(S)\right| - \left|\Sigma^{O_{1/2}}(S)\right| \tag{3}$$

and measure the asymmetry of the interaction by the bias

$$\beta(O_1, O_2; S) = \frac{|\Gamma_1(O_1, O_2; S) - \Gamma_2(O_1, O_2; S)|}{\Gamma_1(O_1, O_2; S) + \Gamma_2(O_1, O_2; S)}. \tag{4}$$

This value is only defined if at least one of the organisms benefits from the interaction. It is 0 if both organisms increase their metabolic capabilities by the same amount of metabolites, and 1 if only one of the partners benefits. For simplicity, the arguments of Γ and β will be omitted if they are unambiguous.

2.3. *Distances between networks*

An intriguing question is whether structural properties such as the degree of similarity of the interacting networks determine the principle capacities for a positive synergetic effect. To study this, we relate the dissimilarity of a pair of organisms with the increase of metabolic capability resulting from cooperation. The increase in capability is quantified by the gain, defined in Eq. (2). To quantify the dissimilarity of two organisms, we introduce two distance measures, one based on differences in the underlying network structure and the other on their phylogenetic distance.

Network distance. We measure the structural distance between two metabolic networks by counting those reactions which occur in only one of the networks. The quantity

$$d_N(O_1, O_2) = |O_1 \cup O_2| - |O_1 \cap O_2| \tag{5}$$

is the Manhattan distance between the two sets of reactions O_1 and O_2, where $|O|$ is the number of biochemical reactions within a network O. Clearly, two identical networks possess zero distance whereas two completely distinct networks possess a distance equal to the sum of the reactions in both single networks. A noteworthy feature of this distance is that, if one network is completely contained in the other, their distance may still be large.

Evolutionary distance. Using the NCBI taxonomy tree [1], we approximate the evolutionary distance of two species by the number of edges on the shortest path between the two organisms. We define the evolutionary distance

$$d_E(O_1, O_2) = \text{number of edges on shortest path connecting } O_1 \text{ and } O_2. \tag{6}$$

This distance measure can only give an approximation of the true evolutionary distance because it depends on the structure of the underlying phylogenetic tree. Moreover, it weighs every edge in this tree equally, while the number of levels may vary substantially from subtree to subtree, often reflecting the thoroughness with which a particular subtree has been studied rather than a true evolutionary distance.

3. Results

3.1. *Metabolic capacities of single organisms*

In the following, we study the ability of organisms to incorporate glucose in their metabolism. For this, we define a seed containing glucose and inorganic material, as described in Sec. 2.1. Because this particular seed contains 92 chemical species, this is also the minimal number of compounds contained in each scope.

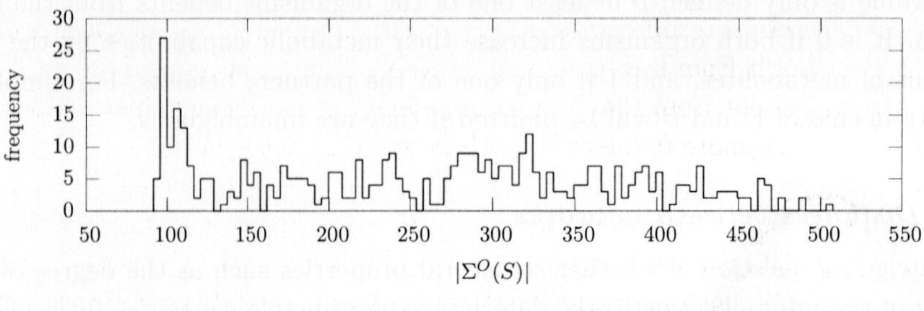

Fig. 1. Histogram of the scope sizes for all considered organisms. The smallest observed scope size is 92, which corresponds to the number of seed metabolites.

Fig. 1 shows a histogram of the metabolic capabilities for all investigated 447 organism specific metabolic networks. A substantial fraction (90) of the considered organisms display a scope size of less than 130, meaning they are capable to produce less than 40 new metabolites. Three of these organisms (*N. equitans* and the two *Phytoplasmae OY* and *AYWP*) are not capable to synthesize any new metabolites. The largest scope size is 508, exhibited by the beta proteobacterium *Burkholderia sp. 383*, meaning it can synthesize 416 new carbon containing metabolites.

3.2. *Interaction of metabolic networks*

We systematically investigate how the synthesizing capacities are increased if two organisms share their metabolic reactions. We first study how even the beneficial effects of the interactions are distributed among the two partners. For this, we calculate the bias values β, defined in Eq. (4), for all 99681 possible network pairs. In 97 cases the bias is not defined ($\Gamma_1 = \Gamma_2 = 0$), meaning that neither partner can increase its biosynthetic capability. Fig. 2 shows a histogram of the bias values β. The pairs in which only one of the partners benefits from the interaction ($\beta = 1$)

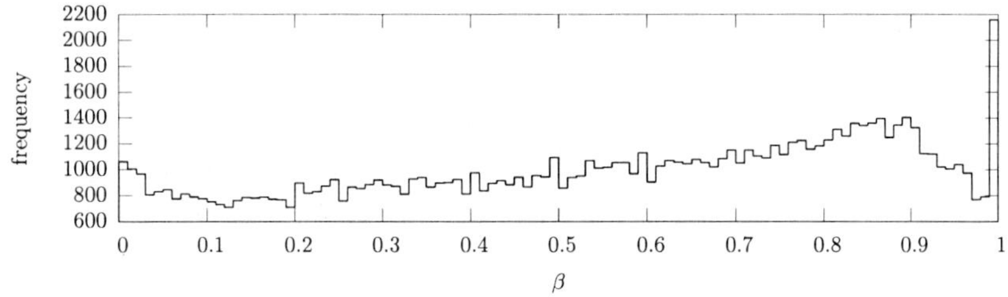

Fig. 2. Histogram of the bias value β as defined in Eq. (4). A value of $\beta = 0$ describes an interaction which is beneficial for both partners, $\beta = 1$ an interaction in which only one partner benefits.

are overrepresented. A closer inspection shows that in most of these cases a small network is almost completely contained in a larger one, thus explaining why the latter cannot increase its capacity due to the interaction. A more detailed analysis will be necessary to determine whether the small sizes reflect the biological reality or whether they result from incomplete annotations. In the majority of pairs (98.5%), both partners benefit from the interaction, where, as a tendency, interactions with a stronger bias are more frequent than those with a weaker bias.

To visualize how the gain Γ, defined in Eq. (2), correlates with the network distance d_N, defined in Eq. (5), we sort the organism pairs by their distance and group them into equidistant bins. The number of organism pairs per bin is plotted in the top panel of Fig. 3, demonstrating that most organisms exhibit a network distance between 400 and 1000. For the gain values in each bin, we determine the 10% quantile, the median, the 90% quantile and the maximum and plot these values versus the network distance in the bottom panel of Fig. 3.

It can be observed that for very small network distances ($d_N < 100$, 0.3% of all organism pairs) the gain is also very small. This is not surprising since similar

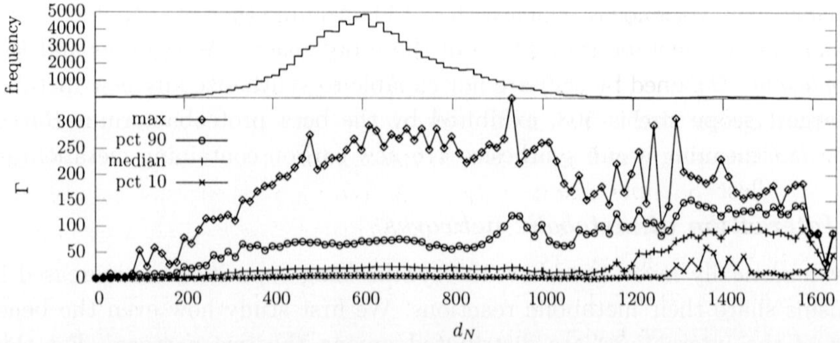

Fig. 3. Top panel: Histogram of the network distances d_N, defined in Eq. (5), for all pairs of networks. Bottom panel: Correlation between the gain Γ, defined in Eq. (2), and the network distance d_N. Plotted are the 10% quantile, the median, the 90% quantile and the maximum gain values as a function of d_N.

networks contain mostly the same pathways. As a consequence, an interaction of such networks is unlikely to yield positive effects.

For network distances $100 < d_N < 400$ (6.6% of pairs), the 90% quantile and the maximum increase strongly with increasing distance. This reflects that a certain degree of dissimilarity is necessary for one network to be able to utilize a subnetwork of the other, in order to increase its capacity.

Remarkably, the quantiles do not change considerably for a large range of network distances ($400 < d_N < 1100$, 91.1% of pairs). This indicates that the difference between two networks alone is not sufficient to predict the increase in biosynthetic capabilities. As long as two networks are not too similar or too distant, particular structural features like the specific occurrence of pathways are apparently more important for the synergetic effect than the degree of dissimilarity of the whole networks. A noticeable characteristic is the increase in the 90% quantile in the range $850 < d_N < 1050$, containing a total of over 8400 network pairs. The most abundant organisms in this region are *C. albicans* and *A. thaliana*, occurring in 386 and 335 pairs, respectively. Only 734 of these 8400 pairs show a gain larger than 100. Surprisingly, in about one third (277) of these pairs, one of the partners is *A. thaliana*, and 239 pairs contain *C. albicans.* The high frequency of pairs containing one of these two organisms together with the fact that pairs involving these organisms tend to produce a high gain explains the observed increase of the 90% quantile.

For even larger network distances ($d_N > 1100$, 2.1% of pairs), the median and 90% quantile significantly increase, whereas the maximum strongly fluctuates. A closer inspection reveals that in 62.8% of the pairs within this distance region, one of the interacting partner is either human (*H. sapiens*), mouse (*M. musculus*), or rat (*R. norvegicus*) and, when considering only those pairs yielding a gain $\Gamma > 50$, the fraction increases to 77.3%. Again, the increase in the quantiles can be explained by a high abundance of organisms which on average yield a high gain.

Fig. 4A depicts the correlation between gain Γ and evolutionary distance d_E, defined by Eq. (6). It corresponds to Fig. 3 with the difference that pairs exhibiting the same evolutionary distance are grouped into bins. The similar appearance of these two figures can be explained by the high correlation of the two distance measures, which are plotted against each other in Fig. 4B.

A remarkable attribute of Fig. 4A is that the maximum is already very pronounced for the smallest possible distance $d_E = 2$. We found that this pair consists of two strains of the gamma proteobacteria *Shewanella* whose network sizes differ significantly by over 400 reactions, thus resulting in a large network distance d_N, which, as outlined above, is required for a high gain. The large network distance hints at incomplete or faulty annotations of the corresponding genomes because it seems unlikely that organisms may develop such a drastic different metabolic network composition during a relatively short evolutionary time span.

4. Discussion and Outlook

Motivated by the observation that no organism exists in complete isolation, but rather exchanges metabolites with organisms belonging to other species, we have provided a conceptional framework to analyze changes in biosynthetic capabilities that result from a cooperation between metabolic networks. In this work, we have measured biosynthetic capabilities using the concept of scopes. A scope describes the maximal synthesizing capacity of a network when it is provided with a specific set of external resources. We have systematically investigated the pairwise interactions of 447 organisms, for which we have retrieved the networks from the KEGG database. We focused on finding correlations between the maximal amount of increase in biosynthetic capabilities and the dissimilarities of the investigated organisms, both with respect to the network structure and to the phylogenetic distance.

We have found that in some cases there is no measurable increase compared to the functioning of the networks in isolation, but for some network pairs the biosynthetic capability increases dramatically due to the mutual cooperation.

Naturally, the obtained results will critically depend on the quality of the underlying

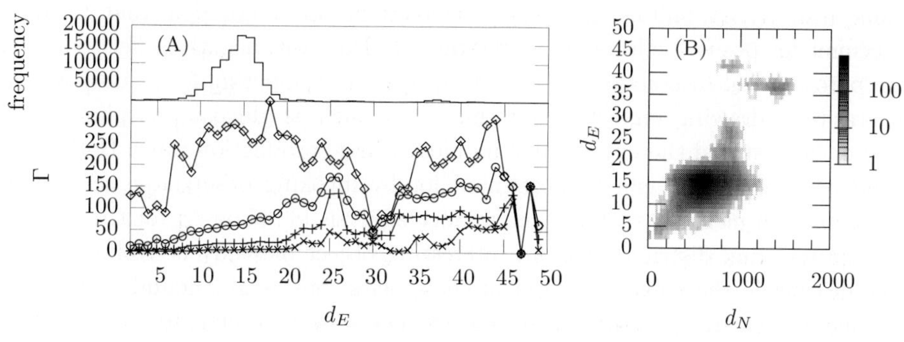

Fig. 4. (A) Same as Fig. 3, but with values plotted against the evolutionary distance d_E, defined by Eq. (6). (B) Correlation of the two distance measures, d_N and d_E.

network data. It is striking that those organisms which, according to our statistical analysis, appear to play an outstanding role, are also among the most thoroughly studied (*H. sapiens*, *R. norvegicus*, *M. musculus* and *A. thaliana*). This observation gives rise to the speculation that the networks of these organisms are far more complete than networks of less intensely studied species, and thus have the potential to infer a stronger synergetic effect in conjunction with other organisms. Erroneous network structures may also arise from reactions in the KEGG database displaying inconsistent stoichiometries. Such reactions were simply ignored in our calculations. In addition, many of the metabolic enzymes included in the database are inferred by sequence analysis. However, homology matches can never provide absolute certainty that an identified gene actually exists in the investigated organism, that it is actively transcribed and translated into the protein, and that the gene product catalyzes the assumed reaction.

Despite the dependence on high quality networks, the here described methodological framework opens a wide field of future investigations. While the systematic approach yields interesting results on a general basis, such as the average increase in biosynthetic capability as a function of network dissimilarity, only the closer investigation of well studied interacting species will provide insight into the specific mechanisms that are responsible for a mutual benefit and therefore into the principles of symbiotic relationships. A particularly interesting field of study will be the symbiosis between plants of the Fabaceae family with Rhizobia, bacteria that possess the ability to fix nitrogen from the air into nitrate or ammonia which is usable by the plant. For an overview of the mechanisms, see e. g. [2].

In real biological systems, the exchange of metabolites between species is limited by the fact that transporters are required to bring the substances into the cells. A more realistic description of interacting networks can be obtained by the inclusion of such transport mechanisms. Whereas this improvement is technically easy to achieve, the actual realization is unfortunately still hindered by the limited knowledge on the transporters and their substrate specificities.

In the presented work, we have applied the method of network expansion to assess biosynthetic capabilities. A major drawback of this method is that it can only account for positive effects of the interaction. However, negative effects may occur in parasite-host interactions where the species are competing for a substrate or the parasite is drawing important intermediates, such as glucose phosphate or ATP, from the host and thereby reduces its biosynthetic production rates. Both, positive and negative effects can be accounted for by invoking other large scale network examination methods such as flux balance analysis [8], allowing for the calculation of optimal flux distributions. An interesting object of study is Wolbachia, which resides inside the cells of several insect species, but is also found in nematodes. While it is clearly parasitic in insects, its relationship to nematodes can rather be described as symbiotic [15]. By comparison of parasites and symbionts, we expect to gain understanding how such mutual interdependencies may have evolved.

5. Acknowledgments

We thank the following organizations for financial support: The International Research Training Group "Genomics and Systems Biology of Molecular Networks" (Christian, N.), the German Research Foundation, in particular the Collaborative Research Center "Theoretical Biology: Robustness, Modularity and Evolutionary Design of Living Systems" (Handorf, T.) and the German Federal Ministry of Education and Research, Systems Biology Research Initiative "GoFORSYS" (Ebenhöh, O.).

References

[1] Benson, D.A., Karsch-Mizrachi, I., Lipman, D.J., Ostell, J., Rapp, B.A., and Wheeler, D.L., GenBank, *Nucleic Acids Res.*, 28(1):15–18, 2000.

[2] Denison, R.F. and Toby, Kiers. E., Why are most rhizobia beneficial to their plant hosts, rather than parasitic?, *Microbes Infect.*, 6(13):1235–1239, 2004.

[3] Edelstein-Keshet, L., *Mathematical Models in Biology*, Society for Industrial and Applied Mathematics, 2005.

[4] Handorf, T. and Ebenhöh, O., Metapath online: A web server implementation of the network expansion algorithm, *Nucleic Acids Res.*, 35(Web Server issue):W613–618, 2007.

[5] Handorf, T., Ebenhöh, O., and Heinrich, R., Expanding metabolic networks: Scopes of compounds, robustness, and evolution, *J. Mol. Evol.*, 61(4):498–512, 2005.

[6] Jeong, H., Tombor, B., Albert, R., Oltvai, Z.N., and Barabasi, A.L., The large-scale organization of metabolic networks, *Nature*, 407(6804):651–654, 2000.

[7] Kanehisa, M., Goto, S., Hattori, M., Aoki-Kinoshita, K.F., Itoh, M., Kawashima, S., Katayama, T., Araki, M., and Hirakawa, M., From genomics to chemical genomics: new developments in KEGG, *Nucleic Acids Res.*, 34(Database issue):354–357, 2006.

[8] Kauffman, K.J., Prakash, P., and Edwards, J.S., Advances in flux balance analysis, *Curr. Opin. Biotechnol*, 14(5):491–496, 2003.

[9] Krieger, C.J., Zhang, P., Mueller, L.A., Wang, A., Paley, S., Arnaud, M., Pick, J., Rhee, S.Y., and Karp, P.D., MetaCyc: a multiorganism database of metabolic pathways and enzymes, *Nucleic Acids Res.*, 32(Database issue):D438–442, 2004.

[10] Papin, J.A., Price, N.D., Wiback, S.J., Fell, D.A., and Palsson, B.O., Metabolic pathways in the post-genome era, *Trends. Biochem. Sci.*, 28(5):250–258, 2003.

[11] Schuster, S., Fell, D.A., and Dandekar, T., A general definition of metabolic pathways useful for systematic organization and analysis of complex metabolic networks, *Nat. Biotechnol.*, 18(3):326–332, 2000.

[12] Schuster, S. and Hilgetag, C., On elementary flux modes in biochemical reaction systems at steady state, *J. Biol. Syst.*, 2(2):165–182, 1994.

[13] Thiele, I., Vo, T.D., Price, N.D., and Palsson, B.O., Expanded metabolic reconstruction of Helicobacter pylori (iIT341 GSM/GPR): an in silico genome-scale characterization of single- and double-deletion mutants, *J. Bacteriol*, 187(16):5818–5830, 2005.

[14] Wagner, A. and Fell, D.A., The small world inside large metabolic networks, *Proc. Biol. Sci.*, 268(1478):1803–1810, 2001.

[15] Werren, J.H., O'Neill, S.L., and Hoffman, A. Eds., *Influential Passengers:inherited microorganisms and arthropod reproduction*, Oxford University Press, 1997.

[16] Wiback, S.J., Mahadevan, R., and Palsson, B.O., Using metabolic flux data to further constrain the metabolic solution space and predict internal flux patterns: the Escherichia coli spectrum, *Biotechnol Bioeng*, 86(3):317–331, 2004.

AUTHOR INDEX